Statistical Inference for Models with Multivariate *t*-Distributed Errors

Statistical Inference for Models with Multivariate t-Distributed Errors

A. K. Md. Ehsanes Saleh
School of Mathematics and Statistics
Carleton University
Ottawa, Canada

M. Arashi
Department of Statistics
School of Mathematical Sciences
Shahrood University
Shahrood, Iran

S. M. M. Tabatabaey
Department of Statistics
Ferdowsi University of Mashhad
Mashhad, Iran

WILEY

Cover image: iStockphoto @ jcarroll-images
Cover design: Wiley

Published by John Wiley & Sons, Inc., Hoboken, New Jersey.
Published simultaneously in Canada.

Library of Congress Cataloging-in-Publication Data:

Saleh, A. K. Md. Ehsanes, author
Statistical inference for models with multivariate t-distributed errors / A.K. Md. Ehsanes Saleh, Department of Mathematics and Statistics, Carleton University, Ottawa, Canada, M. Arashi, Department of Mathematical Sciences, Shahrood University of Technology, Shahrood, Iran, S.M.M. Tabatabaey, Department of Statistics, Ferdowsi University of Mashhad, Mashhad, Iran.
pages cm
Includes bibliographical references and index.
ISBN 978-1-118-85405-1 (hardback)
1. Regression analysis. 2. Multivariate analysis. I. Arashi, M. (Mohammad), 1981– author. II. Tabatabaey, S. M. M., author. III. Title.
QA278.2.S254 2014
519.5'36—dc23 2014007304

Printed in the United States of America.

10 9 8 7 6 5 4 3 2 1

To our wives

SHAHIDARA SALEH

REIHANEH ARASHI

IN LOVING MEMORY
OF PARI TABATABAEY

CONTENTS IN BRIEF

CONTENTS

LIST OF FIGURES

LIST OF TABLES

PREFACE

In problems of statistical inference, it is customary to use normal distribution as the basis of statistical analysis. Many results related to univariate analysis can be extended to multivariate analysis using multidimensional normal distribution. Fisher (1956) pointed out, from his experience with Darwin's data analysis, that a slight change in the specification of the distribution may play havoc on the resulting inferences. To overcome this problem, statisticians tried to broaden the scope of the distributions and achieve reasonable inferential conclusions. Zellner (1976) introduced the idea of using Student's t-distribution, which can accommodate the heavier tailed distributions in a reasonable way and produce robust inference procedures for applications. Most of the research with Student's t-distribution, so far, is focused on the agreement of the results with that of the normal theory. For example, the maximum likelihood estimator of the location parameter agrees with the mean-vector of a normal distribution. Similarly, the likelihood ratio test under the Student's t-distribution has same distribution as the normal distribution under the null hypothesis. This book is an attempt to fill the gap in statistical inference on linear models based on the multivariate t-errors.

This book consists of 12 chapters. Chapter 1 summarizes the results of various models under normal theory with a brief review of the literature. Chapter 2 contains the basic properties of various known distributions and opens the discussion of multivariate t-distribution with its basic properties. We begin Chapter 3 with a discussion of the statistical analysis of a location model from estimation of the intercept. We include the preliminary test and shrinkage type estimators of the location parameter. We also include a discussion of various estimators of variance and their statistical properties. Chapter 4 covers estimation and test of slope and intercept parameters of a simple linear regression model. Chapter 5 is devoted to ANOVA models. Chapter 6 deals with the parallelism model in the same spirit. The multiple regression model is the subject of Chapter 7, and ridge regression is dealt with in Chapter 8. Statistical inference of multivariate models and simple multivariate linear models are discussed in Chapter 9. The Bayesian view point is discussed in multivariate t-models in Chapter 10. The statistical analysis of linear prediction models is included in Chapter 11. The book concludes with Chapter 12, devoted to Stein estimation.

The aim of the book is to provide a clear and balanced introduction to inference techniques using the Student's t-distribution for students and teachers alike, in mathematics, statistics, engineering, and biostatistics programs among other disciplines. Prerequisite for this book is a modest background in statistics, preferably having used textbooks such as Introduction to Probability and Statistics by Rohatgi and Saleh (2001), Introduction to Mathematical Statistics by Hogg, Mckean and Craig (2012), and some exposure to the book Theory of Preliminary Test and Stein-Type Estimation with Applications by Saleh (2006).

After the preliminary chapters, we begin with the location model in Chapter 3, detailing the mathematical development of the theory using multivariate t-distribution as error distribution and we proceed from chapter to chapter raising the level of discussions with various topics of applied statistics.

The content of the book may be covered in two semesters. Various problems are given at the end of every chapter to enhance the knowledge and application of the theory with multivariate t-distribution.

We wish to thank Ms. Mina Noruzirad (Shahrood University) for diligently typesetting the manuscript with care and for reading several chapters and producing most of the graphs and tables that appear in the book. Without her help, the book could not have been completed in time.

A.K.MD.EHSANES SALEH

Carleton University, Canada

M. ARASHI

Shahrood University, Iran

S. M. M. TABATABAEY

Ferdowsi University of Mashhad, Iran

June, 2014

GLOSSARY

ANOVA	analysis of variance
BLF	balanced loss function
BTE	Baranchik-type estimator
C.L.T	central limit theorem
cdf	cumulative distribution function
d.f.	degree of freedom
dim	dimensional
FRV	future regression vector
HPD	highest posterior density
iff	if and only if
JSE	James-Stein estimator
LR	likelihood ratio
LRC	likelihood ratio criterion
LSE	least squares estimator
MLE	maximum likelihood estimator
MRE(.;.)	mean square error based relative efficiency

MSE	mean square error
NCP	natural conjugate prior
p.d.	positive definite
OLS	ordinary least squares
pdf	probability density function
RE	restricted estimator
RLSE	restricted least-square estimator
RMLE	restricted maximum likelihood estimator
RRE	restricted rank estimator
RRE	risk-based relative efficiency
RUE	restricted unbiased estimator
R(.;.)	relative efficiency
r.v.	random variable
PRSE	positive rule Stein-type estimator
PRSRRE	positive-rule Stein-type ridge regression estimator
PTE	preliminary test estimator
PTMLE	preliminary test maximum likelihood estimator
PTLSE	preliminary test least-square estimator
PTRRE	preliminary test ridge regression estimator
RRRE	restricted ridge regression estimator
RSS	residual sum of square
SE	Stein-type estimator
SMLE	Stein-type maximum likelihood estimator
SLSE	Stein-type least-square estimator
SRRE	Stein-type ridge regression estimator
SRV	sum of regression vector
SSE	Stein-type shrinkage estimator
UE	unrestricted estimator
ULSE	unrestricted least-square estimator
UMLE	unrestricted maximum likelihood estimator
URE	unrestricted rank estimator
URRE	unrestricted ridge regression estimator
UUE	unrestricted unbiased estimator

LIST OF SYMBOLS

$\hat{\boldsymbol{\theta}}_n; \hat{\boldsymbol{\beta}}_n$	UE
$\tilde{\boldsymbol{\theta}}_n; \tilde{\boldsymbol{\beta}}_n$	RE
$\hat{\boldsymbol{\theta}}_n^{PT}; \hat{\boldsymbol{\beta}}_n^{PT}$	PTE
$\hat{\boldsymbol{\theta}}_n^{S}; \hat{\boldsymbol{\beta}}_n^{S}$	SE
$\hat{\boldsymbol{\theta}}_n^{S+}; \hat{\boldsymbol{\beta}}_n^{S+}$	PRSE
$b(.)/\boldsymbol{b}(.)$	bias
$R(\cdot)$	risk
Δ^2/Δ_*^2	noncentrality parameter
$\mathcal{L}_n/\mathcal{L}_n^*/\mathcal{L}_n^{**}$	test-statistics
γ_o	d.f.
$\mathbf{V}_n/\mathbf{V}_p$	known scale matrix
$\mathbf{1}_n$	vector of n-tuple of one's
$\varphi(.)$	Borel measurable function
$\phi(.)$	pdf of the standard normal distribution
$\Phi(.)$	cdf of the standard normal distribution
$\lvert a \rvert$	absolute value of the scalar a
$\lVert \boldsymbol{a} \rVert$	norma of $\boldsymbol{a}$

$\|\boldsymbol{A}\|$	determinant of the matrix $\boldsymbol{A}$
$\Gamma(.)$	gamma function
$\Gamma_p(.)$	multivariate gamma function
$B(.,.)$	beta function
$J(t \to z)$	Jacobian of the transformation t to z
$\succeq$	dominates
$\otimes$	Kronecker product
vec	vectorial operator
$Re(a)/Re(\boldsymbol{A})$	real part of $a/\boldsymbol{A}$
Ω_1	unrestricted parameter space
Ω_0	restricted parameter space
$Diag$	diagonal matrix
$Ch_{\max}(\boldsymbol{A})/\lambda_1(\boldsymbol{A})$	largest eigenvalue of the matrix $\boldsymbol{A}$
$Ch_{\min}(\boldsymbol{A})/\lambda_p(\boldsymbol{A})$	smallest eigenvalue of the matrix $\boldsymbol{A}$
$\mathcal{N}(.,.)$	univariate normal distribution
$\mathcal{N}_p(.,.)$	p-variate normal distribution
χ^2_p	chi-square distribution with p d.f.
$\chi^2_p(\Delta^2)$	noncentral chi-square distribution with p d.f.
$M_t^{(1)}(.,.,.)$	Student's t-distribution
$M_t^{(p)}(.,.,.)$	P-variate t-distribution
$IG(.,.)$	inverse gamma distribution
F_{γ_1,γ_2}	F distribution
$F_{\gamma_1,\gamma_2}(\Delta^2)$	noncentral F distribution with noncentrality parameter Δ^2

CHAPTER 1

INTRODUCTION

Outline

1.1 Objective of the Book

The classical theory of statistical analysis is primarily based on the assumption that the errors of various models are normally distributed. The normal distribution is also the basis of the (i) chi-square, (ii) Student's t-, and (iii) F-distributions. Fisher (1956) pointed out that slight differences in the specification of the distribution of the model errors may play havoc on the resulting inferences. To examine the effects on inference, Fisher (1960) analyzed Darwin's data under normal theory and later under a

A. K. Md. Ehsanes Saleh, M. Arashi, S.M.M. Tabatabaey.

Table 1.1 Comparison of quantiles of N(0,1) and Student's t-distribution

CDF	N(0,1)	t_3	t_5	t_6	t_8	t_{10}	t_{15}	t_{20}
.99	2.3263	4.5407	3.3649	3.1430	2.8965	2.7638	2.6025	2.5280
.95	1.6449	2.3534	2.0150	1.9430	1.8595	1.8125	1.7531	1.7247
.90	1.2816	1.6377	1.4759	1.4400	1.3968	1.3722	1.3406	1.3253

symmetric non-normal distribution. Many researchers have since investigated the influence on inference of distributional assumptions differing from normality. Further, it has been observed that most economic and business data, e.g., stock return data, exhibit long-tailed distributions. Accordingly, Fraser and Fick (1975) analyzed Darwin's data and Baltberg and Gonedes (1974) analyzed stock returns using a family of Student's t-distribution to record the effect of distributional assumptions compared to the normal theory analysis. Soon after, Zellner (1976) considered analyzing stock return data by a simple regression model, assuming the error distribution to have a multivariate t-distribution. He revealed the fact that *dependent* but uncorrelated responses can be analyzed by multivariate t-distribution. He discussed differences as well as similarities of the results in both classical and Bayesian contexts for multivariate normal and multivariate t-based models.

Fraser (1979, p. 37) emphasized that the normal distribution is extremely short-tailed and thus *unrealistic* as a sole distribution for variability. He demonstrated the robustness of the Student's t-family as opposed to the normal distribution based on numerical studies. In justifying the appropriateness and the essence of the use of Student's t-distribution, Prucha and Kalajian (1984) pointed out that the normal model based analysis (i) is generally very sensitive to deviations from its assumptions, (ii) places too much weight on outliers, (iii) fails to utilize sample information beyond the first two moments, and (iv) appeals to the central limit theorem for asymptotic approximations.

Table 1.1 shows the differences in quantiles of the standard normal and Student's t-distributions with selected degrees of freedom.

The objective of this book is to present systematic analytical results for various linear models with error distributions belonging to the class of multivariate t. The book will cover results involving (i) estimation, (ii) test of hypothesis, and (iii) improved estimation of location/regression parameters as well as scale parameters.

We shall consider models whose error-distribution is multivariate t. Explicit formulation of *multivariate t-distribution* is given by

$$f\left(\varepsilon \mid \sigma^2\mathbf{V}_n\right) = C\left(\gamma_o\right)\sigma^{-n}\left|\,\mathbf{V}_n\,\right|^{-\frac{1}{2}}\left(1+\frac{1}{\gamma_o\sigma^2}\varepsilon'\mathbf{V}_n^{-1}\varepsilon\right)^{-\frac{1}{2}(\gamma_o+n)}, \quad (1.1.1)$$

where $\sigma > 0$, $C(\gamma_o) = \frac{(\gamma_o)^{\frac{\gamma_o}{2}}\Gamma\left(\frac{\gamma_o+n}{2}\right)}{\pi^{n/2}\Gamma\left(\frac{\gamma_o}{2}\right)}$, $\mathbf{V}_n$ is a known positive definite matrix of rank n and $\gamma_o > 2$. The mean vector and the dispersion matrix of (1.1.1) are

$$E(\varepsilon) = \mathbf{0} \qquad \text{and} \qquad E(\varepsilon\varepsilon') = \frac{\gamma_o\sigma^2}{(\gamma_o - 2)}\mathbf{V}_n. \tag{1.1.2}$$

This distribution will be denoted by $M_t^{(n)}\left(\mathbf{0}, \sigma^2\mathbf{V}_n, \gamma_o\right)$ and will be called the n-dim t-distribution.

In a nutshell, this distribution may be obtained as a mixture of the normal distribution, $\mathcal{N}_n(\mathbf{0}, t^{-1}\sigma^2\mathbf{V}_n)$, and the inverse gamma distribution, $IG(t^{-1}, \gamma_o)$, given by

$$W_o(t) = \frac{1}{\Gamma\left(\frac{\gamma_o}{2}\right)}\, e^{-\frac{\gamma_o t}{2}}\left(\frac{\gamma_o t}{2}\right)^{\frac{\gamma_o}{2}} t^{-1}, \tag{1.1.3}$$

where

$$\Gamma(a) = \int_0^\infty y^{a-1}e^{-y}dy$$

is the gamma function.

1.2 Models under Consideration

In this section, we consider some basic statistical models that are frequently used in applied statistical and econometric analysis with preliminary information with regard to estimation and testing of a hypothesis under normal theory. These will be used to discuss the estimation and test of hypothesis problems under the multivariate t-distribution in later chapters.

1.2.1 Location Model

Consider the location model

$$\boldsymbol{Y} = \theta\mathbf{1}_n + \boldsymbol{\varepsilon}, \tag{1.2.1}$$

where $\boldsymbol{Y} = (Y_1, \ldots, Y_n)'$ is the response vector of n components, $\mathbf{1}_n = (1, \ldots, 1)'$ is a vector of an n-tuple of ones, θ is a scalar parameter, and $\boldsymbol{\varepsilon} = (\varepsilon_1, \ldots, \varepsilon_n)'$ is the error vector of n components having the $\mathcal{N}_n(\mathbf{0}, \sigma^2\mathbf{V}_n)$ -distribution.

An *unrestricted estimator* (UE) of θ is given by the least-square/MLE method as

$$\tilde{\theta}_n = (\mathbf{1}_n'\mathbf{V}_n\mathbf{1}_n)^{-1}\mathbf{1}_n'\mathbf{V}_n^{-1}\boldsymbol{Y}, \tag{1.2.2}$$

while the *unrestricted* unbiased estimator (UUE) of σ^2 is given by

$$S_u^2 = m^{-1}(\boldsymbol{Y} - \tilde{\theta}_n\mathbf{1}_n)'\mathbf{V}_n^{-1}(\boldsymbol{Y} - \tilde{\theta}_n\mathbf{1}_n), \qquad m = n - 1. \tag{1.2.3}$$

Under the normal theory, the exact distribution of $\tilde{\theta}_n - \theta$ is $\mathcal{N}_n(\mathbf{0}, \sigma^2(\mathbf{1}_n'\mathbf{V}_n\mathbf{1}_n)^{-1})$, and the exact distribution of mS_u^2/σ^2 is a chi-square distribution with m degrees of freedom (d.f.).

For testing the null hypothesis $H_o : \theta = \theta_o$ vs $H_A : \theta \neq \theta_o$ the normal theory likelihood ratio (LR)-test statistic for the case where σ^2 is unknown is given by

$$\mathcal{L}_n = \frac{1}{S_u^2}(\mathbf{1}_n'\mathbf{V}_n^{-1}\mathbf{1}_n)(\tilde{\theta}_n - \theta_o)^2. \tag{1.2.4}$$

The exact distribution of $\mathcal{L}_n$ under H_o is the central F-distribution with $(1, m)$ d.f.

1.2.2 Simple Linear Model

Consider a simple linear model

$$\boldsymbol{Y} = \theta\mathbf{1}_n + \beta\boldsymbol{x} + \boldsymbol{\varepsilon} = \boldsymbol{A\eta} + \boldsymbol{\varepsilon}, \quad \boldsymbol{A} = [\mathbf{1}_n, \boldsymbol{x}], \quad \boldsymbol{\eta} = (\theta, \beta)' \tag{1.2.5}$$

where $\boldsymbol{Y} = (Y_1, \ldots, Y_n)'$ is the response vector and $\boldsymbol{x} = (x_1, \ldots, x_n)'$ is a fixed vector of known constants, while $\boldsymbol{\varepsilon} = (\varepsilon_1, \ldots, \varepsilon_n)'$ is the error vector distributed as $\mathcal{N}_n(\mathbf{0}, \sigma^2\mathbf{V}_n)$. Based on the LS or ML principle, the *unrestricted* estimator of $\boldsymbol{\eta} = (\theta, \boldsymbol{\beta})$ is given by

$$\tilde{\boldsymbol{\eta}} = \left(\boldsymbol{A}'\mathbf{V}_n^{-1}\boldsymbol{A}\right)^{-1}\left(\boldsymbol{A}'\mathbf{V}_n^{-1}\mathbf{Y}\right) = \begin{pmatrix} K_1 & K_3 \\ K_3 & K_2 \end{pmatrix}^{-1} \begin{bmatrix} (\mathbf{1}_n'\mathbf{V}_n^{-1}\mathbf{Y}) \\ (\boldsymbol{x}'\mathbf{V}_n^{-1}\mathbf{Y}) \end{bmatrix} = \begin{pmatrix} \tilde{\theta}_n \\ \tilde{\beta}_n \end{pmatrix}, \tag{1.2.6}$$

where $K_1 = (\mathbf{1}_n'\mathbf{V}_n^{-1}\mathbf{1}_n)$, $K_2 = (\boldsymbol{x}'\mathbf{V}_n^{-1}\boldsymbol{x})$, and

$$K_3 = (\mathbf{1}_n'\mathbf{V}_n^{-1}\boldsymbol{x}) = (\boldsymbol{x}'\mathbf{V}_n^{-1}\mathbf{1}_n); \quad \boldsymbol{K} = (\boldsymbol{A}'\mathbf{V}_n^{-1}\boldsymbol{A}), \tag{1.2.7}$$

with the covariance matrix $\sigma^2\boldsymbol{K}^{-1}$, where

$$(\boldsymbol{A}'\mathbf{V}_n^{-1}\boldsymbol{A})^{-1} = \boldsymbol{K}^{-1} = \begin{pmatrix} K_1 & K_3 \\ K_3 & K_2 \end{pmatrix}^{-1} = \frac{1}{K_1K_2 - K_3^2}\begin{pmatrix} K_2 & -K_3 \\ -K_3 & K_1 \end{pmatrix}. \tag{1.2.8}$$

The exact distribution of $\tilde{\boldsymbol{\eta}} - \boldsymbol{\eta}$ follows $\mathcal{N}_2(\mathbf{0}, \sigma^2\boldsymbol{K}^{-1})$. Also, the unbiased estimator of σ^2 is S_u^2 given by

$$S_u^2 = m^{-1}(\boldsymbol{Y} - \boldsymbol{A}\tilde{\boldsymbol{\eta}})'\mathbf{V}_n^{-1}(\boldsymbol{Y} - \boldsymbol{A}\tilde{\boldsymbol{\eta}}); \qquad m = n - 2. \tag{1.2.9}$$

The distribution of mS_u^2/σ^2 follows a chi-square distribution with m d.f.

In order to test the null hypothesis $H_o : \beta = \beta_o$ against an alternative $H_A : \beta \neq \beta_o$, one uses the test statistic $\mathcal{L}_n$, defined by

$$\mathcal{L}_n = \frac{(\tilde{\beta}_n - \beta_o)^2 K_2}{S_u^2}. \tag{1.2.10}$$

Using normal theory, the exact distribution of $\mathcal{L}_n$ under H_o follows the central F-distribution with $(1, m)$ d.f. Similarly, for the test of $H_o : \theta = \theta_o$ against $H_A : \theta \neq \theta_o$ one uses the test-statistic

$$\mathcal{L}_n^* = \frac{(\tilde{\theta}_n - \theta_o)^2 K_1}{S_u^2}. \tag{1.2.11}$$

The exact distribution of $\mathcal{L}_n^*$ under H_o follows the same central F-distribution with $(1, m)$ d.f.

Further, for the test of $H_o : \boldsymbol{\eta} = \boldsymbol{\eta}_o = (\theta_o, \beta_o)$ against $H_A : \boldsymbol{\eta} \neq \boldsymbol{\eta}_o$, one uses the test statistic

$$\mathcal{L}_n^{**} = S_u^{-2} \left[\frac{1}{2} (\tilde{\boldsymbol{\eta}} - \boldsymbol{\eta}_0)' \boldsymbol{K} (\tilde{\boldsymbol{\eta}} - \boldsymbol{\eta}_0) \right]. \tag{1.2.12}$$

The exact distribution of $\mathcal{L}_n^{**}$ under H_o follows the central F-distribution with $(2, m)$ d.f.

1.2.3 ANOVA Model

Suppose that the response vector $\boldsymbol{Y}$ is modeled as

$$\boldsymbol{Y} = \boldsymbol{B}\boldsymbol{\theta} + \boldsymbol{\varepsilon}, \tag{1.2.13}$$

where

$$\boldsymbol{Y} = (Y_{11}, \ldots, Y_{1n_1}; \ldots; Y_{p1}, \ldots, Y_{pn_p})' \tag{1.2.14}$$

$$\boldsymbol{\varepsilon} = (\varepsilon_{11}, \ldots, \varepsilon_{1n_1}; \ldots; \varepsilon_{p1}, \ldots, \varepsilon_{pn_p})' \tag{1.2.15}$$

$$\boldsymbol{B} = \text{Block diagonal matrix} = \text{Diag}(\mathbf{1}_{n_1}, \ldots, \mathbf{1}_{n_p}), \tag{1.2.16}$$

with $\mathbf{1}_{n_i} = (1, \ldots, 1)'$ an n_i-tuple of 1's and

$$\boldsymbol{\theta} = (\theta_1, \ldots, \theta_p)' \quad \text{and} \quad n = n_1 + \ldots + n_p \,. \tag{1.2.17}$$

We assume that

$$\boldsymbol{\varepsilon} \sim \mathcal{N}_n(\mathbf{0}, \sigma^2 \mathbf{V}_n). \tag{1.2.18}$$

According to the LSE/MLE principle, the unrestricted estimators of $\boldsymbol{\theta}$ and σ^2 are given by

$$\tilde{\boldsymbol{\theta}}_n = \boldsymbol{K}^{-1}(\boldsymbol{B}\mathbf{V}_n^{-1}\boldsymbol{Y}), \qquad \boldsymbol{K} = (\boldsymbol{B}'\mathbf{V}_n^{-1}\boldsymbol{B}) \tag{1.2.19}$$

and

$$S_u^2 = m^{-1}(\boldsymbol{Y} - \boldsymbol{B}\tilde{\boldsymbol{\theta}}_n)'\mathbf{V}_n^{-1}(\boldsymbol{Y} - \boldsymbol{B}\tilde{\boldsymbol{\theta}}_n)\,, \qquad m = n - p, \tag{1.2.20}$$

respectively. Moreover, under normal theory, the exact distribution of $\tilde{\boldsymbol{\theta}}_n - \boldsymbol{\theta}$ is $\mathcal{N}_p(\mathbf{0}, \sigma^2 \boldsymbol{K}^{-1})$ and that of mS_u^2/σ^2 follows the chi-square distribution with m d.f. independent of $\tilde{\boldsymbol{\theta}}_n$.

In order to test the null hypothesis $H_o : \boldsymbol{\theta} = \theta_o \mathbf{1}_p$, where θ_o is a scalar and $\mathbf{1}_p$ is a p-vector of 1's, one uses the LR-test statistic, $\mathcal{L}_n$, under normal theory, given by

$$\mathcal{L}_n = \frac{(\tilde{\boldsymbol{\theta}}_n - \hat{\theta}_o \mathbf{1}_p)' \boldsymbol{K} (\tilde{\boldsymbol{\theta}}_n - \hat{\theta}_o \mathbf{1}_p)}{p S_u^2}, \tag{1.2.21}$$

where

$$\hat{\theta}_o = (\mathbf{1}_n' \mathbf{V}_n^{-1} \mathbf{1}_n)^{-1} (\mathbf{1}_n' \mathbf{V}_n^{-1} \boldsymbol{Y}) \tag{1.2.22}$$

is the *restricted estimator* of $\boldsymbol{\theta}_o$ obtained from the null hypothesis $H_o : \boldsymbol{\theta} = \theta_o \mathbf{1}_p$. The exact distribution of $\mathcal{L}_n$ under H_o is the central F-distribution with (p, m) d.f.

1.2.4 Parallelism Model

Consider the n-dim (dimensional) response vector $\boldsymbol{Y}$ modeled as

$$\boldsymbol{Y} = \boldsymbol{B}\boldsymbol{\theta} + \boldsymbol{X}\boldsymbol{\beta} + \boldsymbol{\varepsilon} = \boldsymbol{A}\eta + \varepsilon, \quad \boldsymbol{A} = [\boldsymbol{B}, \boldsymbol{X}], \quad \eta = \begin{pmatrix} \boldsymbol{\theta} \\ \boldsymbol{\beta} \end{pmatrix}, \tag{1.2.23}$$

where

$$\boldsymbol{Y} = (Y_{11}, \ldots, Y_{1n_1}; \ldots; Y_{p1}, \ldots, Y_{pn_p})', \tag{1.2.24}$$

$$\boldsymbol{\varepsilon} = (\varepsilon_{11}, \ldots, \varepsilon_{1n_1}; \ldots; \varepsilon_{p1}, \ldots, \varepsilon_{pn_p})', \tag{1.2.25}$$

$$\boldsymbol{B} = \text{Block diagonal matrix}(\mathbf{1}_{n_1}, \ldots, \mathbf{1}_{n_p}), \tag{1.2.26}$$

$$\boldsymbol{X} = \text{Block diagonal matrix}(\boldsymbol{x}_1, \ldots, \boldsymbol{x}_p), \tag{1.2.27}$$

and

$$\boldsymbol{\theta} = (\theta_1, \ldots, \theta_p)', \qquad \text{and} \qquad \boldsymbol{\beta} = (\beta_1, \ldots, \beta_p)'. \tag{1.2.28}$$

We assume that the error distribution is $\mathcal{N}_n(\mathbf{0}, \sigma^2 \mathbf{V}_n)$, where $n = n_1 + \ldots + n_p$.

The LSE of $\boldsymbol{\theta}$ and $\boldsymbol{\beta}$ are given by

$$\tilde{\boldsymbol{\eta}} = \begin{pmatrix} \tilde{\boldsymbol{\theta}}_n \\ \tilde{\boldsymbol{\beta}}_n \end{pmatrix} = \left(\boldsymbol{A}' \mathbf{V}_n^{-1} \boldsymbol{A}\right)^{-1} \left(\boldsymbol{A}' \mathbf{V}_n^{-1} \boldsymbol{Y}\right) = \begin{pmatrix} \boldsymbol{K}_1 & \boldsymbol{K}_3 \\ \boldsymbol{K}_3 & \boldsymbol{K}_2 \end{pmatrix}^{-1} \begin{pmatrix} \boldsymbol{B}' \mathbf{V}_n^{-1} \boldsymbol{Y} \\ \boldsymbol{X}' \mathbf{V}_n^{-1} \boldsymbol{Y} \end{pmatrix}, \tag{1.2.29}$$

where

$$\boldsymbol{K}_1 = \boldsymbol{B}' \mathbf{V}_n^{-1} \boldsymbol{B}, \quad \boldsymbol{K}_2 = \boldsymbol{X}' \mathbf{V}_n^{-1} \boldsymbol{X}, \quad \text{and} \quad \boldsymbol{K}_3 = (\boldsymbol{B}' \mathbf{V}_n^{-1} \boldsymbol{X}). \tag{1.2.30}$$

The covariance matrix of $\tilde{\boldsymbol{\eta}}$ is given by $\sigma^2 \boldsymbol{K}^{-1}$, where

$$\boldsymbol{K} = (\boldsymbol{A}' \mathbf{V}_n^{-1} \boldsymbol{A})^{-1}. \tag{1.2.31}$$

Normal theory results lead to the distribution of $\tilde{\boldsymbol{\eta}}$ as $\mathcal{N}_{2p}(\boldsymbol{\eta}, \sigma^2 \boldsymbol{K}^{-1})$.

The unbiased estimator of σ^2 is defined by

$$S_u^2 = m^{-1} (\boldsymbol{Y} - \boldsymbol{A}\tilde{\boldsymbol{\eta}})' \mathbf{V}_n^{-1} (\boldsymbol{Y} - \boldsymbol{A}\tilde{\boldsymbol{\eta}}) \,; \qquad (m = n - 2p), \tag{1.2.32}$$

and mS_u^2/σ^2 follows the chi-square distribution with m d.f., independent of $\tilde{\boldsymbol{\eta}}$ under normal theory.

For test of the null hypothesis $H_o : \boldsymbol{\beta} = \boldsymbol{\beta}_o$ vs $H_A : \boldsymbol{\beta} \neq \boldsymbol{\beta}_o$ one uses the test statistic

$$\mathcal{L}_n^* = \frac{(\tilde{\boldsymbol{\beta}}_n - \boldsymbol{\beta}_o)'\boldsymbol{K}_2(\tilde{\boldsymbol{\beta}}_n - \boldsymbol{\beta}_o)}{pS_u^2}. \tag{1.2.33}$$

Under H_o, $\mathcal{L}_n$ follows the central F-distribution with (p, m) d.f. On the other hand, if $\boldsymbol{\beta}_o$ is unknown but equals to $\beta_o\mathbf{1}_p$, β_o scalar and *unknown*, then the common estimator of $\boldsymbol{\beta}_o$ is given by

$$\hat{\boldsymbol{\beta}}_n = \mathbf{1}_p(\mathbf{1}_p'\boldsymbol{X}'\mathbf{V}_n^{-1}\boldsymbol{X}\mathbf{1}_p)^{-1}(\mathbf{1}_p'\boldsymbol{X}'\mathbf{V}_n^{-1}\boldsymbol{Y}), \tag{1.2.34}$$

and the test-statistic is defined by

$$\mathcal{L}_n = \frac{(\tilde{\boldsymbol{\beta}}_n - \hat{\boldsymbol{\beta}}_n)'\boldsymbol{K}_2(\tilde{\boldsymbol{\beta}}_n - \hat{\boldsymbol{\beta}}_n)}{qS_u^2}, \qquad (q = p - 1). \tag{1.2.35}$$

The exact distribution of $\mathcal{L}_n$ under H_o follows the central F-distribution with (q, m) d.f.

1.2.5 Multiple Regression Model

Consider the multiple regression model

$$\boldsymbol{Y} = \boldsymbol{X}\boldsymbol{\beta} + \boldsymbol{\varepsilon}, \tag{1.2.36}$$

where $\boldsymbol{Y} = (Y_1, Y_2, \ldots, Y_n)'$, $\boldsymbol{X}$ is the $n \times p$ matrix of known constants, $\boldsymbol{\beta} = (\beta_1, \beta_2, \ldots, \beta_p)'$, and $\boldsymbol{\varepsilon}$ is the n-vector of errors with distribution $\mathcal{N}_n(\mathbf{0}, \sigma^2\mathbf{V}_n)$. The LSE/MLE of $\boldsymbol{\beta}$ is then given by

$$\tilde{\boldsymbol{\beta}}_n = (\boldsymbol{X}'\mathbf{V}_n^{-1}\boldsymbol{X})^{-1}(\boldsymbol{X}'\mathbf{V}_n^{-1}\boldsymbol{Y}), \tag{1.2.37}$$

where $\tilde{\boldsymbol{\beta}}_n \sim \mathcal{N}_n\left(\boldsymbol{\beta}, \sigma^2(\boldsymbol{X}'\mathbf{V}_n\boldsymbol{X})^{-1}\right)$, and the unbiased estimator of σ^2 is given by

$$S_u^2 = m^{-1}(\boldsymbol{Y} - \boldsymbol{X}\tilde{\boldsymbol{\beta}}_n)'\mathbf{V}_n^{-1}(\boldsymbol{Y} - \boldsymbol{X}\tilde{\boldsymbol{\beta}}_n); \qquad (m = n - p), \tag{1.2.38}$$

where mS_u^2/σ^2 follows the chi-square distribution with m d.f., independent of $\tilde{\boldsymbol{\beta}}_n$ under normal theory.

For the test of null hypothesis, $H_o : \boldsymbol{H}\boldsymbol{\beta} = \boldsymbol{h}$ against the alternative, $H_A : \boldsymbol{H}\boldsymbol{\beta} \neq \boldsymbol{h}$, where $\boldsymbol{H}$ is a $q \times p$ matrix of known constants and $\boldsymbol{h}$ is a q-vector of known constants, the *restricted estimator* of $\boldsymbol{\beta}$ is given by

$$\hat{\boldsymbol{\beta}}_n = \tilde{\boldsymbol{\beta}}_n - \boldsymbol{C}^{*-1}\boldsymbol{H}(\boldsymbol{H}\boldsymbol{C}^{*-1}\boldsymbol{H}')^{-1}(\boldsymbol{H}\tilde{\boldsymbol{\beta}}_n - \boldsymbol{h}), \tag{1.2.39}$$

where $\boldsymbol{C}^* = \boldsymbol{X}'\mathbf{V}_n^{-1}\boldsymbol{X}$. Thus, one uses the test statistic for the test of $H_o : \boldsymbol{H\beta} = \boldsymbol{h}$, given by

$$\mathcal{L}_n = \frac{(\boldsymbol{H}\tilde{\boldsymbol{\beta}}_n - \boldsymbol{h})'(\boldsymbol{H}\boldsymbol{C}^{*-1}\boldsymbol{H}')^{-1}(\boldsymbol{H}\tilde{\boldsymbol{\beta}}_n - \boldsymbol{h})}{q\boldsymbol{S}_u^2}. \tag{1.2.40}$$

The exact distribution of $\mathcal{L}_n$ under $H_o : \boldsymbol{H\beta} = \boldsymbol{h}$ follows the central F-distribution with (q, m) d.f.

1.2.6 Ridge Regression

If in the model (1.2.36), $\boldsymbol{X}'\mathbf{V}_n^{-1}\boldsymbol{X}$ is a near-singular or ill-conditioned matrix that prevents the reasonable inversion of $\boldsymbol{X}'\mathbf{V}_n^{-1}\boldsymbol{X}$, then multicollinearity may be present among the elements of $\boldsymbol{X}'\mathbf{V}_n^{-\frac{1}{2}}$, and $\boldsymbol{X}'\mathbf{V}_n^{-1}\boldsymbol{X}$ has rank less than p or one or more characteristic roots of $\boldsymbol{X}'\mathbf{V}_n^{-1}\boldsymbol{X}$ are small, leading to a large variance of $\tilde{\boldsymbol{\beta}}_n$. Thus, following Hoerl and Kennard (1970), one may define the "Ridge estimator" of $\boldsymbol{\beta}$ as

$$\begin{aligned}\hat{\boldsymbol{\beta}}_{HK} &= \left(\boldsymbol{X}'\mathbf{V}_n^{-1}\boldsymbol{X} + k\boldsymbol{I}_p\right)^{-1}\boldsymbol{X}'\mathbf{V}_n^{-1}\boldsymbol{Y}\,, \quad k > 0,\\ &= \left(\boldsymbol{C}^* + k\boldsymbol{I}_p\right)^{-1}\boldsymbol{X}'\mathbf{V}_n^{-1}\boldsymbol{Y}\,, \quad \boldsymbol{C}^* = \boldsymbol{X}'\mathbf{V}_n^{-1}\boldsymbol{X}.\end{aligned} \tag{1.2.41}$$

Properties of these estimators have been studied by Hoerl and Kennard (1970) for $\mathbf{V}_n = \boldsymbol{I}_n$. A detailed discussion is available in Saleh (2006), among other papers.

1.2.7 Multivariate Model

Let $\boldsymbol{Y}_1, \boldsymbol{Y}_2, \ldots, \boldsymbol{Y}_N$ be N observation vectors of p-dim satisfying the model

$$\boldsymbol{Y}_\alpha = \boldsymbol{\theta} + \boldsymbol{\varepsilon}_\alpha\,, \qquad \alpha = 1, \ldots, N, \tag{1.2.42}$$

where $\boldsymbol{Y}_\alpha = (Y_{\alpha 1}, \ldots, Y_{\alpha p})'$, $\boldsymbol{\theta} = (\theta_1, \ldots, \theta_p)'$ is the location vector parameter, and $\boldsymbol{\varepsilon}_\alpha = (\varepsilon_{\alpha 1}, , \ldots, \varepsilon_{\alpha p})'$. $\{\boldsymbol{\varepsilon}_\alpha | \alpha = 1, 2, \ldots, N\}$ are distributed as $\mathcal{N}_n(\boldsymbol{0}, \boldsymbol{\Sigma})$ for each $\alpha = 1, \ldots, N$.

The point estimator of $\boldsymbol{\theta}$ is $\tilde{\boldsymbol{\theta}}_n = \bar{\boldsymbol{X}}$, and an unbiased estimator of $\boldsymbol{\Sigma}$ is given by

$$\boldsymbol{S}_u = \frac{1}{N-1}\sum_{\alpha=1}^{N}(\boldsymbol{Y}_\alpha - \bar{\boldsymbol{Y}})(\boldsymbol{Y}_\alpha - \bar{\boldsymbol{Y}})'. \tag{1.2.43}$$

The exact distribution of $\tilde{\boldsymbol{\theta}}_N$ is $\mathcal{N}_p(\boldsymbol{0}, \frac{1}{N}\boldsymbol{\Sigma})$, and that of $(N-1)\boldsymbol{S}_u^2 = A$ is the Wishart distribution, given by the probability density function (pdf)

$$\boldsymbol{f}(\boldsymbol{A}|\boldsymbol{\Sigma}) = \frac{|\boldsymbol{\Sigma}|^{-\frac{n}{2}}|\boldsymbol{A}|^{-\frac{1}{2}(n-p-1)}e^{(-\frac{1}{2}tr\boldsymbol{\Sigma}^{-1}A)}}{2^{\frac{np}{2}}\pi^{\frac{p(p-1)}{4}}\prod_{i=1}^{p}\left(\frac{n-i+1}{2}\right)!}\,; \qquad (n = N-1). \tag{1.2.44}$$

For the test of the null hypothesis $H_o : \boldsymbol{\theta} = \boldsymbol{\theta}_o$ one uses the Hotelling's T^2 statistic as

$$\mathcal{L}_n = (N-1)(\tilde{\boldsymbol{\theta}}_n - \boldsymbol{\theta}_o)'\boldsymbol{S}_u^{-1}(\tilde{\boldsymbol{\theta}}_n - \boldsymbol{\theta}_o), \tag{1.2.45}$$

which follows the Hotelling's T^2-distribution with (p, m) d.f., i.e., $\mathcal{L}_n/(N-1)$ is distributed as central F-distribution with (p, m) d.f., where $m = N - p$.

An improved estimator of $\boldsymbol{\theta}$ is known to be

$$\hat{\boldsymbol{\theta}}_N^S = \boldsymbol{\theta}_o + (1 - c\mathcal{L}_n^{-1})(\tilde{\boldsymbol{\theta}}_N - \boldsymbol{\theta}_o) \; ; \qquad c = \frac{(p-2)m}{p(m+2)}. \tag{1.2.46}$$

See for example, James and Stein (1961) and Saleh (2006).

1.2.8 Simple Multivariate Linear Model

Consider the simple linear model

$$\boldsymbol{Y}_\alpha = \boldsymbol{\theta} + \boldsymbol{\beta} x_\alpha + \boldsymbol{\varepsilon}_\alpha \; ; \qquad \alpha = 1, \ldots, N, \tag{1.2.47}$$

where $\boldsymbol{Y}_\alpha = (Y_{\alpha 1}, \ldots, \boldsymbol{Y}_{\alpha p})'$, $\boldsymbol{\theta} = (\theta_1, \ldots, \theta_p)'$, is the *intercept* vector, $\boldsymbol{\beta} = (\beta_1, \ldots, \beta_p)'$ is *slope* vector, $x_\alpha = (x_1, \ldots, x_p)'$ is the p-vector of fixed constants for every α, and $\boldsymbol{\varepsilon}_\alpha = (\varepsilon_{\alpha 1}, \ldots, \varepsilon_{\alpha p})'$ is the vector of errors, having the distribution $\mathcal{N}_n(\mathbf{0}, \boldsymbol{\Sigma})$ for each $\alpha = 1, \ldots, N$.

The point estimators of $\boldsymbol{\theta}$ and $\boldsymbol{\beta}$ are given by

$$\tilde{\boldsymbol{\theta}}_n = \bar{\boldsymbol{Y}} - \tilde{\boldsymbol{\beta}}_N \bar{\boldsymbol{x}}_N, \quad \tilde{\boldsymbol{\beta}}_N = \frac{1}{Q_N} \left\{ \boldsymbol{x}' \begin{pmatrix} \boldsymbol{Y}_1 \\ \vdots \\ \boldsymbol{Y}_N \end{pmatrix} - \frac{1}{N}(\mathbf{1}'_N \boldsymbol{x}) \left(\mathbf{1}'_N \begin{pmatrix} \boldsymbol{Y}_1 \\ \vdots \\ \boldsymbol{Y}_N \end{pmatrix} \right) \right\}, \tag{1.2.48}$$

with $Q_N = \boldsymbol{x}'\boldsymbol{x} - \frac{1}{N}(\mathbf{1}'_N \boldsymbol{x})$.

The point estimator of $\boldsymbol{\Sigma}$ is given by

$$\boldsymbol{S}_u = \frac{1}{N-2} \sum_{\alpha=1} \left\{ (\boldsymbol{Y}_\alpha - \bar{\boldsymbol{Y}}) - \tilde{\boldsymbol{\beta}}_N (\boldsymbol{x}_\alpha - \bar{\boldsymbol{x}}_N) \right\} \left\{ (\boldsymbol{Y}_\alpha - \bar{\boldsymbol{Y}}) - \tilde{\boldsymbol{\beta}}_N (\boldsymbol{x}_\alpha - \bar{\boldsymbol{x}}_N) \right\}' \tag{1.2.49}$$

For the test of null hypothesis $H_o : \boldsymbol{\beta} = \mathbf{0}$ the LR-test is given by

$$\mathcal{L}_n = \frac{m}{p} Q_N \tilde{\boldsymbol{\beta}}'_N \boldsymbol{S}_u^{-1} \tilde{\boldsymbol{\beta}}_N, \tag{1.2.50}$$

which follows the central F-distribution with (p, m) d.f.

1.3 Organization of the Book

This book consists of twelve chapters. Chapter 1 summarizes the results of various models under normal theory with brief review of the literature. Chapter 2 contains the basic properties of various known distributions and opens discussion of multivariate t-distribution with its basic properties. We open up Chapter 3 to discuss the statistical analysis of a location model from estimation of the intercept and slope, to test of hypothesis of the parameters. We also add the preliminary test and shrinkage-type estimators of the three parameters which include the estimation of the scale parameter of the model while Chapter 4 contains similar details of a simple regression model. Chapter 5 is devoted to ANOVA models and discussing on preliminary test and shrinkage-type estimators in multivariate t-distribution and Chapter 6 deals with the parallelism model in the same spirit. Multiple regression model is the content of Chapter 7 and ridge regression is dealt with in Chapter 8. Statistical inference of multivariate models and simple multivariate linear models are discussed in Chapter 9. Bayesian view point is discussed in multivariate t-models in Chapter 10. Finally, we discuss linear prediction models in Chapter 11. We conclude the book with Chapter 12 containing the Stein estimation as complementary results.

1.4 Problems

1. Derive the estimates given in (1.2.2) and (1.2.3), using the LSE method.

2. Derive the test statistic given by (1.2.4), using normal theory.

3. Show that under $H_o : \theta = \theta_o$, (1.2.4) follows the central F-distribution with $(1, m)$ d.f., $m = n - 1$.

4. Derive the estimates given by (1.2.6)–(1.2.9), using LSE method.

5. **(a)** Show that under $H_o : \beta = \beta_o$, the test statistic (1.2.10) follows the central F-distribution with $(1, m)$ d.f.

(b) Show that under $H_o : \theta = \theta_o$, the test statistic (1.2.11) follows the central F-distribution with $(1, m)$ d.f.

(c) Show that under $H_o : \boldsymbol{\eta} = \boldsymbol{\eta}_o$, the test statistic (1.2.12) follows the central F-distribution with $(2, m)$ d.f.

6. Derive the estimators given by (1.2.19)–(1.2.20), using the LSE method.

7. Show that the LR-test statistic (1.2.21) follows the central F-distribution with (p, m) d.f., $m = n - p$.

8. Derive the estimates given by (1.2.29)–(1.2.32) using the LSE method.

9. Show that the test statistic (1.2.33) follows the central F-distribution with (p, m) d.f., $m = n - 2p$.

10. Show that the test statistic (1.2.35) follows central F-distribution with (q, m) d.f., $q = p - 1$, $m = n - p - 1$.

11. Derive the estimators $\tilde{\boldsymbol{\beta}}_n$ (1.2.37) and $\hat{\boldsymbol{\beta}}_n$ (1.2.39) using the LSE method.

12. Derive the LR-test statistic (1.2.40), and show that it follows the central F-distribution with (q, m) d.f. , $m = n - p$.

13. Consider the ridge regression estimate of $\boldsymbol{\beta}$

$$\hat{\boldsymbol{\beta}}_{HK} = \left[\boldsymbol{X}'\boldsymbol{X} + k\boldsymbol{I}_p\right]^{-1} \boldsymbol{X}'\boldsymbol{Y} \ , \qquad \text{when} \quad \mathbf{V}_n = \boldsymbol{I}_n$$

Find the bias and mean-square-error of $\hat{\boldsymbol{\beta}}_{HK}$ and compare it with the unrestricted estimator $\tilde{\boldsymbol{\beta}}_n = (\boldsymbol{X}'\boldsymbol{X})^{-1}\boldsymbol{X}'\boldsymbol{Y}$ of $\boldsymbol{\beta}$.

14. Derive the LR test statistic (1.2.45) and verify its distribution under the null hypothesis.

15. Derive the James-Stain-Saleh estimator given by (1.2.46), and find its bias and MSE-matrix.

16. Verify the estimates of $\boldsymbol{\theta}$ and $\boldsymbol{\beta}$ given by (1.2.48) and derive the test statistic given by (1.2.50), using the LR-principle to test the null hypothesis $H_o : \boldsymbol{\beta} = \mathbf{0}$ for the model (1.2.47).

CHAPTER 2

PRELIMINARIES

Outline

In this chapter, we discuss some basic results on various distributions, particularly the normal, chi-square, Student's t-, multivariate t-distributions.

A. K. Md. Ehsanes Saleh, M. Arashi, S.M.M. Tabatabaey.

2.1 Normal Distribution

The most basic distribution in statistical theory is the normal distribution, $\mathcal{N}(\theta, \sigma^2)$, with pdf

$$f(y; \theta, \sigma^2) = \frac{1}{\sigma\sqrt{2\pi}} e^{\left\{-\frac{1}{2\sigma^2}(y-\theta)^2\right\}}, \quad -\infty < y, \theta < \infty, \ \sigma > 0, \qquad (2.1.1)$$

where θ is the mean and σ^2 is the variance of this distribution.

It is well known that

(i) If Y is $\mathcal{N}(\theta, \sigma^2)$, then $Z = \frac{Y-\theta}{\sigma}$ is $\mathcal{N}(0, 1)$.

(ii) If Y is $\mathcal{N}(\theta, \sigma^2)$ and $\varphi(Y)$ is a differentiable function satisfying $|\varphi'(Y)| < \infty$, then

$$E(\varphi(Y)(Y - \theta)) = \sigma^2 E[\varphi'(Y)]. \qquad (2.1.2)$$

For more information see Stein (1981).

(iii) If Z is $\mathcal{N}(\theta, 1)$, then

$$\begin{aligned} E(|Z|) &= \sqrt{\frac{2}{\pi}} e^{-\frac{\theta^2}{2}} + \theta(2\Phi(\theta) - 1) \\ E\left(\frac{Z}{|Z|}\right) &= 1 - 2\Phi(-\theta), \end{aligned} \qquad (2.1.3)$$

where $\Phi(.)$ is the cumulative distribution function (cdf) of the standard normal (zero mean and unit variance) distribution. For more information see Saleh (2006).

(iv) Noncentral normal distribution.
If Y is $\mathcal{N}(\theta, 1)$, then

$$f(y; \theta) = e^{\left(-\frac{1}{2}\theta^2\right)} \sum_{r \geq 0} \frac{1}{r!} \theta^r y^r f(y; 0), \quad f(y; 0) = \frac{1}{\sqrt{2\pi}} e^{-\frac{y^2}{2}}. \qquad (2.1.4)$$

2.2 Chi-Square Distribution

If Z is $\mathcal{N}(0, 1)$, then Z^2 follows the central chi-square distribution with one degree of freedom (d.f.). However, if Z is $\mathcal{N}(\theta, \sigma^2)$, then $\frac{Z^2}{\sigma^2}$ follows the noncentral chi-square distribution with one d.f. and noncentrality parameter $\frac{\Delta^2}{2}$, with $\Delta^2 = \frac{\theta^2}{\sigma^2}$. The pdf/cdf of this noncentral chi-square variable with one d.f. is given by

$$h(y; \Delta^2) = \sum_{r \geq 0} \frac{1}{r!} e^{-\frac{\Delta^2}{2}} \left(\frac{\Delta^2}{2}\right)^r h_{1+2r}(y; 0), \quad \Delta^2 = \frac{\theta^2}{\sigma^2}, \qquad (2.2.1)$$

where

$$h_{1+2r}(y;0) = \frac{e^{-\frac{y}{2}}}{2\Gamma\left(\frac{1+2r}{2}\right)}\left(\frac{y}{2}\right)^{\frac{1+2r}{2}-1}, \qquad y \geq 0, \tag{2.2.2}$$

and the cdf of the chi-square distribution is given by

$$H(c;\Delta^2) = \sum_{r\geq 0}\frac{1}{r!}e^{-\frac{\Delta^2}{2}}\left(\frac{\Delta^2}{2}\right)^r H_{1+2r}(c;0), \tag{2.2.3}$$

where $H_{1+2r}(c;0)$ is the cdf of a central chi-square distribution with $1+2r$ d.f.

An important identity w.r.t. the cdf is given by

$$H_{\gamma_o+2}(c;0) = -h_{\gamma_o+2}(c;0) + H_{\gamma_o}(c;0). \tag{2.2.4}$$

A central and noncentral chi-square variable will be denoted by $\chi^2_{\gamma_o}$ and $\chi^2_{\gamma_o}(\Delta^2)$, respectively. In statistical theory, moment results involving chi-square variables are important and they are given below.

If $Z \sim \mathcal{N}(\theta, \sigma^2)$ and $\varphi(.)$ is a measurable function of Z^2, then

(i)

$$E\left[Z\varphi(Z^2)\right] = \theta E\left[\varphi(\chi^2_3(\Delta^2))\right],$$

(ii)

$$E\left[Z^2\varphi(Z^2)\right] = \sigma^2 E\left[\varphi(\chi^2_3(\Delta^2))\right] + \theta^2 E\left[\varphi(\chi^2_5(\Delta^2))\right].$$

Further,

(iii)

$$E\left[\chi^2_{\gamma_o}\varphi(\chi^2_{\gamma_o})\right] = \gamma_o E\left[\varphi(\chi^2_{\gamma_o+2})\right],$$

(iv)

$$E\left[\chi^2_{\gamma_o}(\Delta^2)\varphi\left(\chi^2_{\gamma_o}(\Delta^2)\right)\right] = \gamma_o E\left[\varphi\left(\chi^2_{\gamma_o+2}(\Delta^2)\right)\right] + \Delta^2 E\left[\chi^2_{\gamma_o+4}(\Delta^2)\right],$$

(v)

$$\begin{aligned} E\left\{\left[\chi^2_{\gamma_o}(\Delta^2)\right]^2\varphi\left(\chi^2_{\gamma_o}(\Delta^2)\right)\right\} &= \gamma_o(\gamma_o+2)E\left[\varphi\left(\chi^2_{\gamma_o+2}(\Delta^2)\right)\right] \\ &\quad +2(\gamma_o+2)\Delta^2 E\left[\varphi\left(\chi^2_{\gamma_o+6}(\Delta^2)\right)\right] \\ &\quad +\Delta^4 E\left[\varphi\left(\chi^2_{\gamma_o+8}(\Delta^2)\right)\right]. \end{aligned} \tag{2.2.5}$$

If $\mathbf{Z} = (Z_1, \ldots, Z_p)'$ be a $\mathcal{N}_p(\boldsymbol{\theta}, \boldsymbol{I}_p)$ (see (2.5.1)) and $\varphi(\mathbf{Z}'\mathbf{Z})$ is a measurable function of $\mathbf{Z}'\mathbf{Z}$, then

(i)

$$E\left[\mathbf{Z}\varphi(\mathbf{Z}'\mathbf{Z})\right] = \boldsymbol{\theta} E\left[\varphi\left(\chi^2_{p+2}(\Delta^2)\right)\right], \qquad \Delta^2 = \boldsymbol{\theta}'\boldsymbol{\theta},$$

(ii)

$$E\left[\mathbf{Z}\mathbf{Z}'\varphi(\mathbf{Z}'\mathbf{Z})\right] = \boldsymbol{I}_p E\left[\varphi\left(\chi^2_{p+2}(\Delta^2)\right)\right] + \boldsymbol{\theta}\boldsymbol{\theta}' E\left[\varphi\left(\chi^2_{p+4}(\Delta^2)\right)\right],$$

(iii)

$$E\left[\mathbf{Z}'\boldsymbol{A}\mathbf{Z}\varphi(\mathbf{Z}'\mathbf{Z})\right] = \operatorname{tr}(\boldsymbol{A})E\left[\varphi(\chi^2_{p+2}(\Delta^2)\right] + \boldsymbol{\theta}'\boldsymbol{A}\boldsymbol{\theta}E\left[\varphi\left(\chi^2_{p+4}(\Delta^2)\right)\right], \tag{2.2.6}$$

where $\boldsymbol{A}$ is a positive definite matrix.

Further, if $\varphi\left(\chi^2_{p+2}(\Delta^2)\right) = \chi^{-2}_{p+2}(\Delta^2) = \left(\chi^2_{p+2}(\Delta^2)\right)^{-1}$, or $\varphi\left(\chi^2_{p+2}(\Delta^2)\right) = \chi^{-4}_{p+2}(\Delta^2) = \left(\chi^2_{p+2}(\Delta^2)\right)^{-2}$, then

(i)

$$E\left[\chi^{-2}_{p+2}(\Delta^2)\right] = e^{-\frac{\Delta^2}{2}} \sum_{r\geq 0} \frac{1}{r!}\left(\frac{\Delta^2}{2}\right)^r (p+2r)^{-1},$$

(ii)

$$E\left[\chi^{-4}_{p+2}(\Delta^2)\right] = e^{-\frac{\Delta^2}{2}} \sum_{r\geq 0} \frac{1}{r!}\left(\frac{\Delta^2}{2}\right)^r (p+2r)^{-1}(p-2+2r)^{-1},$$

(iii)

$$\Delta^2 E\left[\chi^{-2}_{p+2}(\Delta^2)\right] = 1 - (p-2)E\left[\chi^{-2}_{p}(\Delta^2)\right],$$

(iv)

$$\Delta^2 E\left[\chi^{-4}_{p+4}(\Delta^2)\right] = E\left[\chi^{-2}_{p+2}(\Delta^2)\right] - (p-2)E\left[\chi^{-2}_{p+2}(\Delta^2)\right],$$

(v)

$$2E\left[\chi^{-4}_{p+2}(\Delta^2)\right] = E\left[\chi^{-2}_{p}(\Delta^2)\right] - E\left[\chi^{-2}_{p+2}(\Delta^2)\right]. \tag{2.2.7}$$

For details, see Judge and Bock (1978) and Saleh (2006).

2.3 Student's t-Distribution

It is well known that if $Z \sim \mathcal{N}(0, \sigma^2)$ and U is independent of Z such that $\frac{\gamma_o U}{\sigma^2}$ is a chi-square variable with γ_o degrees of freedom, then $u = \frac{Z}{\sqrt{U}}$ follows the Student's t-distribution with γ_o d.f. The pdf of this t-statistic is given by

$$f(u) = \frac{1}{\sqrt{\gamma_o}B(\frac{1}{2}, \frac{\gamma_o}{2})}\left(1 + \frac{1}{\gamma_o}u^2\right)^{\frac{1}{2}(\gamma_o+1)}, \qquad -\infty < u < \infty. \tag{2.3.1}$$

The distribution $f(u)$ of u may be obtained as the expected value of the pdf of $\mathcal{N}(0, t^{-1})$, which is the conditional distribution of u given t with respect to the inverse gamma distribution, $IG\left(t^{-1}, \gamma_o\right)$, with the pdf given by

$$W_o(t) = \frac{1}{\Gamma\left(\frac{\gamma_o}{2}\right)} e^{-\frac{\gamma_o t}{2}} \left(\frac{\gamma_o t}{2}\right)^{\frac{\gamma_o}{2}} t^{-1}. \tag{2.3.2}$$

Then,

$$\begin{aligned}
f(u) &= \int_0^\infty W_o(t)\mathcal{N}(0, t^{-1})dt \\
&= E_t\left\{\left(\frac{t}{2\pi}\right)^{\frac{1}{2}} e^{-\frac{u^2 t}{2}}\right\}, \qquad -\infty < u < \infty \qquad (2.3.3)\\
&= \frac{1}{\sqrt{\pi}\Gamma\left(\frac{\gamma_o}{2}\right)} \left(\frac{\gamma_o}{2}\right)^{\frac{\gamma_o}{2}} \int_0^\infty t^{\frac{1}{2}(\gamma_o+1)-1} e^{-t\left(\frac{u^2}{2}+\frac{\gamma_o}{2}\right)} dt \\
&= \frac{1}{\sqrt{\pi}\Gamma\left(\frac{\gamma_o}{2}\right)} \left(\frac{\gamma_o}{2}\right)^{\frac{\gamma_o}{2}} \left(\frac{u^2}{2}+\frac{\gamma_o}{2}\right)^{-\frac{1}{2}(\gamma_o+1)} \int_0^\infty y^{\frac{1}{2}(\gamma_o+1)-1} e^{-y} dy \\
&= \frac{\Gamma\left(\frac{\gamma_o+1}{2}\right)}{\sqrt{\gamma_o}\Gamma\left(\frac{1}{2}\right)\Gamma\left(\frac{\gamma_o}{2}\right)} \left(1+\frac{1}{\gamma_o}u^2\right)^{-\frac{1}{2}(\gamma_o+1)} \\
&= \frac{1}{\sqrt{\gamma_o}B(\frac{1}{2}, \frac{\gamma_o}{2})} \left(1+\frac{1}{\gamma_o}u^2\right)^{-\frac{1}{2}(\gamma_o+1)}, \qquad -\infty < u < \infty \qquad (2.3.4)
\end{aligned}$$

since $\Gamma\left(\frac{1}{2}\right) = \sqrt{\pi}$.

This distribution will be denoted by $M_t^{(1)}(0, 1, \gamma_o)$.

Clearly, $E(u) = 0$ and $Var(u) = \frac{\gamma_o}{\gamma_o - 2}$. Further, the odd central moments are zero while the even central moments are given by

$$\mu_{2r}(u) = \gamma_o^r \left\{\frac{1, 3, 5, \ldots, (2r-1)}{(\gamma_o - 2r), (\gamma_o - 2r + 2) \ldots, (\gamma_o - 2)}\right\}, \quad \gamma_o > 2r. \tag{2.3.5}$$

Now consider the distribution of Z to be $\mathcal{N}(\theta, t^{-1})$. The distribution of tZ^2 is then the *noncentral chi-square distribution* with one d.f. having the pdf

$$h_1\left(\chi^2(\Delta_t^2)\right) = \sum_{r\geq 0} \frac{1}{r!} e^{-\frac{\Delta_t^2}{2}} \left(\frac{\Delta_t^2}{2}\right)^r h_{1+2r}(\chi^2; 0), \quad \Delta_t^2 = t\theta^2, \tag{2.3.6}$$

where $h_{1+2r}(x; 0)$ is the pdf of a central chi-square variable with $(1+2r)$ d.f. The cdf of this random variable (r.v.) is given by

$$H_1\left(x; \Delta_t^2\right) = \sum_{r\geq 0} \frac{1}{r!} e^{-\frac{\Delta_t^2}{2}} \left(\frac{\Delta_t^2}{2}\right)^r H_{1+2r}(x; 0), \tag{2.3.7}$$

where $H_{1+2r}(x; 0)$ is the cdf of a central chi-square variable with $(1+2r)$ d.f.

Let $\varphi(.)$ be a measurable function, then

$$E_N\left[\varphi\left(\chi^2_\gamma(\Delta^2_t)\right)\right]=\sum_{r\geq 0}\frac{1}{r!}\,e^{-\frac{\Delta^2_t}{2}}\left(\frac{\Delta^2_t}{2}\right)^r E_N\left[\varphi\left(\chi^2_{\gamma+2r}(0)\right)\right] \tag{2.3.8}$$

based on (2.3.6), where E_N denotes getting expectation w.r.t. the normal theory, i.e., $\mathcal{N}(\theta,t^{-1})$.

In this regard, if t^{-1} follows the *inverse gamma distribution* (2.3.2), then the exact distribution of tZ^2 is given by the expectation of $h_1\left(\chi^2(\Delta^2_t)\right)$ or $H_1\left(x;\Delta^2_t\right)$ using (2.3.3). Thus, the *unconditional* distribution of tZ^2 is given by the pdf and cdf, respectively, as

$$h_1^{(2)}\left(\chi^2(\Delta^2)\right)=\sum_{r\geq 0}K_r^{(0)}(\Delta^2)h_{1+2r}(\chi^2;0)\,, \tag{2.3.9}$$

and

$$H_1^{(2)}\left(x;\Delta^2\right)=\sum_{r\geq 0}K_r^{(0)}(\Delta^2)H_{1+2r}(x;0)\,;\quad \Delta^2=\frac{(\gamma_o-2)\theta^2}{\gamma_o}, \tag{2.3.10}$$

where the mixing distribution of r is given by

$$K_r^{(0)}(\Delta^2)=\frac{\Gamma\left(\frac{\gamma_o}{2}+r\right)}{\Gamma(r+1)\Gamma\left(\frac{\gamma_o}{2}\right)}\frac{\left(\frac{\Delta^2}{\gamma_o-2}\right)^r}{\left(1+\frac{\Delta^2}{\gamma_o-2}\right)^{\frac{\gamma_o}{2}+r}} \tag{2.3.11}$$

for $\gamma_o\geq 3$. Let us define

$$E^{(2-h)}\left[\varphi\left(\chi^2_\gamma(\Delta^2)\right)\right]=E_t\left\{\left(t^{-1}\right)^h E_N\left[\varphi\left(\chi^2_\gamma(\Delta^2_t)\right)\right]\right\}. \tag{2.3.12}$$

Then we may write

$$E^{(2-h)}\left[\varphi\left(\chi^2_\gamma(\Delta^2)\right)\right]=\sum_{r\geq 0}K_r^{(h)}(\Delta^2)E_N\left[\varphi\left(\chi^2_{\gamma+2r}(0)\right)\right], \tag{2.3.13}$$

where

$$K_r^{(h)}(\Delta^2)=\left(\frac{\gamma_o}{2}\right)^h\frac{\Gamma\left(\frac{\gamma_o}{2}+r-h\right)}{\Gamma(r+1)\Gamma\left(\frac{\gamma_o}{2}\right)}\frac{\left(\frac{\Delta^2}{\gamma_o-2}\right)^r}{\left(1+\frac{\Delta^2}{\gamma_o-2}\right)^{\frac{\gamma_o}{2}+r-h}};\quad h=0,1. \tag{2.3.14}$$

Specifically, using the fact that $\Delta^2_t=\frac{\gamma_o}{\gamma_o-2}t\Delta^2$

$$\begin{aligned}K_r^{(h)}(\Delta^2) &= \int_0^\infty t^{-h}W_o(t)\,\frac{1}{r!}e^{-\frac{\Delta^2_t}{2}}\left(\frac{\Delta^2_t}{2}\right)^r dt\\ &= \frac{1}{\Gamma(r+1)\Gamma\left(\frac{\gamma_o}{2}\right)}\left(\frac{\gamma_o}{2}\right)^{\frac{\gamma_o}{2}}\left(\frac{\gamma_o}{\gamma_o-2}\frac{\Delta^2}{2}\right)^r\end{aligned}$$

$$\times \int_0^\infty t^{\frac{\gamma_o}{2}+r-h-1} e^{-\frac{\gamma_o t}{2}\left(1+\frac{\Delta^2}{\gamma_o-2}\right)} dt.$$

Make the transformation $u = \frac{\gamma_o t}{2}\left(1+\frac{\Delta^2}{\gamma_o-2}\right)$ with the Jacobian $J(t \to u) = 1/\frac{\gamma_o}{2}\left(1+\frac{\Delta^2}{\gamma_o-2}\right)$ to get

$$\begin{aligned} K_r^{(h)}(\Delta^2) &= \frac{1}{\Gamma(r+1)\Gamma\left(\frac{\gamma_o}{2}\right)}\left(\frac{\gamma_o}{2}\right)^h \frac{\left(\frac{\Delta^2}{\gamma_o-2}\right)^r}{\left(1+\frac{\Delta^2}{\gamma_o-2}\right)^{\frac{\gamma_o}{2}+r-h}} \\ &\quad \times \int_0^\infty u^{\frac{\gamma_o}{2}+r-h-1} e^{-u} du \\ &= \left(\frac{\gamma_o}{2}\right)^h \frac{\Gamma\left(\frac{\gamma_o}{2}+r-h\right)}{\Gamma(r+1)\Gamma\left(\frac{\gamma_o}{2}\right)} \frac{\left(\frac{\Delta^2}{\gamma_o-2}\right)^r}{\left(1+\frac{\Delta^2}{\gamma_o-2}\right)^{\frac{\gamma_o}{2}+r-h}}. \end{aligned}$$

Now, we have the following theorem similar to the equations (2.2.5)–(2.2.6).

Theorem 2.3.1. *If $Z \sim M_t^{(1)}(\theta, 1, \gamma_o)$ and $\varphi(Z^2)$ is a measurable function, then*

(i)

$$E\left[\varphi(Z^2)\right] = E^{(2)}\left[\varphi\left(\chi_1^2(\Delta^2)\right)\right]$$

(ii)

$$E\left[Z\varphi(Z^2)\right] = \theta E^{(2)}\left[\varphi\left(\chi_3^2(\Delta^2)\right)\right]$$

(iii)

$$E\left[Z^2\varphi(Z^2)\right] = E^{(1)}\left[\varphi\left(\chi_3^2(\Delta^2)\right)\right] + \theta^2 E^{(2)}\left[\varphi\left(\chi_5^2(\Delta^2)\right)\right]$$

Proof:

(i)

$$E\left[\varphi(Z^2)\right] = E_t E_N\left[\varphi(tZ^2)\right] = E^{(2)}\left[\varphi\left(\chi_1^2(\Delta^2)\right)\right] \qquad \textit{by } (2.3.13).$$

(ii)

$$\begin{aligned} E\left[Z\varphi(Z^2)\right] &= E_t\left\{\theta E_N\left[\varphi(tZ^2)\right]\right\} \\ &= \theta E_t\left\{E_N\left[\varphi\left(\chi_3^2(\Delta_t^2)\right)\right]\right\} \qquad \textit{by } (2.2.7) \\ &= \theta E^{(2)}\left[\varphi\left(\chi_3^2(\Delta^2)\right)\right] \qquad \textit{by } (2.3.13). \end{aligned}$$

(iii)

$$
\begin{aligned}
E\left[Z^2\varphi(Z^2)\right] &= E_t\left\{t^{-1}E_N\left[\varphi\left(\chi_3^2(\Delta_t^2)\right)\right]+\theta^2E_N\left[\varphi\left(\chi_5^2(\Delta^2)\right)\right]\right\} \\
&\qquad by\ (2.2.5) \\
&= E^{(1)}\left[\varphi\left(\chi_3^2(\Delta^2)\right)\right]+\theta^2E^{(2)}\left[\varphi\left(\chi_5^2(\Delta^2)\right)\right] \\
&\qquad by\ (2.2.5)
\end{aligned}
$$

The expressions inside $\{.\}$ *are obtained from (2.2.5)–(2.2.6).*

2.4 F-Distribution

Let $Z \sim \mathcal{N}(0,1)$ and mU be the central chi-square variable with m d.f. Then it is well known that $F = \frac{Z^2}{U}$ follows the central F-distribution with $(1,m)$ d.f. In other words, u^2 in the other section follows the central F-distribution with $(1,m)$ d.f.

Now, consider two independent chi-square variables $\chi^2_{\gamma_1}$ and $\chi^2_{\gamma_1}$ with γ_1 and γ_2 d.f.s respectively. Then, the ratio $F = \frac{\chi^2_{\gamma_1}/\gamma_1}{\chi^2_{\gamma_2}/\gamma_2}$ follows the central F-distribution with (γ_1,γ_2) d.f. The pdf of F is given by

$$
g_{\gamma_1,\gamma_2}(x) = \frac{\left(\frac{\gamma_1}{\gamma_2}\right)^{\frac{1}{2}\gamma_1}}{B\left(\frac{\gamma_1}{2},\frac{\gamma_2}{2}\right)}\frac{F^{\frac{1}{2}\gamma_1-1}}{\left(1+\frac{\gamma_1}{\gamma_2}F\right)^{\frac{1}{2}(\gamma_1+\gamma_2)}}; \qquad 0<F<\infty \tag{2.4.1}
$$

and the cdf of F is

$$
\begin{aligned}
G_{\gamma_1,\gamma_2}(x) &= \frac{\left(\frac{\gamma_1}{\gamma_2}\right)^{\frac{1}{2}\gamma_1}}{B\left(\frac{\gamma_1}{2},\frac{\gamma_2}{2}\right)}\int_0^x\frac{F^{\frac{1}{2}\gamma_1-1}\,dF}{\left(1+\frac{\gamma_1}{\gamma_2}F\right)^{\frac{1}{2}(\gamma_1+\gamma_2)}} \\
&= I_y\left(\frac{1}{2}\gamma_1,\frac{1}{2}\gamma_2\right), \qquad y=\frac{\gamma_1x}{\gamma_2+\gamma_1x},
\end{aligned} \tag{2.4.2}
$$

where $I_y(a,b)$ is the incomplete beta (Pearson's regularized incomplete beta) function.

Further, consider two independent chi-square variables, where one is a noncentral chi-square variable $\chi^2_{\gamma_1}(\Delta^2)$ with γ_1 d.f. and noncentral parameter Δ^2, and the other, $\chi^2_{\gamma_2}$, is a central chi-square variable with γ_2 d.f. Then, $F_{\gamma_1,\gamma_2}(\Delta^2) = \frac{\chi^2_{\gamma_1}(\Delta^2)/\gamma_1}{\chi^2_{\gamma_2}/\gamma_2}$ follows the noncentral F-distribution with (γ_1,γ_2) d.f. and noncentrality parameter Δ^2. The cdf of $F_{\gamma_1,\gamma_2}(\Delta^2)$ is given by

$$G_{\gamma_1,\gamma_2}(x;\Delta^2) = \sum_{r\geq 0} \frac{e^{\frac{\Delta^2}{2}}}{r!}\left(\frac{\Delta^2}{2}\right)^r I_y\left(\frac{1}{2}\gamma_1 + r; \frac{1}{2}\gamma_2\right), \quad y = \frac{\gamma_1 x}{\gamma_2 + \gamma_1 x} \tag{2.4.3}$$

with pdf

$$g_{\gamma_1,\gamma_2}(F(\Delta^2)) = \sum_{r\geq 0} \frac{e^{\frac{\Delta^2}{2}}}{r!}\left(\frac{\Delta^2}{2}\right)^r \frac{\left(\frac{\gamma_1}{\gamma_2}\right)^{\frac{1}{2}\gamma_1+r} F^{\frac{1}{2}\gamma_1-1}}{B\left(\frac{1}{2}\gamma_1+r,\frac{1}{2}\gamma_2\right)\left(1+\frac{\gamma_1}{\gamma_2}F\right)^{\frac{1}{2}(\gamma_1+\gamma_2)+r}}. \tag{2.4.4}$$

Let $\varphi(.)$ be measurable function of $F_{\gamma_1,\gamma_2}(\Delta^2)$, then

$$E\left[\varphi\left(F_{\gamma_1,\gamma_2}(\Delta^2)\right)\right] = \sum_{r\geq 0} \frac{e^{\frac{\Delta^2}{2}}}{r!}\left(\frac{\Delta^2}{2}\right)^r E_N\left[\varphi\left(\frac{\gamma_1+2r}{\gamma_1}F_{\gamma_1+2r,\gamma_2}(0)\right)\right]. \tag{2.4.5}$$

Now, if $Z \sim \mathcal{N}(\theta, t^{-1})$, then tZ^2 follows the noncentral chi-square distribution with one d.f. and noncentrality parameter $\Delta_t^2 = t\theta^2$. Further, let mU be a central chi-square with m d.f. Then, $\frac{tZ^2}{U}$ follows the noncentral F-distribution with $(1,m)$ d.f. and noncentrality parameter Δ_t^2. Thus, for a measurable function $\varphi(.)$, we have

$$E_N\left[\varphi\left(\frac{tZ^2}{U}\right)\right] = \sum_{r\geq 0} \frac{1}{r!}e^{\frac{\Delta_t^2}{2}}\left(\frac{\Delta_t^2}{2}\right)^r E_N\left[\varphi\left((1+2r)F_{1+2r,m}(0)\right)\right] \tag{2.4.6}$$

by (2.4.5).

Further, if $Z \sim M_t^{(1)}(\theta, 1, \gamma_o)$, then

$$\begin{aligned} E\left[\varphi\left(\frac{Z^2}{U}\right)\right] &= E^{(2)}\left[\varphi\left(F_{1,m}(\Delta^2)\right)\right] \\ &= \sum_{r\geq 0} K_r^{(0)}(\Delta^2) E_N\left[\varphi\left((1+2r)F_{1+2r,m}(0)\right)\right], \end{aligned} \tag{2.4.7}$$

where

$$E^{(2-h)}\left[\varphi\left(F_{1,m}(\Delta^2)\right)\right] = \sum_{r\geq 0} K_r^{(h)}(\Delta^2) E_N\left[\varphi\left((1+2r)F_{1+2r,m}(0)\right)\right], \; h = 0, 1. \tag{2.4.8}$$

See (2.3.13)–(2.3.14) for details with $\Delta^2 = \frac{(\gamma_o-2)\theta^2}{\gamma_o}$.

Thus, we have

(i)

$$E\left[Z\varphi\left(\frac{Z^2}{U}\right)\right] = \theta E^{(2)}\left[\varphi\left(3F_{3,m}(\Delta^2)\right)\right] \tag{2.4.9}$$

(ii)

$$E\left[Z^2\varphi\left(\frac{Z^2}{U}\right)\right] = E^{(1)}\left[\varphi\left(3F_{3,m}\left(\Delta^2\right)\right)\right] + \theta^2 E^{(2)}\left[\varphi\left(5F_{5,m}\left(\Delta^2\right)\right)\right], \tag{2.4.10}$$

similar to (2.3.15). Here, mU is a central chi-square variable with m d.f.

If $G_{q+i,m+j}\left(c;\Delta_t^2\right)$ denote the cdf of a noncentral F-distribution with $(q+i, m+j)$ d.f. with noncentrality parameter $\Delta_t^2 = t\theta^2$, then we write

$$E_t\left\{\left(t^{-1}\right)^h G_{q+i,m+j}\left(x;\Delta_t^2\right)\right\} = G_{q+i,m+j}^{(2-h)}\left(x;\Delta^2\right), \quad h = 0, 1, \tag{2.4.11}$$

where

$$G_{q+i,m+j}^{(2-h)}\left(l_c;\Delta^2\right) = \sum_{r\geq 0} K_r^{(h)}(\Delta^2) I_{l_c}\left[\frac{1}{2}(q+i)+r, \frac{1}{2}(m+j)\right], l_c = \frac{c(m+j)}{m(q+i)}. \tag{2.4.12}$$

2.5 Multivariate Normal Distribution

In the multivariate setup, multivariate normal distribution, $\mathcal{N}_p(\boldsymbol{\theta}, \boldsymbol{\Sigma})$, $\boldsymbol{\theta} \in \mathbb{R}^p$, and $\boldsymbol{\Sigma} \in \mathcal{S}(p)$ (space of all positive definite matrices of order p), is the basic distribution on which all statistical inference depends. The pdf of a $\mathcal{N}_p(\boldsymbol{\theta}, \boldsymbol{\Sigma})$ is given by

$$f\left(\boldsymbol{Y}|\boldsymbol{\theta}, \boldsymbol{\Sigma}\right) = \frac{1}{(2\pi)^{p/2}|\boldsymbol{\Sigma}|^{\frac{1}{2}}} e^{\left\{-\frac{1}{2}(\boldsymbol{Y}-\boldsymbol{\theta})'\boldsymbol{\Sigma}^{-1}(\boldsymbol{Y}-\boldsymbol{\theta})\right\}}, \qquad \boldsymbol{Y} \in \mathbb{R}^p. \tag{2.5.1}$$

Then,

$$E(\boldsymbol{Y}) = \boldsymbol{\theta} \quad \text{and} \quad E\left[(\boldsymbol{Y}-\boldsymbol{\theta})(\boldsymbol{Y}-\boldsymbol{\theta})'\right] = \boldsymbol{\Sigma}. \tag{2.5.2}$$

Further, the distribution of $\boldsymbol{Y}'\boldsymbol{\Sigma}^{-1}\boldsymbol{Y}$ follows the noncentral chi-square distribution with p d.f. and noncentrality parameter $\frac{\Delta^2}{2}$,where $\Delta^2 = \boldsymbol{\theta}'\boldsymbol{\Sigma}^{-1}\boldsymbol{\theta}$. The point estimator of $\boldsymbol{\theta}$ and $\boldsymbol{\Sigma}$ based on a sample $\boldsymbol{Y}_1, \ldots, \boldsymbol{Y}_N$ of size N are the sample mean vector $\bar{\boldsymbol{Y}}$ and $\boldsymbol{S}_u$;

$$\boldsymbol{S}_u = \frac{1}{N-1}\sum_{\alpha=1}^{N}\left(\boldsymbol{Y}_\alpha - \bar{\boldsymbol{Y}}\right)\left(\boldsymbol{Y}_\alpha - \bar{\boldsymbol{Y}}\right)'. \tag{2.5.3}$$

It is known that $\bar{\boldsymbol{Y}}$ and $\boldsymbol{S}_u$ are independent and

$$\bar{\boldsymbol{Y}} \sim \mathcal{N}_p(\boldsymbol{\theta}, \frac{1}{N}\boldsymbol{\Sigma}) \quad \text{and} \quad \boldsymbol{S}_u \sim W_p(\boldsymbol{\Sigma}, N-1), \tag{2.5.4}$$

where W_p stands for the Wishart distribution with $N-1$ d.f.

In order to test the null hypothesis $H_o : \boldsymbol{\theta} = \boldsymbol{\theta}_o$ vs $H_o : \boldsymbol{\theta} \neq \boldsymbol{\theta}_o$ when $\boldsymbol{\Sigma}$ is unknown, one uses Hotelling's T^2-statistic

$$\mathcal{L}_N = N(\bar{\boldsymbol{Y}} - \boldsymbol{\theta}_o)' \boldsymbol{S}_u^{-1} (\bar{\boldsymbol{Y}} - \boldsymbol{\theta}_o), \tag{2.5.5}$$

where $\left(\frac{\mathcal{L}_N}{N-1}\right)\left(\frac{N-p}{p}\right)$ is distributed as central F-distribution with (p, m) d.f., where $m = N - p$.

An improved estimator of $\boldsymbol{\theta}$ under a quadratic loss function,

$$L(\boldsymbol{\theta}^*; \boldsymbol{\theta}) = (\boldsymbol{\theta}^* - \boldsymbol{\theta})' Q (\boldsymbol{\theta}^* - \boldsymbol{\theta}),$$

may be obtained as

$$\hat{\boldsymbol{\theta}}_N = \tilde{\boldsymbol{\theta}}_N - k(\tilde{\boldsymbol{\theta}}_N - \boldsymbol{\theta}_o)\mathcal{L}_N \,, \quad k = \frac{(p-2)m}{p(m+2)}, \tag{2.5.6}$$

where $\tilde{\boldsymbol{\theta}}_N = \bar{\boldsymbol{Y}}$ and $m = N - p$, $n = N - 1$. See, for example, Saleh (2006) and James and Stein (1961).

Similarly, under the loss of the type

$$\boldsymbol{L}_1(\boldsymbol{\Sigma}, \boldsymbol{W}) = tr(\boldsymbol{W}\boldsymbol{\Sigma}^{-1} - \boldsymbol{I})^2 \tag{2.5.7}$$

or

$$\boldsymbol{L}_2(\boldsymbol{\Sigma}, \boldsymbol{W}) = tr(\boldsymbol{W}\boldsymbol{\Sigma}^{-1}) - log|\boldsymbol{W}\boldsymbol{\Sigma}^{-1}| - p \tag{2.5.8}$$

the improved estimator of $\boldsymbol{\Sigma}$ is of the form

$$a(N-1)\boldsymbol{S}_u, \qquad \text{where} \quad a = (N+p)^{-1} \tag{2.5.9}$$

may be obtained following James and Stein (1961).

In the sequel, we may need the generalized *Stein's identity*, namely, if $\boldsymbol{Y} = (Y_1, \ldots, Y_p)' \sim \mathcal{N}_p(\boldsymbol{\theta}, \sigma^2 \boldsymbol{I}_p)$ and if a function $\boldsymbol{\varphi}(\boldsymbol{Y}) = (\varphi_1(\boldsymbol{Y}), \ldots, \varphi_p(\boldsymbol{Y}))'$ is partially differentiable, that $\left|\frac{\partial \varphi_j(\boldsymbol{Y})}{\partial Y_j}\right| < \infty$, $j = 1, \ldots, p$ and $\varphi_j(.)$ are continuous functions of all vectors $\boldsymbol{Y}_{[j]} = (Y_1, \ldots, Y_{j-1}, Y_{j+1}, \ldots, Y_p)'$, then

$$E\left\{[\boldsymbol{\varphi}(\boldsymbol{Y})]'(\boldsymbol{Y} - \boldsymbol{\theta})\right\} = \sigma^2 E\left[\sum_{j=1}^{p} \frac{\partial \varphi_j(\boldsymbol{Y})}{\partial Y_j}\right] \quad \text{provided} \quad E\left\|\boldsymbol{\varphi}(\boldsymbol{Y})\right\|^2 < \infty. \tag{2.5.10}$$

2.6 Multivariate t-Distribution

In this section, we discuss the p-dim multivariate t-distribution denoted by $M_t^{(p)}(\boldsymbol{\eta}, \sigma^2 \mathbf{V}_p, \gamma_o)$, defined by

$$f(\boldsymbol{Y}|\mathbf{V}_p, \gamma_o) = \frac{\Gamma\left(\frac{\gamma_o + p}{2}\right)}{(\pi\gamma_o)^{\frac{p}{2}}\,\Gamma\left(\frac{\gamma_o}{2}\right)\sigma^p|\mathbf{V}_p|^{\frac{1}{2}}} \left\{1 + \frac{1}{\gamma_o\sigma^2}(\boldsymbol{Y} - \boldsymbol{\eta})'\mathbf{V}_p^{-1}(\boldsymbol{Y} - \boldsymbol{\eta})\right\}^{-\frac{1}{2}(\gamma_o + p)}, \boldsymbol{Y} \in \mathbb{R}^p \tag{2.6.1}$$

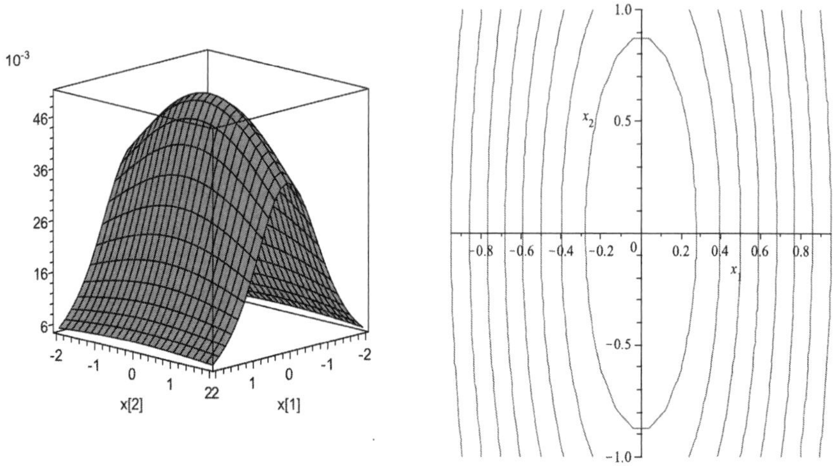

Figure 2.1 Bivariate t-distribution

with γ_o degrees of freedom (also referred to as the *shape* parameter), which preserves, diminishes, or increases the peakedness of the distribution as γ_o varies. Figure 2.1 shows the density and contour plots of bivariate t-distribution. This distribution and the multivariate normal one belong to a more general class of distributions, namely "elliptically contoured distributions" (ECDs), for which the kernel of any distribution in the class is a function of a quadratic form.

The mean of $\boldsymbol{Y}$ is $\boldsymbol{\eta}$ while the Cov (covariance) matrix is given by

$$E\left[(\boldsymbol{Y}-\boldsymbol{\eta})(\boldsymbol{Y}-\boldsymbol{\eta})'\right] = \frac{\gamma_o\sigma^2}{\gamma_o - 2}\mathbf{V}_p. \tag{2.6.2}$$

If $p = 1$, $\boldsymbol{\eta} = \mathbf{0}$, and $\mathbf{V}_p = 1$ (scalar), (2.5.1) reduces to the univariate Student's t-distribution. If $p = 2$, $\boldsymbol{\eta} = \mathbf{0}$, and $\mathbf{V}_2 = \boldsymbol{I}_2$, then (2.5.1) reduces to a slight modification of the bivariate surface of Pearson's (1923) distribution. If $\frac{1}{2}(\gamma_o + p) = m$, an integer (2.6.1) reduces to the p-variate Pearson's type VII distribution. Further, as $\gamma_o \to \infty$, (2.6.1) tends to the p-variate normal distribution, while $\gamma_o = 1$ gives the Cauchy distribution.

According to the weight representation of ECDs due to Chu (1973), the p-variate t-pdf at (2.6.1) may be written as

$$f\left(\mathbf{Y}|\mathbf{V}_p, \gamma_o\right) = \int W_o(t)\mathcal{N}_p(\boldsymbol{\eta}, t^{-1}\sigma^2\mathbf{V}_p)dt, \tag{2.6.3}$$

where $\mathcal{N}_p(\boldsymbol{\eta}, t^{-1}\sigma^2\mathbf{V}_p)$ stands for the pdf of a p-dim normal distribution with mean $\boldsymbol{\eta}$ and Cov matrix $t^{-1}\sigma^2\mathbf{V}_p$ and $W_o(t)$ is the inverse gamma distribution with

$$\begin{cases} W_o(t) = \frac{1}{\Gamma(\frac{\gamma_o}{2})}\left(\frac{\gamma_o t}{2}\right)^{\frac{\gamma_o}{2}} e^{-\frac{\gamma_o t}{2}} t^{-1}, \qquad 0 < t < \infty \\ \text{and} \\ \kappa^{(h)} = E\left(t^{-h}\right) = \int \left(\frac{1}{t}\right)^h W_o(t)dt = \left(\frac{\gamma_o}{2}\right)^h \left(\frac{\Gamma(\frac{\gamma_o}{2}-h)}{\Gamma(\frac{\gamma_o}{2})}\right). \end{cases} \tag{2.6.4}$$

The multivariate t-distribution, $M_t^{(p)}(\boldsymbol{\eta}, \sigma^2\mathbf{V}_p, \gamma_o)$ is a generalization of the univariate Student's t-distribution and the basis of many statistical inferences. Thus, it has the potential to be a broad statistical model for application. It is a promising distribution to broaden the multivariate analysis and offers a viable alternative to real-world data, particularly because its tails are more realistic. Recent applications of $M_t^{(p)}(\boldsymbol{\eta}, \sigma^2\mathbf{V}_p, \gamma_o)$ in novel areas such as cluster analysis, discriminant analysis, multiple regressions, robust projection indices, and missing data imputations are evidence of its influence in improving statistical analysis. Further, $M_t^{(p)}(\boldsymbol{\eta}, \sigma^2\mathbf{V}_p, \gamma_o)$ has played a crucial role in Bayesian analysis for several decades, as a prior distribution in various physical, engineering, and financial phenomenon closely associated with this distribution to generate meaningful posterior distributions.

In this section, we discuss many characteristic properties of this distribution for application in the chapters that follow. An extensive overview on multivariate t-distribution is provided in Kotz and Nadarajah (2004).

2.6.1 Expected Values of Functions of $M_t^{(p)}(\boldsymbol{\eta}, \sigma^2\mathbf{V}_p, \gamma_o)$-Variables

Let $\boldsymbol{Y}$ be distributed as $M_t^{(p)}(\boldsymbol{\eta}, \sigma^2\mathbf{V}_p, \gamma_o)$ and $\varphi(\boldsymbol{Y})$ be a measurable function. Then, it can be seen that

$$E\left[\varphi(\boldsymbol{Y})\right] = \int_0^\infty W_o(t) E_N\left[\varphi(\boldsymbol{Y})|t\right] dt, \tag{2.6.5}$$

where $W_o(t)$ is given by (2.6.4) and E_N meets our expectations with respect to $\mathcal{N}(\boldsymbol{\eta}, t^{-1}\sigma^2\mathbf{V}_p)$. Explicitly,

$$E\left[\varphi(\boldsymbol{Y})\right] = \frac{1}{\Gamma\left(\frac{\gamma_o}{2}\right)} \int_0^\infty \left(\frac{\gamma_o t}{2}\right)^{\frac{\gamma_o}{2}} e^{-\frac{\gamma_o t}{2}} t^{-1} E_N\left[\varphi(\boldsymbol{Y})|t\right] dt. \tag{2.6.6}$$

If $\varphi(\boldsymbol{Y}) = \boldsymbol{Y}$, then $E(\boldsymbol{Y}) = \boldsymbol{\eta} \int W_o(t)dt = \boldsymbol{\eta}$ since $W_o(t)$ is a pdf.

If $\varphi(\boldsymbol{Y}) = \boldsymbol{Y}\boldsymbol{Y}'$, then

$$\begin{aligned} E(\boldsymbol{Y}\boldsymbol{Y}') &= \int W_o(t) E_N[\boldsymbol{Y}\boldsymbol{Y}'|t]dt \\ &= \int W_o(t)(t^{-1}\sigma^2\mathbf{V}_p + \boldsymbol{\eta}\boldsymbol{\eta}')dt \\ &= \boldsymbol{\eta}\boldsymbol{\eta}' + \sigma^2\mathbf{V}_p \int t^{-1}W_o(t)dt \\ &= \boldsymbol{\eta}\boldsymbol{\eta}' + \sigma_\varepsilon^2\mathbf{V}_p, \qquad \sigma_\varepsilon^2 = \frac{\gamma_o\sigma^2}{\gamma_o - 2}. \end{aligned} \tag{2.6.7}$$

Thus, the covariance matrix of the $M_t^{(p)}(\boldsymbol{\eta}, \sigma^2\mathbf{V}_p, \gamma_o)$ variable is $\sigma_\varepsilon^2\mathbf{V}_p$.

If $\varphi_{\boldsymbol{u}}(\boldsymbol{Y}) = e^{i\boldsymbol{u}'\boldsymbol{Y}}$ for some vector $\boldsymbol{u} \neq \mathbf{0}$, then we obtain the *characteristic function* of $M_t^{(p)}(\boldsymbol{\eta}, \sigma^2\mathbf{V}_p, \gamma_o)$ as

$$\begin{aligned} E(\varphi_{\boldsymbol{u}}(\boldsymbol{Y})) &= E\left[e^{i\boldsymbol{u}'\boldsymbol{Y}}\right] = \int W_o(t)\left[e^{i\boldsymbol{u}'\boldsymbol{\eta} - \frac{t^{-1}\sigma^2}{2}(\boldsymbol{u}'\mathbf{V}_p\boldsymbol{u})}\right]dt \\ &= e^{i\boldsymbol{u}'\boldsymbol{\eta}} \int W_o(t) e^{-\frac{t^{-1}\sigma^2}{2}(\boldsymbol{u}'\mathbf{V}_p\boldsymbol{u})}dt \\ &= e^{i\boldsymbol{u}'\boldsymbol{\eta}} \frac{1}{\Gamma\left(\frac{\gamma_o}{2}\right)} \int e^{-\frac{\gamma_o t}{2}\left\{1 + \frac{\sigma^2}{\gamma_o t^2}(\boldsymbol{u}'\mathbf{V}_p\boldsymbol{u})\right\}} \left(\frac{\gamma_o t}{2}\right)^{\frac{\gamma_o}{2}} \frac{dt}{t} \\ &= \frac{e^{i\boldsymbol{u}'\boldsymbol{\eta}}}{2^{\frac{\gamma_o}{2}-1}\Gamma\left(\frac{\gamma_o}{2}\right)} \left\|\left(\gamma_o\sigma^2\mathbf{V}_p\right)^{\frac{1}{2}}\boldsymbol{u}\right\|^{\frac{\gamma_o}{2}} K_{\frac{\gamma_o}{2}}\left(\left\|\left(\gamma_o\sigma^2\mathbf{V}_p\right)^{\frac{1}{2}}\boldsymbol{u}\right\|^{\frac{1}{2}}\right), \end{aligned} \tag{2.6.8}$$

where $K_{\frac{\gamma_o}{2}}$ is the Macdonald function of order $\frac{\gamma_o}{2}$ and argument $\left\|\left(\gamma_o\sigma^2\mathbf{V}_p\right)^{\frac{1}{2}}\boldsymbol{u}\right\|$. See Kotz and Nadarajah (2004).

2.6.2 Sampling Distribution of Quadratic Forms

Now, consider the distribution of the quadratic form under $M_t^{(p)}(\boldsymbol{\eta}, \sigma^2\mathbf{V}_p, \gamma_o)$.

First, consider the distribution of $(\boldsymbol{Y}'(t^{-1}\mathbf{V}_p)^{-1}\boldsymbol{Y})$ when $\boldsymbol{Y} \sim \mathcal{N}_p(\boldsymbol{\eta}, t^{-1}\sigma^2\mathbf{V}_p)$. In this case, the distribution of $(\boldsymbol{Y}'(t^{-1}\sigma^2\mathbf{V}_p)^{-1}\boldsymbol{Y})$ is given by the pdf

$$h_p\left(\chi^2\left(\Delta_t^2\right)\right) = \sum_{r \geq 0} \frac{e^{-\frac{\Delta_t^2}{2}}}{r!}\left(\frac{\Delta_t^2}{2}\right)^r h_{p+2r}(\chi^2; 0), \quad \Delta_t^2 = \frac{\boldsymbol{\eta}'\mathbf{V}_p^{-1}\boldsymbol{\eta}}{\sigma^2 t^{-1}}, \tag{2.6.9}$$

where $h_{p+2r}(\chi^2; 0)$ is the pdf of a central chi-square distribution with $p + 2r$ d.f. Therefore, integrating (2.6.9) over t using the inverse gamma distribution given by

(2.6.4), we obtain the pdf under $M_t^{(p)}(\boldsymbol{\eta}, \sigma^2\mathbf{V}_p, \gamma_o)$ as

$$h_p^{(2)}\left(\chi^2\left(\Delta_t^2\right)\right) = \sum_{r\geq 0} K_r^{(0)}\left(\Delta^2\right) h_{p+2r}(\chi^2; 0); \quad \Delta^2 = \frac{(\boldsymbol{\eta}'\mathbf{V}_p^{-1}\boldsymbol{\eta})}{\sigma_\varepsilon^2}; \sigma_\varepsilon^2 = \frac{\gamma_o\sigma^2}{\gamma_o - 2}, \tag{2.6.10}$$

with

$$K_r^{(0)}\left(\Delta^2\right) = \frac{\Gamma\left(\frac{\gamma_o}{2} + r\right)}{\Gamma(r+1)\Gamma\left(\frac{\gamma_o}{2}\right)} \frac{\left(\frac{\Delta^2}{\gamma_o - 2}\right)^r}{\left(1 + \frac{\Delta^2}{\gamma_o - 2}\right)^{\frac{\gamma_o}{2} + r}}. \tag{2.6.11}$$

Consequently, the cdf of $(\mathbf{Y}'(t^{-1}\sigma^2\mathbf{V}_p)^{-1}\mathbf{Y})$ under $M_t^{(p)}(\boldsymbol{\eta}, \sigma^2\mathbf{V}_p, \gamma_o)$ is given by

$$H_p^{(2)}\left(x, \Delta^2\right) = \sum_{r\geq 0} K_r^{(0)}\left(\Delta^2\right) H_{p+2r}(x; 0), \tag{2.6.12}$$

where $H_{p+2r}(x; 0)$ is the cdf of the central chi-square variable with $p + 2r$ d.f.

Let $\varphi(.)$ be a measurable function of $\mathbf{Y}'(t^{-1}\sigma^2\mathbf{V}_p)^{-1}\mathbf{Y}$; then, under normal theory, we have

$$E_N\left[\varphi(\mathbf{Y}'(t^{-1}\sigma^2\mathbf{V}_p)^{-1}\mathbf{Y})\right] = \sum_{r\geq 0} \frac{e^{-\frac{\Delta_t^2}{2}}}{r!}\left(\frac{\Delta_t^2}{2}\right)^r E_N\left[\varphi(\chi_{p+2r}^2(0))\right] \tag{2.6.13}$$

based on (2.3.6). Then, using (2.3.12) we have

$$E^{(2-h)}\left[\varphi\left(\chi_p^2(\Delta^2)\right)\right] = E_t\left\{\left(t^{-1}\right)^h E_N\left[\varphi(\chi_p^2(\Delta_t^2))\right]\right\}, \quad h = 0, 1. \tag{2.6.14}$$

Then it follows that for $h = 0, 1$, we may write

$$E^{(2-h)}\left[\varphi(\chi_p^2\left(\Delta^2\right))\right] = \sum_{r\geq 0} K_r^{(h)}\left(\Delta^2\right) E_N\left[\varphi(\chi_{p+2r}^2(0))\right], \tag{2.6.15}$$

where

$$K_r^{(h)}\left(\Delta^2\right) = \left(\frac{\gamma_o}{2}\right)^h \frac{\Gamma\left(\frac{\gamma_o}{2} + r - h\right)}{\Gamma(r+1)\Gamma\left(\frac{\gamma_o}{2}\right)} \frac{\left(\frac{\Delta^2}{\gamma_o - 2}\right)^r}{\left(1 + \frac{\Delta^2}{\gamma_o - 2}\right)^{\frac{\gamma_o}{2} + r - h}}. \tag{2.6.16}$$

As a consequence, the next theorem follows.

Theorem 2.6.1. *If* $\mathbf{Y} \sim M_t^{(p)}(\boldsymbol{\eta}, \sigma^2\mathbf{V}_p, \gamma_o)$ *and* $\varphi(.)$ *is a measurable function, then*

(i)

$$E\left[\varphi(\mathbf{Y}'\mathbf{V}_p^{-1}\mathbf{Y})\right] = E^{(2)}\left[\varphi\left(\chi_p^2\left(\Delta^2\right)\right)\right] \tag{2.6.17}$$

(ii)

$$E\left[\boldsymbol{Y}\varphi(\boldsymbol{Y}'\mathbf{V}_p^{-1}\boldsymbol{Y})\right] = \boldsymbol{\eta}E^{(2)}\left[\varphi\left(\chi^2_{p+2}\left(\Delta^2\right)\right)\right]$$

(iii)

$$\begin{aligned} E\left[\boldsymbol{Y}\boldsymbol{Y}'\varphi(\boldsymbol{Y}'\mathbf{V}_p^{-1}\boldsymbol{Y})\right] &= \sigma^2\mathbf{V}_pE^{(1)}\left[\varphi\left(\chi^2_{p+2}\left(\Delta^2\right)\right)\right] \\ &\quad + (\boldsymbol{\eta}\boldsymbol{\eta}')E^{(2)}\left[\varphi\left(\chi^2_{p+4}\left(\Delta^2\right)\right)\right] \end{aligned}$$

(iv)

$$\begin{aligned} E\left[\boldsymbol{Y}'\boldsymbol{A}\boldsymbol{Y}\varphi(\boldsymbol{Y}'\mathbf{V}_p^{-1}\boldsymbol{Y})\right] &= \operatorname{tr}\left(\boldsymbol{A}\mathbf{V}_p\right)E^{(1)}\left[\varphi\left(\chi^2_{p+2}\left(\Delta^2\right)\right)\right] \\ &\quad +(\boldsymbol{\eta}'\boldsymbol{A}\boldsymbol{\eta})E^{(2)}\left[\varphi\left(\chi^2_{p+4}\left(\Delta^2\right)\right)\right] \end{aligned}$$

using equations (2.6.15) and (2.2.6).

As an example, if $\varphi(\boldsymbol{Y}'\mathbf{V}_p^{-1}\boldsymbol{Y}) = (\boldsymbol{Y}'\mathbf{V}_p^{-1}\boldsymbol{Y})^{-1}$, then

$$\begin{aligned} E\left[\boldsymbol{Y}(\boldsymbol{Y}'\mathbf{V}_p\boldsymbol{Y})^{-1}\right] &= \boldsymbol{\eta}E^{(2)}\left[\varphi\left(\chi^{-2}_{p+2}\left(\Delta^2\right)\right)\right] \\ &= \boldsymbol{\eta}\left\{\frac{2}{(\gamma_o-2)}\sum_{r\geq 0}\frac{\left(\frac{\Delta^2}{\gamma_o-2}\right)^r}{B\left(r+1,\frac{\gamma_o}{2}-1\right)}\frac{(p+2r)^{-1}}{\left(1+\frac{\Delta^2}{\gamma_o-2}\right)^{\frac{1}{2}(\gamma_o+2r)}}\right\}; \end{aligned} \tag{2.6.18}$$

Similarly, if $\varphi(\boldsymbol{Y}'\mathbf{V}_p\boldsymbol{Y}) = I(\boldsymbol{Y}'\mathbf{V}_p^{-1}\boldsymbol{Y} < c)$, then

(i)

$$E\left[\boldsymbol{Y}I(\boldsymbol{Y}'\mathbf{V}_p\boldsymbol{Y} < c)\right] = \boldsymbol{\eta}H^{(2)}_{p+2}(c;\Delta^2), \tag{2.6.19}$$

where $I(A)$ is the indicator function of the set A and

$$\begin{aligned} E\left[I\left(\chi^2_{p+2}\left(\Delta^2\right) < c\right)\right] &= H^{(2)}_{p+2}(c;\Delta^2) \\ &= \frac{2}{(\gamma_o-2)}\sum_{r\geq 0}\frac{\left(\frac{\Delta^2}{\gamma_o-2}\right)^r}{B\left(r+1,\frac{\gamma_o}{2}-1\right)}\frac{H_{p+2+2r}(c;0)}{\left(1+\frac{\Delta^2}{\gamma_o-2}\right)^{\frac{1}{2}(\gamma_o+2r)}}. \end{aligned} \tag{2.6.20}$$

(ii)

$$\begin{aligned} E\left[\boldsymbol{YY}'I(\boldsymbol{Y}'\mathbf{V}_p^{-1}\boldsymbol{Y})<c)\right] &= \mathbf{V}_pE^{(1)}[I(\chi^2_{p+2}(\Delta^2)<c)] \\ &\quad +\boldsymbol{\eta\eta}'E^{(2)}[I(\chi^2_{p+4}(\Delta^2)<c)] \\ &= \mathbf{V}_pH^{(1)}_{p+2}(c;\Delta^2)+\boldsymbol{\eta\eta}'H^{(2)}_{p+2}(c;\Delta^2) \end{aligned} \tag{2.6.21}$$

(iii)

$$E\left[(\boldsymbol{Y}'\mathbf{V}_p^{-1}\boldsymbol{Y})I(\boldsymbol{Y}'\mathbf{V}_p^{-1}\boldsymbol{Y}<c)\right]=pH^{(1)}_{p+2}(c;\Delta^2)+\Delta^2H^{(2)}_{p+2}(c;\Delta^2) \tag{2.6.22}$$

with

$$H^{(h)}_{\gamma}(c;\Delta^2)=\sum K_r^{(0)}H_{\gamma+2r}(c,0)=E_t^{(2-h)}\left\{\left(t^{-1}\right)^hH_\gamma(c,\Delta_t)\right\} \tag{2.6.23}$$

Furthermore, we may have to use the following results:

(iv)

$$E^{(2-h)}\left[\chi^{-2}_{p+q}\left(\Delta^2\right)\right]=\sum_{r\geq 0}K_r^{(h)}\left(\Delta^2\right)(p+q-2+2r)^{-1} \tag{2.6.24}$$

and

(v)

$$E^{(2-h)}\left[\chi^{-4}_{p+q}\left(\Delta^2\right)\right]=\sum_{r\geq 0}K_r^{(h)}\left(\Delta^2\right)(p+q-2+2r)^{-1}(p+q-4+2r)^{-1} \tag{2.6.25}$$

for $h=0,1$ in the sequel.

Next, we consider the sampling distribution of the ratio of two independent quadratic forms.

Let $\boldsymbol{Y}'(t^{-1}\sigma^2\mathbf{V}_p)^{-1}\boldsymbol{Y}$ be the quadratic form as before where $\boldsymbol{Y}\sim\mathcal{N}_p(\boldsymbol{\eta},t^{-1}\sigma^2\mathbf{V}_p)$ and mU be a central chi-square variable with m d.f. Then, $F=\frac{\mathbf{Y}'(t^{-1}\sigma^2\mathbf{V}_p)^{-1}\mathbf{Y}}{pU}$ follows the noncentral F-distribution with (p,m) d.f. and noncentrality parameter $\Delta_t^2=\frac{\boldsymbol{\eta}'\mathbf{V}_p^{-1}\boldsymbol{\eta}}{\sigma^2t^{-1}}$ given by

$$g_{p,m}\left(F_{p,m}\left(\Delta_t^2\right)\right)=\sum_{r\geq 0}\frac{e^{-\frac{\Delta_t^2}{2}}}{r!}\left(\frac{\Delta_t^2}{2}\right)^r\frac{\left(\frac{p}{m}\right)^{\frac{p}{2}}F^{\frac{1}{2}p+r-1}}{B\left(\frac{1}{2}p+r;\frac{1}{2}m\right)\left(1+\frac{p}{m}F\right)^{\frac{1}{2}[(p+m)+2r]}} \tag{2.6.26}$$

with the cdf

$$G_{p,m}\left(x;\Delta_t^2\right)=\sum_{r\geq 0}\frac{e^{-\frac{\Delta_t^2}{2}}}{r!}\left(\frac{\Delta_t^2}{2}\right)^r\frac{I_y\left(\frac{1}{2}(p+2r);\frac{m}{2}\right)}{B\left(r+1,\frac{m}{2}\right)} \tag{2.6.27}$$

with $y = \frac{px}{m+px}$. Then, integrating over t using $IG\left(\gamma_o, t^{-1}\right)$-distribution, we obtain the cdf of F as $G^{(2)}_{p,m}\left(c, \Delta^2\right)$, where

$$G^{(2-h)}_{p,m}\left(c, \Delta^2\right) = \sum_{r\geq 0} K^{(h)}_r\left(\Delta^2\right) I_y\left(\frac{1}{2}(p+2r); \frac{m}{2}\right), \quad h = 0, 1 \tag{2.6.28}$$

having the pdf

$$g^{(2)}_{p,m}\left(F, \Delta^2\right) = \sum_{r\geq 0} K^{(0)}_r\left(\Delta^2\right) \frac{\left(\frac{p}{m}\right)^{\frac{p}{2}+r} F^{\frac{1}{2}(p+2r)-1}}{B\left(\frac{1}{2}(p+2r); \frac{m}{2}\right)\left(1+\frac{p}{m}F\right)^{\frac{1}{2}(p+m)+r}}. \tag{2.6.29}$$

If $\varphi(.)$ is a measurable function of $F_{p,m}\left(\Delta^2_t\right)$, then

$$E\left[\varphi\left(F_{p,m}\left(\Delta^2_t\right)\right)\right] = \sum_{r\geq 0} \frac{e^{-\frac{\Delta^2_t}{2}}}{r!}\left(\frac{\Delta^2_t}{2}\right)^r E_N\left[\varphi\left(F_{p+2r,m}(0)\right)\right]. \tag{2.6.30}$$

Thus, we have the following theorem:

Theorem 2.6.2. *If* $\mathbf{Y} \sim M^{(p)}_t(\boldsymbol{\eta}, \sigma^2\mathbf{V}_p, \gamma_o)$ *and* $\varphi(.)$ *is a measurable function, then*

(i)

$$E\left[\varphi\left(F_{p,m}\left(\Delta^2\right)\right)\right] = E^{(2)}\left[\varphi\left(F_{p,m}\left(\Delta^2\right)\right)\right], \tag{2.6.31}$$

where

$$\begin{aligned} E^{(2-h)}\left[\varphi\left(F_{p,m}\left(\Delta^2\right)\right)\right] &= E_t\left\{\left(t^{-1}\right) E_N\left[\varphi\left(F_{p,m}(\Delta^2_t)\right)\right]\right\} \\ &= \sum_{r\geq 0} K^{(h)}_r\left(\Delta^2\right) E_N\left[\varphi\left(F_{p+2r,m}(0)\right)\right] \end{aligned}$$

(ii)

$$E\left[\mathbf{Y}\varphi\left(F_{p,m}\left(\Delta^2\right)\right)\right] = \boldsymbol{\eta} E^{(1)}\left[\varphi\left(\frac{p+2r}{p} F_{p+2r,m}\left(\Delta^2\right)\right)\right] \tag{2.6.32}$$

(iii)

$$\begin{aligned} E\left[\mathbf{Y}\mathbf{Y}'\varphi\left(F_{p,m}\left(\Delta^2\right)\right)\right] &= \sigma^2\mathbf{V}_p E^{(1)}\left[\varphi\left(\frac{p+2+2r}{p} F_{p+2+2r,m}\left(\Delta^2\right)\right)\right] \\ &\quad + (\boldsymbol{\eta}\boldsymbol{\eta}')E^{(2)}\left[\varphi\left(\frac{p+4+2r}{p} F_{p+4+2r,m}\left(\Delta^2\right)\right)\right] \end{aligned}$$

(iv)

$$E\left[\boldsymbol{Y}'\boldsymbol{A}\boldsymbol{Y}\varphi\left(F_{p,m}\left(\Delta^2\right)\right)\right] = \sigma^2 \operatorname{tr}\left(\boldsymbol{A}\right) E^{(1)}\left[\varphi\left(\frac{p+2+2r}{p}F_{p+2+2r,m}\left(\Delta^2\right)\right)\right] + (\boldsymbol{\eta}\boldsymbol{A}\boldsymbol{\eta}')E^{(2)}\left[\varphi\left(\frac{p+4+2r}{p}F_{p+4+2r,m}\left(\Delta^2\right)\right)\right] \quad (2.6.33)$$

with $\Delta = \left(\boldsymbol{\eta}'\mathbf{V}_p^{-1}\boldsymbol{\eta}\right)\sigma_{\varepsilon}^{-2}$.

Example 2.6.1. *If* $\varphi\left(F_{p,m}\left(\Delta^2\right)\right) = I\left(F_{p,m}\left(\Delta^2\right) < c\right)$, *then*

(i)

$$E\left[\boldsymbol{Y} I\left(F_{p,m}\left(\Delta^2\right) < c\right)\right] = \boldsymbol{\eta} G^{(2)}_{p+2,m}\left(c, \Delta^2\right), \quad (2.6.34)$$

where $G^{(2)}_{p+2,m}\left(c,\Delta^2\right)$ *is given by*

$$\frac{2}{(\gamma_o-2)}\sum_{r\geq 0}\frac{\left(\frac{\Delta^2}{\gamma_o-2}\right)^r}{B\left(r+1,\frac{\gamma_o}{2}-1\right)}\frac{I_y\left(\frac{1}{2}\left(p+2r\right);\frac{m}{2}\right)}{\left(1+\frac{\Delta^2}{\gamma_o-2}\right)^{\frac{1}{2}(\gamma_o+2r)}}\ ; \quad y=\frac{pc}{m+pc}. \quad (2.6.35)$$

(ii)

$$\begin{aligned} E\left[\boldsymbol{Y}\boldsymbol{Y}' I\left(F_{p,m}\left(\Delta^2\right) < c\right)\right] &= \sigma^2\mathbf{V}_p E^{(1)}\left[I\left(F_{p+2,m}\left(\Delta_t^2\right) < \frac{p}{p+2}c\right)\right] \\ &\quad + (\boldsymbol{\eta}\boldsymbol{\eta}')E^{(2)}\left[I\left(F_{p+4,m}\left(\Delta_t^2\right) < \frac{p}{p+4}c\right)\right] \\ &= \sigma_{\varepsilon}^2\mathbf{V}_p G^{(1)}_{p+2,m}\left(c,\Delta^2\right) + \boldsymbol{\eta}\boldsymbol{\eta}' G^{(2)}_{p+4,m}\left(c,\Delta^2\right). \end{aligned} \quad (2.6.36)$$

where for $h = 0, 1$ *we have*

$$\begin{aligned} G^{(2-h)}_{p+s,m}\left(c,\Delta^2\right) &= E^{(2-h)}\left[I\left(F_{p+s,m}\left(\Delta_t^2\right) < c\right)\right] \\ &= \sum_{r\geq 0} K_r^{(h)}\left(\Delta^2\right) I_y\left(\frac{1}{2}(p+2r);\frac{m}{2}\right) \end{aligned} \quad (2.6.37)$$

with $y = \frac{pc}{m+pc}$ *and* $K_r^{(h)}\left(\Delta^2\right)$ *given by* (2.6.16).

(iii)

$$\begin{aligned} E\left[(\boldsymbol{Y}'\mathbf{V}_p^{-1}\boldsymbol{Y})I\left(\frac{\boldsymbol{Y}'\mathbf{V}_p^{-1}\boldsymbol{Y}}{\sigma_\varepsilon^2 pU}<c\right)\right] &= p\,E^{(1)}\left[I(F_{p,m}(\Delta^2)<c)\right] \\ &\quad +\Delta^2 E^{(1)}\left[I(F_{p,m}(\Delta^2)<c)\right] \\ &= p\,G^{(1)}_{p+2,m}\left(\frac{pc}{p+2},\Delta^2\right) \\ &\quad +\Delta^2 G^{(2)}_{p+4,m}\left(\frac{pc}{p+2},\Delta^2\right) \end{aligned} \tag{2.6.38}$$

using (2.6.39).

Example 2.6.2. *If* $\varphi\left(F_{p+s,m}\left(\Delta^2\right)\right)=F^{-1}_{p+s,m}\left(\Delta^2\right)I\left(F_{p+s,m}\left(\Delta^2\right)<c\right)$*, then*

$$\begin{aligned} & E^{(2-h)}\left[F^{-1}_{p+s,m}\left(\Delta^2\right)I\left(F_{p+s,m}\left(\Delta^2\right)<c\right)\right] \\ = & \textstyle\sum_{r\geq 0} K_r^{(h)}\left(\Delta^2\right)m(p+s)(p+s-2+2r)^{-1}I_y\left(\frac{1}{2}(p+s+2r);\frac{1}{2}(m+2)\right) \end{aligned} \tag{2.6.39}$$

with $y=\frac{(p+s)c}{m+(p+s)c}$.

2.6.3 Distribution of Linear Functions of t-Variables

Let $\boldsymbol{Y}$ follow the distribution $M_t^{(p)}(\boldsymbol{\eta},\sigma^2\mathbf{V}_p,\gamma_o)$ and $\mathbf{Z}=\boldsymbol{H}\boldsymbol{Y}$, where $\boldsymbol{H}$ is a $q\times p$ matrix of linear formulations. What would be the distribution of $\mathbf{Z}$? We use the method of characteristic function to build the distribution of $\mathbf{Z}$.

The characteristic function of the distribution of $\boldsymbol{Y}$ is given by

$$\begin{aligned} \phi_{\mathbf{Z}}(\boldsymbol{u}) &= E\left[e^{i\boldsymbol{u}'\mathbf{Z}}\right]=E\left[e^{i\boldsymbol{u}'\boldsymbol{H}\boldsymbol{Y}}\right], \quad \boldsymbol{u} \text{ is a } q\text{-vector} \\ &= e^{i\boldsymbol{u}'\boldsymbol{H}\boldsymbol{\eta}}\int W_o(t)\,e^{-\frac{\sigma^2}{2t}\boldsymbol{u}'\boldsymbol{H}'\mathbf{V}_p\boldsymbol{H}\boldsymbol{u}}dt \end{aligned} \tag{2.6.40}$$

so that we have the following theorem.

Theorem 2.6.3. *The pdf of* $\mathbf{Z}$ *is given by*

$$\wp(\mathbf{Z}) = \Im^{-1}\left[e^{i\boldsymbol{u}'\boldsymbol{H}\boldsymbol{\eta}}\int W_o(t)N_q\left\{\boldsymbol{H}\boldsymbol{\eta},t^{-1}\sigma^2(\boldsymbol{H}\mathbf{V}_p\boldsymbol{H}')\right\}dt\right],$$

where $\Im^{-1}$ is the inverse Fourier operator. Then

$$
\begin{aligned}
\wp(\mathbf{Z}) &= \int W_o(t)\Im^{-1}\left\{e^{i\boldsymbol{u}'\boldsymbol{H\eta}-\frac{\sigma^2}{2t}\left[\boldsymbol{u}'(\boldsymbol{H}\mathbf{V}_p\boldsymbol{H}')\boldsymbol{u}\right]}\right\}dt \\
&= \int W_o(t)N_q\left\{\boldsymbol{H\eta}, t^{-1}\sigma^2(\boldsymbol{H}\mathbf{V}_p\boldsymbol{H}')\right\}dt \\
&= \frac{\Gamma\left(\frac{\gamma_o+q}{2}\right)}{(\pi\gamma_o)^{\frac{q}{2}}\Gamma\left(\frac{\gamma_o}{2}\right)\sigma^q\left|\boldsymbol{H}\mathbf{V}_p\boldsymbol{H}'\right|^{\frac{1}{2}}} \\
&\quad\times\left\{1+\frac{1}{\gamma_o\sigma^2}\left[(\mathbf{Z}-\boldsymbol{H\eta})'(\boldsymbol{H}\mathbf{V}_p\boldsymbol{H}')^{-1}(\mathbf{Z}-\boldsymbol{H\eta})\right]\right\}^{-\frac{1}{2}(\gamma_o+q)}.
\end{aligned}
\tag{2.6.41}
$$

Hence, $\mathbf{Z}$ *follows* $M_t^{(q)}(\boldsymbol{H\eta}, \sigma^2\left(\boldsymbol{H}\mathbf{V}_p\boldsymbol{H}'\right), \gamma_o)$*-distribution.*

Corollary 2.6.3.1.

$$E\left(\mathbf{ZZ}'\right) = \frac{\gamma_o\sigma^2}{\gamma_o-2}(\boldsymbol{H}\mathbf{V}_p\boldsymbol{H}')$$

Corollary 2.6.3.2. *If* $\boldsymbol{Y}$ *is partitioned as* $(\boldsymbol{Y}_1', \boldsymbol{Y}_2')'$, $\boldsymbol{\eta} = (\boldsymbol{\eta}_1', \boldsymbol{\eta}_2')'$ *and the covariance matrix* $\mathbf{V}_p$ *is partitioned as*

$$\mathbf{V}_p = \begin{pmatrix} \mathbf{V}_{11} & \mathbf{V}_{12} \\ \mathbf{V}_{21} & \mathbf{V}_{22} \end{pmatrix},$$

then,

$$
\begin{aligned}
\wp(\boldsymbol{Y}_1) &= \frac{\Gamma\left(\frac{\gamma_o+p_1}{2}\right)}{(\pi\gamma_o)^{\frac{p_1}{2}}\Gamma\left(\frac{\gamma_o}{2}\right)\sigma^{p_1}\left|\mathbf{V}_{11}\right|^{\frac{1}{2}}} \\
&\quad\times\left\{1+\frac{1}{\gamma_o\sigma^2}\left[(\boldsymbol{Y}_1-\boldsymbol{\eta}_1)'\mathbf{V}_{11}^{-1}(\boldsymbol{Y}_1-\boldsymbol{\eta}_1)\right]\right\}^{-\frac{1}{2}(\gamma_o+p_1)},
\end{aligned}
\tag{2.6.42}
$$

where p_1 *is dimension of* $\boldsymbol{Y}_1$, $E(\boldsymbol{Y}_1) = \boldsymbol{\eta}_1$ *and* $Cov(\boldsymbol{Y}_1) = \frac{\gamma_o\sigma^2}{\gamma_o-2}\mathbf{V}_{11}$.

Corollary 2.6.3.3.

$$E\left(\boldsymbol{Y}_1|\boldsymbol{Y}_2\right) = \boldsymbol{\eta}_1 + \mathbf{V}_{12}\mathbf{V}_{22}^{-1}(\boldsymbol{Y}_2-\boldsymbol{\eta}_2) = \boldsymbol{\eta}_{1.2}$$

$$Cov\,(\boldsymbol{Y}_1|\boldsymbol{Y}_2) = \frac{\gamma_o\sigma^2}{(\gamma_o - 2)}\mathbf{V}_{11.2}\,, \quad \mathbf{V}_{11.2} = \mathbf{V}_{11} - \mathbf{V}_{12}\mathbf{V}_{22}^{-1}\mathbf{V}_{21} \tag{2.6.43}$$

and

$$\begin{aligned}\wp(\boldsymbol{Y}_1|\boldsymbol{Y}_2) &= \frac{\Gamma\left(\frac{\gamma_o+p_1}{2}\right)}{(\pi\gamma_o)^{\frac{p_1}{2}}\,\Gamma\left(\frac{\gamma_o}{2}\right)\sigma^{p_1}\,|\mathbf{V}_{11.2}|^{\frac{1}{2}}} \\ &\times \left\{1 + \frac{1}{\gamma_o\sigma^2}\left[(\boldsymbol{Y}_1 - \boldsymbol{\eta}_{1.2})'\mathbf{V}_{11.2}^{-1}(\boldsymbol{Y}_1 - \boldsymbol{\eta}_{1.2})\right]\right\}^{-\frac{1}{2}(\gamma_o+p_1)}.\end{aligned}$$

2.7 Problems

1. Show that

$$|\alpha\boldsymbol{A} + (1-\alpha)\boldsymbol{B}| \geq |\boldsymbol{A}|^{\alpha}|\boldsymbol{B}|^{1-\alpha},$$

where $\boldsymbol{A}_n$ and $\boldsymbol{B}_n$ are p.d. (positive definite) matrices and $0 < \alpha < 1$.

2. If $\boldsymbol{A}$ is a symmetric matrix and $\boldsymbol{B} = \boldsymbol{C}^{-1}\boldsymbol{A}\boldsymbol{C}$ for some nonsingular matrix $\boldsymbol{C}$, then $\boldsymbol{A}$ and $\boldsymbol{B}$ have the same eigenvalues with the same multiplicities. Further, if $\boldsymbol{x}$ is an eigenvector of $\boldsymbol{A}$ for eigenvalue λ, then $\boldsymbol{C}^{-1}\boldsymbol{x}$ is an eigenvector of $\boldsymbol{B}$.

3. Let $\boldsymbol{A}$ be symmetric matrix and $\boldsymbol{B}$ be a p.d. matrix. Show that $\max_{\boldsymbol{x}} \frac{\boldsymbol{x}'\boldsymbol{A}\boldsymbol{x}}{\boldsymbol{x}'\boldsymbol{B}\boldsymbol{x}}$ is given by the largest eigenvalue of $\boldsymbol{B}^{-1}\boldsymbol{A}$.
Hint: $\max_{\boldsymbol{x}} \frac{\boldsymbol{x}'\boldsymbol{A}\boldsymbol{x}}{\boldsymbol{x}'\boldsymbol{B}\boldsymbol{x}} = \max_{\boldsymbol{x}} \boldsymbol{x}'\boldsymbol{A}\boldsymbol{x}$ subject to $(\boldsymbol{x}'\boldsymbol{B}\boldsymbol{x} = 1)$.

4. Show that

(i) $|\boldsymbol{S}_{\boldsymbol{a}}|$ is minimized when $\boldsymbol{a} = \bar{\boldsymbol{x}}$, and

(ii) $tr(\boldsymbol{S}_{\boldsymbol{a}})$ is minimized when $\boldsymbol{a} = \bar{\boldsymbol{x}}$,

where $\bar{\boldsymbol{x}} = \frac{1}{n}\sum_{i=1}^{n}\boldsymbol{x}_i$ and $\boldsymbol{S}_{\boldsymbol{a}} = \frac{1}{n}\sum_{i=1}^{n}(\boldsymbol{x}_i - \boldsymbol{a})(\boldsymbol{x}_i - \boldsymbol{a})'$.

5. Suppose $X \sim \chi_p^2(\Delta^2)$ and $Y \sim \chi_k^2$. Show that for some constant c

(i) $Pr(X > c) \geq Pr(Y > c)$.

(ii) $Pr(X > c)$ is an increasing function.

6. Let $X_i \sim \chi_{p_i}^2(\Delta_i^2)$, $i = 1, \cdots, k$, and suppose that X_is are all independent. Then, show that

$$X = \sum_{i=1}^{k} X_i \sim \chi_p^2\left(\sum_{i=1}^{k}\Delta_i^2\right), \qquad p = \sum_{i=1}^{k} p_i.$$

7. Prove the identity (2.2.4).

8. Let $F \sim F_{\gamma_1,\gamma_2}(\Delta^2)$, where γ_1 is even. Show that

$$Pr\left(F < \frac{X_1}{\gamma_1}\right) = Pr\left(X_1 - X_2 \geq \frac{\gamma_1}{2}\right),$$

where X_1 and X_2 are independent, with X_1 having a Poisson distribution with mean equal to $\frac{\Delta^2}{2}$ and X_2 having a negative binomial distribution.

9. Prove that if $\boldsymbol{y} \sim \mathcal{N}_p(\boldsymbol{\mu}, \boldsymbol{I}_p)$, then

(i) $E\left[\frac{1}{\boldsymbol{y}'\boldsymbol{y}}\right] = E[\frac{1}{p-2+2K}]$,

(ii) $E\left[\frac{(\boldsymbol{y}-\boldsymbol{\mu})'\boldsymbol{y}}{\boldsymbol{y}'\boldsymbol{y}}\right] = (p-2)E\left[\frac{1}{p-2+2K}\right]$,

where K is a random variable having a Poisson distribution with mean equal to $\frac{\boldsymbol{\mu}'\boldsymbol{\mu}}{2}$.

10. Show that for any vector $\boldsymbol{a}$ and matrix $\boldsymbol{\Sigma} > 0$,

$$(2\pi)^{\frac{p}{2}}|\boldsymbol{\Sigma}|^{\frac{1}{2}}e^{\frac{1}{2}\,\boldsymbol{a}'\boldsymbol{\Sigma}\boldsymbol{a}} = \int_{-\infty}^{\infty}\cdots\int_{-\infty}^{\infty}\exp\left[-\frac{1}{2}\,\boldsymbol{x}'\boldsymbol{\Sigma}^{-1}\boldsymbol{x} + \boldsymbol{a}'\boldsymbol{x}\right]\,d\boldsymbol{x},$$

where $\boldsymbol{x} = (x_1, \cdots, x_p)$.

11. Let $\boldsymbol{X} \sim \mathcal{N}_p(\boldsymbol{\mu}, \Sigma)$. Show that the characteristic function of $\boldsymbol{X}'\boldsymbol{A}\boldsymbol{X}$ has the following form:

$$|\boldsymbol{I}_p - 2it\boldsymbol{\Sigma}\boldsymbol{A}|^{-\frac{1}{2}}\exp\left[it\boldsymbol{\mu}'\boldsymbol{A}(\boldsymbol{I}_p - 2it\boldsymbol{A})^{-1}\boldsymbol{\mu}\right].$$

12. Derive the test statistic given by (2.5.5) using normal theory and verify its distribution under the null hypothesis.

13. Show that the estimator given by (2.5.6) outperforms $\tilde{\boldsymbol{\theta}}_N$.

14. Show that the estimator given by (2.5.9) outperforms $\boldsymbol{S}_u$ w.r.t. the losses given by (2.5.7) and (2.5.8).

15. Consider the estimator $\boldsymbol{S}_u$ given by (2.5.3). Prove that

(i) If $p > 1$ and $N > p + 2$, then the estimator

$$\hat{\boldsymbol{\Sigma}}_1(\alpha) = \frac{\alpha}{N-1}\,\boldsymbol{S}_u + \frac{(1-\alpha)p}{(N-1)tr(\boldsymbol{S}_u^{-1})}\,\boldsymbol{I}_p$$

has smaller risk than $\boldsymbol{S}_u$ with respect to the entropy loss given by (2.5.8), provided

$$\frac{-(N-1)p + \left[(N-1)^2p^2 + 8(N-1)p(p-1)\right]^{\frac{1}{2}}}{4(p-1)} \leq \alpha < 1.$$

(ii) If $p > 3$ and $N - 1 > p + 3 + 12/(p - 3)$, then the estimator

$$\hat{\boldsymbol{\Sigma}}_2(\alpha) = \frac{\alpha}{N+p}\,\boldsymbol{S}_u + \frac{(1-\alpha)p}{(N+p)tr(\boldsymbol{S}_u^{-1})}\,\boldsymbol{I}_p$$

has smaller risk than $\boldsymbol{S}_u$ with respect to the quadratic loss given by (2.5.7), provided $\frac{b_1}{a_1} < \alpha < 1$ where

$$\begin{aligned} a_1 &= \frac{[p(N-p+2) - 2(N+1)](N-p)}{N+p} + N - 1, \\ b_1 &= \frac{p(N-p+2)(N-p)}{N+p} - N + 2p + 1. \end{aligned}$$

16. Prove that if given $\boldsymbol{\Sigma}$, $\boldsymbol{Y}$ is distributed according to $\boldsymbol{Y} \sim \mathcal{N}_p(\boldsymbol{\theta}, \gamma_o\boldsymbol{\Sigma})$ and $\boldsymbol{\Sigma} \sim$ inverse-Wishart $(\Psi^{-1}, \gamma_o + p - 1)$, then $\boldsymbol{Y} \sim M_t^{(p)}(\boldsymbol{\theta}, \Psi, \gamma_o)$.

17. Verify equations given by (2.6.9) and (2.6.10).

18. Prove Theorems 2.6.1 and 2.6.2.

19. Let $\boldsymbol{Y} \sim M_t^{(p)}(\boldsymbol{\eta}, \sigma^2\mathbf{V}_p, \gamma_o)$. Prove Corollaries 2.6.3.2 and 2.6.3.3 and show that the kurtosis parameter is equal to $\kappa = \frac{2}{\gamma_o - 4}$.

20. Let $\boldsymbol{x} \sim M_t^{(p)}(\mathbf{0}, \mathbf{V}_p, \gamma_o)$, where $\mathbf{V}_p$ is diagonal. Prove that if $x_1, \cdots, x_p$ are all independent, then $\boldsymbol{x}$ is normal.

21. Show that if $\boldsymbol{X}$ has the density $g(\boldsymbol{x}'\boldsymbol{x})$, then $\boldsymbol{Y} = \boldsymbol{\Gamma}\boldsymbol{X}$, where $\boldsymbol{\Gamma}'\boldsymbol{\Gamma} = \boldsymbol{I}$ has the density $g(\boldsymbol{y}'\boldsymbol{y})$.

22. Suppose $\boldsymbol{X} = (X_1, \cdots, X_k)'$ has *pdf*

$$f(\boldsymbol{x}) = \frac{1}{2\pi}\,e^{-(\boldsymbol{x}'\boldsymbol{x})^{\frac{1}{2}}}, \qquad \boldsymbol{x} \in \mathbb{R}^k.$$

Let $\boldsymbol{A}_k$ denote a nonsingular matrix, and $\boldsymbol{m} \in \mathbb{R}^k$ is a fixed vector. Show that the *pdf* of $\boldsymbol{Y} = \boldsymbol{m} + \boldsymbol{A}\boldsymbol{X}$ is given by

$$f(\boldsymbol{y}) = \frac{1}{2\pi\,|\boldsymbol{A}\boldsymbol{A}'|^{\frac{1}{2}}}\,e^{-(\boldsymbol{y}-\boldsymbol{m})(\boldsymbol{A}\boldsymbol{A}')^{-1}(\boldsymbol{y}-\boldsymbol{m})}, \qquad \boldsymbol{y} \in \mathbb{R}^k.$$

23. Let the density of $\boldsymbol{X}$ be $f(\boldsymbol{x}) = k$ for all $\boldsymbol{x}'\boldsymbol{x} \leq p + 2$ and 0 elsewhere. Prove that

$$k = \frac{\Gamma\left(\frac{p}{2} + 1\right)}{[(p+2)\pi]^{\frac{p}{2}}},$$

and, moreover, show that $E(\boldsymbol{X}) = \mathbf{0}$ and $E(\boldsymbol{X}\boldsymbol{X}') = \boldsymbol{I}_p$.

24. Let $\boldsymbol{X} = \begin{pmatrix} X_1 \\ X_2 \end{pmatrix}$ denote a bivariate random vector with *pdf*

$$f(\boldsymbol{x}) = \frac{1}{2\pi} \exp\left[-(\boldsymbol{x}'\boldsymbol{x})^{\frac{1}{2}}\right], \qquad \boldsymbol{x} \in \mathbb{R}^2.$$

Show that the two variables X_1 and X_2 are independent.

25. Let $3x_1^2 + 2x_2^2 - 2\sqrt{2}x_1x_2 = 1$.

(i) Write the quadratic form in matrix notation.

(ii) Draw the graph of the quadratic form.

CHAPTER 3

LOCATION MODEL

Outline

In this chapter, we discuss some basic results on the location model under the assumption that the error-vector is distributed according to the multivariate t-distribution.

Statistical Inference for Models with Multivariate t-Distributed Errors, First Edition.

A. K. Md. Ehsanes Saleh, M. Arashi, S.M.M. Tabatabaey.

3.1 Model Specification

Consider the location model with the response vector $\boldsymbol{Y} = (Y_1, \cdots, Y_n)'$ such that it satisfies the relation

$$\boldsymbol{Y} = \theta \mathbf{1}_n + \boldsymbol{\varepsilon}, \tag{3.1.1}$$

where $\mathbf{1}_n = (1, \cdots, 1)'$ is an n-tuple of 1's, and the error vector $\boldsymbol{\varepsilon}$ follows the multivariate $M_t^{(n)}(\mathbf{0}, \sigma^2 \mathbf{V}_n, \gamma_o)$ with pdf

$$f(\boldsymbol{\varepsilon}) = \frac{(\gamma_o)^{\frac{\gamma_o}{2}} \Gamma\left(\frac{\gamma_o+n}{2}\right)}{(\pi\sigma^2)^{n/2} \Gamma\left(\frac{\gamma_o}{2}\right)} |\mathbf{V}_n|^{-\frac{1}{2}} \left(1 + \frac{1}{\gamma_o \sigma^2} \boldsymbol{\varepsilon}' \mathbf{V}_n^{-1} \boldsymbol{\varepsilon}\right)^{-\frac{1}{2}(\gamma_o+n)},$$

where $\sigma > 0$, $\mathbf{V}_n$ is a positive definite matrix of rank n and $\gamma_o > 2$.

The mean of $\boldsymbol{\varepsilon}$ is the zero-vector and the covariance-matrix of $\boldsymbol{\varepsilon}$ is

$$E(\boldsymbol{\varepsilon}'\boldsymbol{\varepsilon}) = \frac{\gamma_o \sigma^2}{\gamma_o - 2} \mathbf{V}_n = \sigma_{\boldsymbol{\varepsilon}}^2 \mathbf{V}_n, \qquad \sigma_{\boldsymbol{\varepsilon}}^2 = \frac{\gamma_o \sigma^2}{\gamma_o - 2}. \tag{3.1.2}$$

The plan of this chapter is as follows. In sections 2 and 3, unbiased estimators of θ and $\sigma_{\boldsymbol{\varepsilon}}^2$ are proposed along with the test statistic for testing the hypothesis $H_0 : \theta = \theta_0$. In addition, we propose some improved estimates of location parameter. Section 4 contains some important theorems for the bias and MSE expressions of the proposed estimators of θ and their mathematical characteristics. A complete discussion on the performance of the estimators is also included in this section. Various estimators of $\sigma_{\boldsymbol{\varepsilon}}^2$ as well as their performance are the subjects discussed in section 5.

3.2 Unbiased Estimates of θ and σ^2 and Test of Hypothesis

In this section, we propose the unbiased estimators for the location parameter θ as well as of the variance $\sigma_{\boldsymbol{\varepsilon}}^2$, based on LSE method, by minimizing the following criterion w.r.t. θ:

$$(\boldsymbol{Y} - \theta \mathbf{1}_n)' \mathbf{V}_n^{-1} (\boldsymbol{Y} - \theta \mathbf{1}_n). \tag{3.2.1}$$

The unbiased estimator of θ is then given by

$$\tilde{\theta}_n = K_1^{-1} \mathbf{1}_n' \mathbf{V}_n^{-1} \boldsymbol{Y}, \qquad K_1 = (\mathbf{1}_n' \mathbf{V}_n^{-1} \mathbf{1}_n). \tag{3.2.2}$$

Define

$$S_u^2 = \frac{1}{m} (\boldsymbol{Y} - \tilde{\theta}_n \mathbf{1}_n)' \mathbf{V}_n^{-1} (\boldsymbol{Y} - \tilde{\theta}_n \mathbf{1}_n), \qquad m = n - 1. \tag{3.2.3}$$

The next theorem gives the distributions of $\tilde{\theta}_n$ and S_u^2.

Theorem 3.2.1. *Suppose* $\boldsymbol{Y} \sim M_t^{(n)}(\theta \mathbf{1}, \sigma^2 \mathbf{V}_n, \gamma_o)$, *then the distribution of* $\tilde{\theta}_n$ *is given by the pdf*

$$\frac{\Gamma\left(\frac{\gamma_o+1}{2}\right)}{(\gamma_o \pi)^{\frac{1}{2}} \Gamma\left(\frac{\gamma_o}{2}\right)} \sqrt{\frac{K_1}{\sigma^2}} \left\{1 + \left[\frac{K_1(\tilde{\theta}_n - \theta)^2}{\gamma_o \sigma^2}\right]\right\}^{-\frac{1}{2}(\gamma_o+1)}.$$

Also, the distribution of S_u^2 is the central F-distribution with (m, γ_o) d.f., given by the pdf

$$\left(\frac{m}{\gamma_o\sigma^2}\right)\frac{1}{B\left(\frac{m}{2},\frac{\gamma_o}{2}\right)}\left(\frac{mS_u^2}{\gamma_o\sigma^2}\right)^{\frac{1}{2}m-1}\left(1+\frac{mS_u^2}{\gamma_o\sigma^2}\right)^{-\frac{m+\gamma_o}{2}},$$

where $m = n-1$.

Proof: Using the representation (2.6.3), the distribution of $\tilde{\theta}_n$ can be written as

$$f_{\tilde{\theta}_n}(x) = \int_0^\infty W_o(t) f^*_{\tilde{\theta}_n}(x)dt,$$

where $f^*_{\tilde{\theta}_n}(.)$ is the *pdf* of $\tilde{\theta}_n$, where $\tilde{\theta}_n \sim \mathcal{N}(\theta, t^{-1}\sigma^2 K_1^{-1})$. Thus, the distribution of $\tilde{\theta}_n$ under multivariate t is given by

$$\begin{aligned} f_{\tilde{\theta}_n}(x) &= \int_0^\infty \frac{t^{-1}}{\Gamma\left(\frac{\gamma_o}{2}\right)}\left(\frac{\gamma_o t}{2}\right)^{\frac{\gamma_o}{2}} e^{-\frac{\gamma_o t}{2}}\frac{\sqrt{K_1}}{\sigma\sqrt{2\pi}}e^{-\frac{K_1 t}{2\sigma^2}(x-\theta)^2}dt \\ &= \frac{\sqrt{K_1}}{\sigma\sqrt{2\pi}\Gamma\left(\frac{\gamma_o}{2}\right)}\left(\frac{\gamma_o}{2}\right)^{\frac{\gamma_o}{2}}\int_0^\infty t^{\frac{1}{2}(\gamma_o-1)}e^{-\frac{t}{2\sigma^2}\left[\gamma_o\sigma^2+K_1(x-\theta)^2\right]}dt. \end{aligned}$$

Now make the transformation $2\sigma^2 z = t\left[\gamma_o\sigma^2 + K_1(x-\theta)^2\right]$ with the given Jacobian $J(t \to z) = z\sigma^2/\left[\gamma_o\sigma^2 + K_1(x-\theta)^2\right]$ to get

$$\begin{aligned} f_{\tilde{\theta}_n}(x) &= \frac{\sqrt{K_1}}{\sigma\sqrt{2\pi}\Gamma\left(\frac{\gamma_o}{2}\right)}\left(\frac{\gamma_o}{2}\right)^{\frac{\gamma_o}{2}}\left(\frac{2\sigma^2}{\gamma_o\sigma^2+K_1(x-\theta)^2}\right)^{\frac{1}{2}(\gamma_o+1)} \\ &\quad\times\int_0^\infty z^{\frac{1}{2}(\gamma_o+1)-1}e^{-z}dz \\ &= \frac{\sqrt{K_1}}{\sigma\sqrt{2\pi}\Gamma\left(\frac{\gamma_o}{2}\right)}\left(\frac{\gamma_o}{2}\right)^{\frac{\gamma_o}{2}}\left(\frac{2\sigma^2}{\gamma_o\sigma^2+K_1(x-\theta)^2}\right)^{\frac{1}{2}(\gamma_o+1)}\Gamma\left(\frac{\gamma_o+1}{2}\right) \\ &= \frac{\Gamma\left(\frac{\gamma_o+1}{2}\right)}{(\gamma_o\pi)^{\frac{1}{2}}\Gamma\left(\frac{\gamma_o}{2}\right)}\sqrt{\frac{K_1}{\sigma^2}}\left\{1+\left[\frac{K_1(x-\theta)^2}{\gamma_o\sigma^2}\right]\right\}^{-\frac{1}{2}(\gamma_o+1)}, \end{aligned}$$

which is the desired result.

For the distribution of S_u^2, define $\mathbf{Z}_1 = \mathbf{V}_n^{-1/2}(\boldsymbol{Y} - \tilde{\theta}_n\mathbf{1})$, then under the assumption $\boldsymbol{Y} \sim \mathcal{N}_n(\theta\mathbf{1}, t^{-1}\sigma^2\mathbf{V}_n)$ it follows $t^{1/2}\sigma^{-1}(\boldsymbol{I}_n - \boldsymbol{A})^{-1/2}\mathbf{Z}_1 \sim \mathcal{N}_n(\mathbf{0}, \boldsymbol{I}_n)$, where $\boldsymbol{A} = K_1^{-1}\mathbf{V}_n^{-1/2}\mathbf{1}\mathbf{1}'\mathbf{V}_n^{-1/2}$ is a symmetric idempotent matrix and so is $(\boldsymbol{I}_n - \boldsymbol{A})$. Then, it follows $rank(\boldsymbol{I}_n - \boldsymbol{A}) = tr(\boldsymbol{I}_n - \boldsymbol{A}) = n-1$. Therefore, we obtain

$$mS_u^2|t = \mathbf{Z}_1'(\boldsymbol{I}_n - \boldsymbol{A})^{-1/2}(\boldsymbol{I}_n - \boldsymbol{A})(\boldsymbol{I}_n - \boldsymbol{A})^{-1/2}\mathbf{Z}_1 \sim \sigma^2 t^{-1}\chi^2_m. \quad (3.2.4)$$

Thus, integrating w.r.t. the weight function $W_o(.)$, the distribution of S_u^2 under normal theory gives the pdf

$$\begin{aligned} f_{S_u^2}(x) &= \frac{m}{(2\sigma^2)\Gamma\left(\frac{m}{2}\right)}\left(\frac{mx}{2\sigma^2}\right)^{\frac{m}{2}-1}\int_0^\infty t^{\frac{m}{2}}e^{-\frac{mtx}{2\sigma^2}}W_o(t)dt \\ &= \frac{m}{(2\sigma^2)\Gamma\left(\frac{m}{2}\right)\Gamma\left(\frac{\gamma_o}{2}\right)}\left(\frac{\gamma_o}{2}\right)^{\frac{\gamma_o}{2}}\left(\frac{mx}{2\sigma^2}\right)^{\frac{1}{2}m-1} \\ &\quad\times\int_0^\infty t^{\frac{m+\gamma_o}{2}-1}e^{-\frac{t}{2\sigma^2}[\gamma_o\sigma^2+mx]}dt. \end{aligned}$$

As before, make the transformation $2\sigma^2 z = t\left[\gamma_o\sigma^2 + K_1(x-\theta)^2\right]$, to obtain

$$\begin{aligned} f_{S_u^2}(x) &= \frac{m}{(2\sigma^2)\Gamma\left(\frac{m}{2}\right)\Gamma\left(\frac{\gamma_o}{2}\right)}\left(\frac{\gamma_o}{2}\right)^{\frac{\gamma_o}{2}}\left(\frac{mx}{2\sigma^2}\right)^{\frac{1}{2}m-1}\left(\frac{2\sigma^2}{\gamma_o\sigma^2+mx}\right)^{\frac{m+\gamma_o}{2}} \\ &\quad\times\int_0^\infty z^{\frac{m+\gamma_o}{2}-1}e^{-z}dz \\ &= \left(\frac{m}{\gamma_o\sigma^2}\right)\frac{\Gamma\left(\frac{m+\gamma_o}{2}\right)}{\Gamma\left(\frac{m}{2}\right)\Gamma\left(\frac{\gamma_o}{2}\right)}\left(\frac{mx}{\gamma_o\sigma^2}\right)^{\frac{1}{2}m-1}\left(1+\frac{mx}{\gamma_o\sigma^2}\right)^{-\frac{m+\gamma_o}{2}} \end{aligned}$$

that completes the proof.
From Theorem 3.2.1, we have $\text{Var}(\tilde{\theta}_n) = \sigma_\varepsilon^2 K_1^{-1}$.
Consequently, we obtain

$$\begin{aligned} &\text{(i)} \quad E\left(S_u^2\right) = \frac{\gamma_o\sigma^2}{\gamma_o-2} = \sigma_\varepsilon^2 = \sigma^2\kappa^{(1)},\ \kappa^{(1)} = \frac{\gamma_o}{\gamma_o-2}, \gamma_o > 2, \\ &\text{(ii)} \quad \text{Var}\left(S_u^2\right) = \frac{2\left(\kappa^{(1)}\right)^2}{m}\sigma^4. \end{aligned}$$

For the test of the null hypothesis $H_0 : \theta = \theta_0$ vs $H_A : \theta \neq \theta_0$, we use the following theorem.

Theorem 3.2.2. *Let the unrestricted and restricted parameter spaces be defined by*

$$\begin{aligned} \Omega_1 &= \{(\theta,\sigma,\mathbf{V}_n) : \theta\in\mathbb{R}, \sigma\in\mathbb{R}^+, \mathbf{V}_n > 0\}, \textit{and}, \\ \Omega_0 &= \{(\theta,\sigma,\mathbf{V}_n) : \theta = \theta_0, \theta_0\in\mathbb{R}, \sigma\in\mathbb{R}^+, \mathbf{V}_n > 0\}, \end{aligned}$$

respectively. Then, the LR criterion (LRC) for testing the hypothesis $H_0 : \theta = \theta_0$ is given by

$$\mathcal{L}_n = K_1\frac{(\theta-\theta_0)^2}{S_u^2} \tag{3.2.5}$$

and it has the following generalized noncentral F distribution with pdf given by

$$g_{1,m}^{*}(\mathcal{L}_n) = \sum_{r\geq 0} \frac{\left(\frac{1}{m}\right)^{\frac{1}{2}(1+2r)} \mathcal{L}_n^{\frac{1}{2}(2r-1)} K_r^{(0)}(\Delta^2)}{r!\, B\left(\frac{2r+1}{2}, \frac{m}{2}\right)\left(1+\frac{1}{m}\mathcal{L}_n\right)^{\frac{1}{2}(1+2r+\gamma_o)}}, \tag{3.2.6}$$

where $\Delta^2 = K_1(\theta-\theta_0)^2/\sigma_{\varepsilon}^2$, *and*

$$\begin{aligned} K_r^{(0)}(\Delta^2) &= \int_0^{\infty} \frac{e^{-\frac{\Delta_t^2}{2}}}{r!}\left(-\frac{\Delta_t^2}{2}\right)^r W_o(t)dt, \quad \Delta_t^2 = tK_1\left(\frac{\theta-\theta_0}{\sigma}\right)^2 \\ &= \frac{\Gamma\left(\frac{\gamma_o}{2}+r\right)}{\Gamma(r+1)\Gamma\left(\frac{\gamma_o}{2}\right)} \frac{\left(\frac{\Delta^2}{\gamma_o-2}\right)^r}{\left(1+\frac{\Delta^2}{\gamma_o-2}\right)^{\frac{\gamma_o}{2}+r}}. \end{aligned} \tag{3.2.7}$$

Proof: For the test of the null hypothesis $H_0 : \theta = \theta_0$ versus $H_A : \theta \neq \theta_0$, firstly, let

$$\begin{aligned} \tilde{\sigma}_{\varepsilon}^2 &= \frac{1}{m}(\boldsymbol{Y}-\tilde{\theta}_n\mathbf{1})'\mathbf{V}_n^{-1}(\boldsymbol{Y}-\tilde{\theta}_n\mathbf{1}), \quad \text{and} \\ \hat{\sigma}_{\varepsilon}^2 &= \frac{1}{m+1}(\boldsymbol{Y}-\theta_0\mathbf{1})'\mathbf{V}_n^{-1}(\boldsymbol{Y}-\theta_0\mathbf{1}). \end{aligned} \tag{3.2.8}$$

From definition of LR test,

$$\Lambda = \frac{L_0}{L_1},$$

where L_i, $i = 0, 1$, is the largest value that the likelihood function takes in the region Ω_i, $i = 0, 1$. Then,

$$\begin{aligned} L_0 &= \frac{\Gamma\left(\frac{n+\gamma_o}{2}\right)}{(\pi\gamma_o)^{\frac{n}{2}}\Gamma\left(\frac{\gamma_o}{2}\right)(\hat{\sigma}_{\varepsilon}^2)^{\frac{n}{2}}}\left[1+\frac{(\boldsymbol{Y}-\theta_0\mathbf{1})'\mathbf{V}_n^{-1}(\boldsymbol{Y}-\theta_0\mathbf{1})}{\gamma_o\hat{\sigma}_{\varepsilon}^2}\right]^{-\frac{1}{2}(n+\gamma_o)} \\ &= \frac{\Gamma\left(\frac{n+\gamma_o}{2}\right)}{(\pi\gamma_o)^{\frac{n}{2}}\Gamma\left(\frac{\gamma_o}{2}\right)(\hat{\sigma}_{\varepsilon}^2)^{\frac{n}{2}}}\left[1+\frac{m+1}{\gamma_o}\right]^{-\frac{1}{2}(n+\gamma_o)} \end{aligned}$$

and

$$\begin{aligned} L_1 &= \frac{\Gamma\left(\frac{n+\gamma_o}{2}\right)}{(\pi\gamma_o)^{\frac{n}{2}}\Gamma\left(\frac{\gamma_o}{2}\right)(\tilde{\sigma}_{\varepsilon}^2)^{\frac{n}{2}}}\left[1+\frac{(\boldsymbol{Y}-\tilde{\theta}_n\mathbf{1})'\mathbf{V}_n^{-1}(\boldsymbol{Y}-\tilde{\theta}_n\mathbf{1})}{\gamma_o\tilde{\sigma}_{\varepsilon}^2}\right]^{-\frac{1}{2}(n+\gamma_o)} \\ &= \frac{\Gamma\left(\frac{n+\gamma_o}{2}\right)}{(\pi\gamma_o)^{\frac{n}{2}}\Gamma\left(\frac{\gamma_o}{2}\right)(\tilde{\sigma}_{\varepsilon}^2)^{\frac{n}{2}}}\left[1+\frac{m}{\gamma_o}\right]^{-\frac{1}{2}(n+\gamma_o)} \end{aligned}$$

giving

$$\begin{aligned} \Lambda &= a\left[\frac{(\boldsymbol{Y}-\tilde{\theta}_n\mathbf{1})'\mathbf{V}_n^{-1}(\boldsymbol{Y}-\tilde{\theta}_n\mathbf{1})}{(\boldsymbol{Y}-\theta_0\mathbf{1})'\mathbf{V}_n^{-1}(\boldsymbol{Y}-\theta_0\mathbf{1})}\right]^{\frac{n}{2}} \\ &= a\left(\frac{mS_u^2}{mS_u^2+K_1(\tilde{\theta}_n-\theta_0)^2}\right)^{\frac{n}{2}} \end{aligned}$$

$$= a\left(\frac{1}{1+\frac{1}{m}\mathcal{L}_n}\right)^{\frac{n}{2}},$$

where

$$a = \left(1+\frac{1}{m}\right)^{\frac{n}{2}}\left(\frac{m+\gamma_o}{m+1+\gamma_o}\right)^{\frac{1}{2}(n+\gamma_o)}.$$

Therefore,

$$\Lambda^{\frac{2}{n}} = a^{\frac{2}{n}}\left(\frac{1}{1+\frac{1}{m}\mathcal{L}_n}\right).$$

Λ is a decreasing function of $\mathcal{L}_n$. Thus, $\Lambda < \Lambda_0$ is equivalent to $\mathcal{L}_n > F_\gamma$ for a certain constant F_γ. Hence, $\mathcal{L}_n$ is the LR test for testing the underlying null hypothesis. For its non-null distribution, we note that under the normality assumption (conditional distribution) $\varepsilon|t \sim \mathcal{N}_n(\mathbf{0}, \sigma^2 t^{-1}\mathbf{V})$

$$\mathcal{L}_n = \frac{K_1(\tilde{\theta}_n - \theta_0)^2}{S_u^2}$$

follows the noncentral F-distribution with $(1, m)$ d.f. and noncentrality parameter $\Delta_t^2 = \frac{tK_1(\theta-\theta_0)^2}{\sigma^2}$. Then, integrating over t w.r.t. the measure W_o, we obtain (3.2.6).

Corollary 3.2.2.1. *Under H_0, the pdf of $\mathcal{L}_n$ is given by*

$$g^*_{1,m}(\mathcal{L}_n) = \frac{\left(\frac{1}{m}\right)^{\frac{1}{2}}\mathcal{L}_n^{\frac{1}{2}-1}}{B\left(\frac{1}{2},\frac{m}{2}\right)\left(1+\frac{1}{m}\mathcal{L}_n\right)^{\frac{1}{2}(m+1)}},$$

which is the central F-distribution with $(1, m)$ d.f.

Corollary 3.2.2.2. *The power function at γ-level of significance based on the statistic $\mathcal{L}_n$ is given by $1 - \mathcal{G}_{1,m}(l_\gamma; \Delta^2)$, where*

$$\mathcal{G}_{1,m}(l_\gamma; \Delta^2) = \sum_{r\geq 0}\frac{1}{r!}K_r^{(0)}(\Delta^2) I_x\left[\frac{1}{2}(1+2r), \frac{m}{2}\right], \tag{3.2.9}$$

where $I_x(.,.)$ is the incomplete Beta function, $x = \frac{l_\gamma}{m+l_\gamma}$, $l_\gamma = F_{1,m}(\gamma)$, and $\mathcal{G}_{1,m}(l_\gamma; \Delta^2)$ is the generalized noncentral F cumulative distribution function.

3.3 Estimators

In addition to $\tilde{\theta}_n$, we present four more estimators of θ. First, we consider the case when it is a priori suspected that θ **may be** equal to θ_0. In this case, following Han and Bancroft (1968) and Saleh (2006), we define the estimators given below:

(i) restricted estimator (RE) of θ is

$$\hat{\theta}_n^{RE} = \tilde{\theta}_n - k_0(\tilde{\theta}_n - \theta_0),\ 0 < k_0 < 1, \tag{3.3.1}$$

(ii) preliminary test estimator (PTE) of θ is given by

$$\hat{\theta}_n^{PT} = \tilde{\theta}_n - k_0(\tilde{\theta}_n - \theta_0)I(\mathcal{L}_n < c_\alpha), \tag{3.3.2}$$

where $I(A)$ is the indicator function of the set A and c_α is the α-level critical value of the F-distribution with $(1, m)$ d.f.
(iii) Stein-type estimator (SE) of θ is given by

$$\hat{\theta}_n^{S} = \tilde{\theta}_n - \frac{c_0 k_0(\tilde{\theta}_n - \theta_0)S_u}{\sqrt{K_1}|\tilde{\theta}_n - \theta_0|},\ \ c_0 > 0,\ 0 < k_0 < 1. \tag{3.3.3}$$

It is a continuous version of PTE.
(iv) Shrinkage-type estimator of θ given by

$$\hat{\theta}_n(c) \quad = \quad c\tilde{\theta}_n, \quad c > 0, \tag{3.3.4}$$

where we only consider two cases relevant to our study, namely,

(a) if $c = \frac{K_1}{K_1+h}$, then $\hat{\theta}_n(h) = \frac{K_1}{K_1+h}\tilde{\theta}_n$, $h > 0$,

(b) if $c = \frac{K_1+d}{K_1+1}$, then $\hat{\theta}_n(d) = \frac{K_1+d}{K_1+1}\tilde{\theta}_n$, $0 < d < 1$.

Theorem 3.3.1. *Suppose* $\mathbf{Y} \sim M_t^{(n)}(\theta\mathbf{1}, \sigma^2\mathbf{V}_n, \gamma_o)$, *then the distribution of* $(\tilde{\theta}_n - \theta_0)^2$ *is given by* $\sigma^2 K_1^{-1} h_1^{(1)}(\chi^2(\Delta^2))$, *where*

$$h_1^{(2-h)}(\chi^2(\Delta^2)) = \sum_{r\geq 0} K_r^{(h)}(\Delta^2)h_{1+2r}(\chi^2; 0), \quad h = 0, 1.$$

Proof: We known that $(\tilde{\theta}_n - \theta_0)^2|t \sim \sigma^2 t^{-1}K_1^{-1}\chi_1^2(\Delta_t^2)$. Integrating w.r.t. W_o gives the density function of $(\tilde{\theta}_n - \theta_0)^2$ as

$$f_{(\tilde{\theta}_n-\theta_0)^2}(x) \quad = \quad \sigma^2 K_1^{-1}\int_0^\infty t^{-1}W_o(t)h_1\,(x)\,dt,$$

where $h_1\,(.)$ is the density function of $\chi_1^2\,(\Delta_t^2)$. Now, making use of equation (2.3.6), and the fact that $\Delta_t^2 = \frac{\gamma_o}{\gamma_o - 2}t\Delta^2$, we get

$$\begin{aligned} f_{(\tilde{\theta}_n-\theta_0)^2}(x) &= \sigma^2 K_1^{-1}\sum_{r\geq 0}\frac{1}{r!}h_{1+2r}(x;0)\frac{1}{\Gamma\left(\frac{\gamma_o}{2}\right)}\left(\frac{\gamma_o}{\gamma_o-2}\frac{\Delta^2}{2}\right)^r\left(\frac{\gamma_o}{2}\right)^{\frac{\gamma_o}{2}} \\ &= \int_0^\infty t^{r+\frac{\gamma_o}{2}-2}e^{-t\left(\frac{\gamma_o\Delta^2}{2(\gamma_o-2)}+\frac{\gamma_o}{2}\right)}dt \end{aligned}$$

$$
\begin{aligned}
&= \sigma^2 K_1^{-1} \sum_{r\geq 0} \frac{1}{r!} h_{1+2r}(x;0) \frac{1}{\Gamma\left(\frac{\gamma_o}{2}\right)} \left(\frac{\gamma_o}{\gamma_o-2}\frac{\Delta^2}{2}\right)^r \left(\frac{\gamma_o}{2}\right)^{\frac{\gamma_o}{2}} \\
&\quad \times \left(\frac{\gamma_o \Delta^2}{2(\gamma_o-2)} + \frac{\gamma_o}{2}\right)^{-\left(r+\frac{\gamma_o}{2}-1\right)} \Gamma\left(r+\frac{\gamma_o}{2}-1\right) \\
&= \sigma^2 K_1^{-1} \sum_{\geq 0} \left(\frac{\gamma_o}{2}\right) \frac{\Gamma\left(\frac{\gamma_o}{2}+r-1\right)}{\Gamma(r+1)\Gamma\left(\frac{\gamma_o}{2}\right)} \frac{\left(\frac{\Delta^2}{\gamma_o-2}\right)^r}{\left(1+\frac{\Delta^2}{\gamma_o-2}\right)^{\frac{\gamma_o}{2}+r-1}} h_{1+2r}(x;0) \\
&= \sigma^2 K_1^{-1} \sum_{r\geq 0} K_r^{(1)}(\Delta^2) h_{1+2r}(x;0) \\
&= \sigma^2 K_1^{-1} h_1^{(1)}(x).
\end{aligned}
$$

The proof is complete.

3.4 Bias and MSE Expressions of the Location Estimators

In the following, we present the expressions for bias and MSE of the estimators under study classified into theorems.

Theorem 3.4.1. *If $\mathbf{Y} \sim M_t^{(n)}(\theta, \sigma^2 \mathbf{V}_n, \gamma_o)$, then the bias expressions of $\tilde{\theta}_n$, $\hat{\theta}_n^{RE}$, $\hat{\theta}_n^{PT}$, and $\hat{\theta}_n^S$ are given by*

$$
\begin{aligned}
&\text{(i)} && b_1(\tilde{\theta}_n) = 0 && (3.4.1) \\
&\text{(ii)} && b_2(\hat{\theta}_n^{RE}) = -k_0(\theta-\theta_0) = -k_0\sigma_{\varepsilon}\Delta, \quad \Delta = (\theta-\theta_0)\sigma_{\varepsilon}^{-1} \\
&\text{(iii)} && b_3(\hat{\theta}_n^{PT}) = -k_0\sigma_{\varepsilon}\Delta G_{3,m}^{(2)}(\ell_\alpha;\Delta^2), \quad \ell_\alpha = \frac{c_\alpha}{m+c_\alpha} \\
&\text{(iv)} && b_4(\hat{\theta}_n^S) = -\frac{c_0 k_0 c_n \sigma_{\varepsilon}}{\sqrt{K_1}} E_t[2\Phi(\Delta_t)-1], \quad \Delta_t = \sqrt{t}(\theta-\theta_0)\sigma^{-1} \\
&\text{(va)} && b_5(\hat{\theta}_n(h)) = -\frac{h}{K_1+h}\theta \\
&\text{(vb)} && b_5(\hat{\theta}_n(d)) = -\frac{(1-d)}{K_1+1}\theta,
\end{aligned}
$$

where

$$
c_n = \sqrt{\frac{2}{n-1}\frac{\Gamma(\frac{n}{2})}{\Gamma(\frac{n-1}{2})}}, \quad E_N\left[\text{sign}(Z)\right] = 1-2\Phi(-\Delta_t), \;\; \text{sign}(Z) = \frac{Z}{|Z|}
$$

and

$$G_{q,m}^{(2-h)}(\ell_\alpha;\Delta^2) = \sum_{r\geq 0} K_r^{(h)}(\Delta^2) I_{l_\alpha}\left[\frac{q}{2}+r,\frac{m}{2}\right]$$
$$= \sum_{r\geq 0}\left(\frac{\gamma_o}{2}\right)^h \frac{\Gamma\left(\frac{\gamma_o}{2}+r-h\right)}{\Gamma(r+1)\Gamma\left(\frac{\gamma_o}{2}\right)} \frac{\left(\frac{\Delta^2}{\gamma_o-2}\right)^r}{\left(1+\frac{\Delta^2}{\gamma_o-2}\right)^{\frac{\gamma_o}{2}+r-h}} I_{l_\alpha}\left[\frac{q}{2}+r,\frac{m}{2}\right]. \quad (3.4.2)$$

For proof see exercise 3.6.4.

Theorem 3.4.2. *If* $\mathbf{Y} \sim M_t^{(n)}(\theta, \sigma^2\mathbf{V}_n, \gamma_o)$*, then the MSE expressions of* $\tilde{\theta}_n$, $\hat{\theta}_n^{RE}$, $\hat{\theta}_n^{PT}$*, and* $\hat{\theta}_n^{S}$ *are given by*

$$\text{(i)} \qquad M_1(\tilde{\theta}_n) = \frac{\sigma_\varepsilon^2}{K_1} \quad (3.4.3)$$

$$\text{(ii)} \qquad M_2(\hat{\theta}_n^{RE}) = \frac{\sigma_\varepsilon^2}{K_1}\{(1-k_0)^2 + k_0^2\Delta^2\}$$

$$\text{(iii)} \qquad M_3(\hat{\theta}_n^{PT}) = \frac{\sigma_\varepsilon^2}{K_1}\Big\{1 - k_0(2-k_0)G_{3,m}^{(1)}(\ell_\alpha;\Delta^2) + k_0\Delta^2[2G_{3,m}^{(2)}(\ell_\alpha;\Delta^2) - (2-k_0)G_{5,m}^{(2)}(\ell_\alpha;\Delta^2)]\Big\}$$

$$\text{(iv)} \qquad M_4(\hat{\theta}_n^{S}) = \frac{\sigma_\varepsilon^2}{K_1}\left(1 - \frac{2}{\pi}c_n^2\left[2\left(1+\frac{\Delta^2}{\gamma_o-2}\right)^{-\frac{\gamma_o}{2}} - 1\right]\right),$$

$$\text{(va)} \qquad M_5(\hat{\theta}_n(h)) = \frac{\sigma_\varepsilon^2}{(K_1+h)^2}\left[K_1 + h^2\frac{\theta^2}{\sigma_\varepsilon^2}\right],$$

$$\text{(vb)} \qquad M_5(\hat{\theta}_n(d)) = \frac{(K_1+d)^2}{(K_1+1)^2}\frac{\sigma_\varepsilon^2}{K_1} + \frac{(1-d)^2}{(K_1+1)^2}\theta^2 = \frac{\sigma_\varepsilon^2}{K_1(K_1+1)^2}\left[(K_1+d)^2 + (1-d)^2K_1\frac{\theta^2}{\sigma_\varepsilon^2}\right]. \quad (3.4.4)$$

Proof: See exercise 3.7.5 for (i)-(iii) and (va)-(vb). We only prove (iv) as follows.
We have

$$\begin{aligned} E[\hat{\theta}_n^S - \theta]^2 &= E\left[(\tilde{\theta}_n - \theta) - \frac{c_0k_0S_u(\tilde{\theta}_n - \theta_0)}{\sqrt{K_1}|\tilde{\theta}_n - \theta_0|}\right]^2 \\ &= E\left[(\tilde{\theta}_n - \theta)^2 + \frac{c_0^2k_0^2S_u^2}{K_1} - \frac{2c_0k_0S_u(\tilde{\theta}_n - \theta)(\tilde{\theta}_n - \theta_0)}{\sqrt{K_1}|\tilde{\theta}_n - \theta_0|}\right] \\ &= \frac{\sigma_\varepsilon^2}{K_1} + \frac{c_0^2k_0^2\sigma_\varepsilon^2}{K_1} - \frac{2c_0k_0c_n\sigma_\varepsilon^2}{K_1}\left[E_t\left\{\sqrt{\frac{2}{\pi}}e^{-\frac{\Delta_t^2}{2}}\right\}\right]. \end{aligned} \quad (3.4.5)$$

Choosing k_0c_0 as $k_0c_0^*$ to minimize $M_4(\tilde{\theta}_n^S)$ given by

$$\begin{aligned}
k_0c_0^* &= c_n\sqrt{\frac{2}{\pi}}\, E_t\left[e^{-\frac{\Delta_t^2}{2}}\right] \\
&= c_n\sqrt{\frac{2}{\pi}}\int_0^\infty e^{-\frac{\Delta_t^2}{2}}W_o(t)dt \\
&= c_n\sqrt{\frac{2}{\pi}}\frac{1}{\Gamma\left(\frac{\gamma_o}{2}\right)}\left(\frac{\gamma_o}{2}\right)^{\frac{\gamma_o}{2}}\int_0^\infty t^{\frac{\gamma_o}{2}-1}e^{-\left(\frac{(\theta-\theta_0)^2+\gamma_o\sigma^2}{2\sigma^2}\right)t}dt \\
&= c_n\sqrt{\frac{2}{\pi}}\left[1+\frac{\Delta^2}{\gamma_o-2}\right]^{-\frac{\gamma_o}{2}},
\end{aligned}$$

which depends on Δ^2. We want $k_0c_0^*$ independent of Δ^2. Thus choosing $k_0c_0^* = c_n\sqrt{\frac{2}{\pi}}$ we obtain

$$M_4(\hat{\theta}_n^S) = \frac{\sigma_\varepsilon^2}{K_1}\left(1-\frac{2}{\pi}c_n^2\left[2\left(1+\frac{\Delta^2}{\gamma_o-2}\right)^{-\frac{\gamma_o}{2}}-1\right]\right).$$

3.4.1 Analysis of the Estimators of Location Parameter

In this section, we provide the analysis of the various estimators of θ considered in section 3.3, namely, (a) the unrestricted estimator, $\tilde{\theta}_n$, (b) the restricted estimator, $\hat{\theta}_n^{RE}$, (c) the preliminary test estimator, $\hat{\theta}_n^{PT}$, (d) the Stein-type estimator, $\hat{\theta}_n^{SE}$, and (e) the shrinkage-type estimator, $\hat{\theta}_n(c)$.

Comparison of $\hat{\theta}_n^{RE}$ and $\tilde{\theta}_n$: The bias of $\tilde{\theta}_n$ is zero and the absolute bias of $\hat{\theta}_n^{RE}$ is $k_0\sigma_\varepsilon\Delta$. At $\Delta = 0$, both are unbiased but as Δ moves away from the origin, bias of $\hat{\theta}_n$ is unbounded. As regards the MSE of the estimators, we have the MSE-difference given by

$$M_1(\tilde{\theta}_n) - M_2(\hat{\theta}_n^{RE}) = \frac{\sigma_\varepsilon^2}{K_1}\left\{1-(1-k_0)^2-k_0^2\Delta^2\right\} \gtreqless 0 \tag{3.4.6}$$

whenever

$$\Delta^2 \gtreqless (2k_0^{-1}-1), \quad 0 < k_0 \leq 1.$$

Thus, if $\Delta^2 \leq (2k_0^{-1}-1)$, then $\hat{\theta}_n^{RE}$ is better than $\tilde{\theta}_n$ and if $\Delta^2 > (2k_0^{-1}-1)$, then $\tilde{\theta}_n$ dominates $\hat{\theta}_n^{RE}$. The relative efficiency of the estimator $\hat{\theta}_n^{RE}$ is

$$\text{RE}(\hat{\theta}_n^{RE} : \tilde{\theta}_n) = [(1-k_0)^2 + k_0^2\Delta^2]^{-1}, \tag{3.4.7}$$

which is a decreasing function of Δ^2.

Comparison of $\hat{\theta}_n^{PT}$, $\hat{\theta}_n^{RE}$ and $\tilde{\theta}_n$: The absolute bias of $\hat{\theta}_n^{PT}$, $\hat{\theta}_n^{RE}$, and $\tilde{\theta}_n$ are given by $k_0\sigma_\varepsilon\Delta G_{3,m}^{(2)}(l_\alpha;\Delta^2)$, $k_0\sigma_\varepsilon\Delta$, and 0, respectively. Thus, we can order the absolute biases as

$$0 \leq k_0\sigma_\varepsilon\Delta G_{3,m}^{(2)}(l_\alpha;\Delta^2) \leq k_0\sigma_\varepsilon\Delta \quad \forall\, \Delta.$$

If $\Delta^2 = 0$, then they are unbiased. As regards MSE for the estimators, we have

$$M_1(\tilde{\theta}_n) - M_3(\hat{\theta}_n^{PT}) =$$
$$\frac{\sigma_\varepsilon^2}{K_1}\left\{k_0(2-k_0)G_{3,m}^{(1)}(\ell_\alpha;\Delta^2) - k_0\Delta^2\left[2G_{3,m}^{(2)}(\ell_\alpha;\Delta^2) - (2-k_0)G_{5,m}^{(2)}(\ell_\alpha;\Delta^2)\right]\right\}.$$

Thus, MSE-difference is $\gtreqless 0$ whenever

$$\Delta^2 \gtreqless \frac{(2k_0^{-1}-1)G_{3,m}^{(1)}(\ell_\alpha;\Delta^2)}{[2G_{3,m}^{(2)}(\ell_\alpha;\Delta^2) - (2-k_0)G_{5,m}^{(2)}(\ell_\alpha;\Delta^2)]}. \tag{3.4.8}$$

The relative efficiency of $\hat{\theta}_n^{PT}$ w.r.t. $\tilde{\theta}_n$ is given by

$$\begin{aligned} E(\alpha,\Delta^2) = \mathrm{RE}(\hat{\theta}_n^{PT};\tilde{\theta}_n) &= \Big[1 - k_0(2-k_0)G_{3,m}^{(1)}(\ell_\alpha;\Delta^2) \\ &\quad + k_0\Delta^2\{2G_{3,m}^{(2)}(\ell_\alpha;\Delta^2) \\ &\quad -(2-k_0)G_{5,m}^{(2)}(\ell_\alpha;\Delta^2)\}\Big]^{-1}. \end{aligned} \tag{3.4.9}$$

Note that

(i) If $\Delta^2 = 0$, then it reduces to $[1 - k_0(2-k_0)G_{3,m}^{(1)}(\ell_\alpha;0)]^{-1} \geq 1$.

(ii) If $\Delta^2 \to \infty$, then, $\mathrm{RE}(\hat{\theta}_n^{PT};\tilde{\theta}_n) \to 1$.

(iii) The $\mathrm{RE}(\hat{\theta}_n^{PT};\tilde{\theta}_n)$ crosses the 1-line in the interval $\left(1 - \frac{1}{2}k_0,\, \frac{1}{k_0} - \frac{1}{2}\right)$.

(iv) $\mathrm{RE}(\hat{\theta}_n^{PT};\tilde{\theta}_n)$ equals $[1 - k_0(2-k_0)G_{3,m}^{(1)}(\ell_\alpha;0)]^{-1}$ at $\Delta^2 = 0$, then drops monotonically crossing the 1-line in the interval $(1 - \frac{1}{2}k_0,\, \frac{1}{k_0} - \frac{1}{2})$ keeping to a minimum, then increases towards the 1-line. Suppose $\mathrm{RE}(\alpha,\Delta^2)$ shows the relative efficiency of $\hat{\theta}_n^{PT}$ and $\tilde{\theta}_n$ as a function of α and Δ^2. Thus, an optimum α-level MSE is obtained by solving the equation for $\alpha \in \mathcal{A} = \{\alpha|\mathrm{RE}(\alpha,\Delta^2) \geq E_0\}$

$$\min_{\Delta^2} \mathrm{RE}(\alpha,\Delta^2) = E(\alpha,\Delta_0^2(\alpha)) = E_0, \tag{3.4.10}$$

where E_0 is a prefixed or guaranteed relative efficiency.

Comparison of $\tilde{\theta}_n$ and $\hat{\theta}_n^S$: The bias expression is given by

$$\begin{aligned} b_4(\hat{\theta}_n^S) &= \left\{-\frac{c_0k_0c_n\sigma_\varepsilon}{\sqrt{K_1}}\right\} E_t[2\Phi(\Delta_t) - 1] \\ &= \left\{-\frac{c_0k_0c_n\sigma_\varepsilon}{\sqrt{K_1}}\right\} \int_0^\infty [2\Phi(\Delta_t) - 1]W_o(t)dt. \end{aligned}$$

Table 3.1 Upper bound (U.B.) of Δ^2 for which $\hat{\theta}_n^S$ outperforms $\tilde{\theta}_n$

γ_o	3	4	5	6	7	8	9	10
U. B. of Δ^2	0.5874	0.8284	0.9585	1.0396	1.0950	1.1352	1.1657	1.1895

As $\Delta_t \to 0$, $|b_4(\hat{\theta}_n^S)| \to 0$ and as $\Delta_t^1 \to \infty$, $|b_4(\hat{\theta}_n^S)| = \frac{c_0 k_0 c_n \sigma_\varepsilon}{\sqrt{K_1}}$. The absolute bias is a nondecreasing function of Δ_t. Thus, near the origin the bias is smallest and becomes largest when $\Delta_t \to \infty$.

As regards the MSE comparison, the relative efficiency RE$(\hat{\theta}_n^S;\tilde{\theta}_n)$ is given by

$$\left\{1 - \frac{2}{\pi}c_n^2\left[2\left(1+\frac{\Delta^2}{\gamma_o-2}\right)^{-\frac{\gamma_o}{2}} - 1\right]\right\}^{-1}, \tag{3.4.11}$$

which is a decreasing function of Δ^2. At $\Delta = 0$, the RE$(\hat{\theta}_n^S;\tilde{\theta}_n) = [1-\frac{2}{\pi}c_n^2]^{-1} \geq 1$ and as $\Delta^2 \to \infty$, RE$(\hat{\theta}_n^S;\tilde{\theta}_n) \to 1$, and it crosses the 1-line at $\Delta^2 = (\gamma_o - 2)\left(2^{\frac{2}{\gamma_o}} - 1\right)$ $(\gamma_o \geq 3)$.

Thus, at $\Delta^2 = 0$, RE$(\hat{\theta}_n^S;\tilde{\theta}_n)$ has maximum value $[1-\frac{2}{\pi}c_n^2]^{-1} \geq 1$ then drops to a minimum value $[1+\frac{2}{\pi}c_n^2]^{-1}$ as $\Delta^2 \to \infty$. The loss of efficiency is $1-[1+\frac{2}{\pi}c_n^2]^{-1}$ while the gain in efficiency is $[1-\frac{2}{\pi}c_n^2]^{-1}$. Further, $\hat{\theta}_n^S$ performs better than $\tilde{\theta}_n$ in the interval $0 \leq \Delta^2 \leq (\gamma_o-2)\left(2^{\frac{2}{\gamma_o}}-1\right)$ and outside of this interval $\tilde{\theta}_n$ performs better.

Some values of $(\gamma_o-2)\left(2^{\frac{2}{\gamma_o}}-1\right)$ for $\gamma_o = 3,\ldots,10$ are provided in Table 3.1. Table 3.2 shows the efficiency of SE for different values Δ^2 and degrees of freedom.

Note that $\hat{\theta}_n^S$ does not depend on the level of significance, whereas $\hat{\theta}_n^{PT}$ does. Thus, the minimum guaranteed efficiency of $\hat{\theta}_n^S$ is $[1+\frac{2}{\pi}c_n^2]^{-1}$, while that of $\hat{\theta}_n^{PT}$ varies with the level of significance. Further, the range of better performance of $\hat{\theta}_n^{PT}$ is $0 \leq \Delta^2 \leq 1$, while that of $\hat{\theta}_n^S$ is $0 \leq \Delta^2 \leq (\gamma_o-2)\left(2^{\frac{2}{\gamma_o}}-1\right)$, which is more than that of $\hat{\theta}_n^{PT}$.

Tables 3.3–3.5 list maximum and minimum efficiencies of SE and efficiency of PTE at Δ_0 for preselected αs and different degrees of freedom.

Comparison of $\tilde{\theta}_n$ and $\hat{\theta}_n(c)$: Note that

$$M_5(\hat{\theta}_n(h)) = \frac{\sigma_\varepsilon^2 K_1}{(K_1+h)^2} + \frac{h^2\theta^2}{(K_1+h)^2}. \tag{3.4.12}$$

If $h = 0$, then the MSE equals σ_ε^2/K_1 and if $h \to \infty$, MSE tends to θ^2. The first term is monotically decreasing in h and the second term is monotically increasing in

Table 3.2 Efficiency of SE for $\Delta_1^2 = 0$, $\Delta_2^2 = 0.5$, $\Delta_3^2 = (\gamma_0 - 2)(2^{\frac{2}{\gamma_0}} - 1)$, $\Delta_4^2 = 100$, and $\Delta_5^2 = \infty$

γ_0	n	10	15	20	25	30	35	40
3	E_{Δ_1}	2.514359	2.59271	2.632221	2.656027	2.671938	2.683321	2.691869
	E_{Δ_2}	1.056412	1.057603	1.058177	1.058515	1.058738	1.058896	1.059014
	E_{Δ_3}	1	1	1	1	1	1	1
	E_{Δ_4}	0.6245715	0.6199272	0.6177145	0.6164205	0.6155714	0.6149716	0.6145252
	E_{Δ_5}	0.6241089	0.6194624	0.6172487	0.615954	0.6151046	0.6145044	0.6140579
4	E_{Δ_1}	2.514359	2.59271	2.632221	2.656027	2.671938	2.683321	2.691869
	E_{Δ_2}	1.202848	1.207737	1.210106	1.211504	1.212426	1.213079	1.213567
	E_{Δ_3}	1	1	1	1	1	1	1
	E_{Δ_4}	0.6241089	0.6196437	0.6174304	0.616136	0.6152867	0.6146866	0.6142402
	E_{Δ_5}	0.6241089	0.6194624	0.6172487	0.615954	0.6151046	0.6145044	0.6140579
5	E_{Δ_1}	2.514359	2.59271	2.632221	2.656027	2.671938	2.683321	2.691869
	E_{Δ_2}	1.277231	1.284337	1.287788	1.289826	1.291172	1.292126	1.292839
	E_{Δ_3}	1	1	1	1	1	1	1
	E_{Δ_4}	0.6241769	0.6195306	0.6173171	0.6160225	0.6151731	0.614573	0.6141265
	E_{Δ_5}	0.6241089	0.6194624	0.6172487	0.615954	0.6151046	0.6145044	0.6140579
6	E_{Δ_1}	2.514359	2.59271	2.632221	2.656027	2.671938	2.683321	2.691869
	E_{Δ_2}	1.322266	1.330825	1.334987	1.337447	1.339072	1.340225	1.341086
	E_{Δ_3}	1	1	1	1	1	1	1
	E_{Δ_4}	0.6241356	0.6194892	0.6172755	0.6159809	0.6151315	0.6145314	0.6140849
	E_{Δ_5}	0.6241089	0.6194624	0.6172487	0.615954	0.6151046	0.6145044	0.6140579
10	E_{Δ_1}	2.514359	2.59271	2.632221	2.656027	2.671938	2.683321	2.691869
	E_{Δ_2}	1.403114	1.414492	1.42004	1.423323	1.425493	1.427033	1.428184
	E_{Δ_3}	1	1	1	1	1	1	1
	E_{Δ_4}	0.62411	0.6194634	0.6172497	0.6159551	0.6151056	0.6145055	0.6140589
	E_{Δ_5}	0.6241089	0.6194624	0.6172487	0.615954	0.6151046	0.6145044	0.6140579

Table 3.3 Maximum and minimum efficiencies of SE and efficiency of PTE at Δ_0 for selected α and $\gamma_0 = 3$

α	n	10	15	20	25	30	35	40
	$E^{\max}$	2.514359	2.59271	2.632221	2.656027	2.671938	2.683321	2.691869
	$E^{\min}$	0.6241089	0.6194624	0.6172487	0.615954	0.6151046	0.6145044	0.6140579
0.05	E_{Δ_0}	0.9958017	0.9980362	0.9988126	0.9991851	0.9993968	0.9995305	0.9996212
	E_0	0.6348965	0.6305902	0.6285252	0.6273136	0.6265174	0.625954	0.6255345
	Δ_0	11.5097	11.2982	11.2104	11.1625	11.1323	11.1116	11.0965
0.15	E_{Δ_0}	0.9976015	0.9988476	0.9992947	0.9995125	0.9996375	0.999717	0.9997711
	E_0	0.6351983	0.6307656	0.6286483	0.6274085	0.6265942	0.6260189	0.6255905
	Δ_0	11.2856	11.1729	11.1239	11.0965	11.0791	11.0669	11.0580
0.25	E_{Δ_0}	0.9985554	0.9992965	0.9995667	0.9996995	0.999776	0.9998248	0.9998582
	E_0	0.6353392	0.6308507	0.6287092	0.6274558	0.626633	0.6260516	0.625619
	Δ_0	11.1844	11.1133	11.0817	11.0639	11.0525	11.0445	11.0386
0.35	E_{Δ_0}	0.9991417	0.9995783	0.9997393	0.9998187	0.9998647	0.9998941	0.9999142
	E_0	0.6354272	0.6309048	0.6287481	0.6274863	0.6266581	0.6260729	0.6256373
	Δ_0	11.1223	11.0758	11.0549	11.043	11.0354	11.0300	11.0261
0.45	E_{Δ_0}	0.9995137	0.9997596	0.999851	0.9998962	0.9999224	0.9999392	0.9999507
	E_0	0.6354877	0.6309424	0.6287753	0.6275076	0.6266757	0.6260878	0.6256503
	Δ_0	11.0801	11.0499	11.0363	11.0285	11.0234	11.0199	11.0173

Table 3.4 Maximum and minimum efficiencies of SE and efficiency of PTE at Δ_0 for selected α and $\gamma_0 = 5$

α	n	10	15	20	25	30	35	40
	$E^{\max}$	2.514359	2.59271	2.632221	2.656027	2.671938	2.683321	2.691869
	$E^{\min}$	0.6241089	0.6194624	0.6172487	0.615954	0.6151046	0.6145044	0.6140579
0.05	E_{Δ_0}	0.9807127	0.9902479	0.993891	0.9957219	0.9967917	0.9974798	0.9979529
	E_0	0.6778234	0.6794704	0.6803586	0.6809112	0.681285	0.6815539	0.6817584
	Δ_0	4.3785	4.1003	3.9784	3.9099	3.8661	3.8357	3.8133
0.15	E_{Δ_0}	0.988064	0.9939851	0.9962378	0.9973675	0.9980269	0.9984506	0.9987418
	E_0	0.6840562	0.683701	0.6835545	0.6834769	0.6834267	0.6833931	0.6833689
	Δ_0	4.0874	3.9266	3.8547	3.8139	3.7877	3.7694	3.7559
0.25	E_{Δ_0}	0.9925038	0.9962312	0.997645	0.9983531	0.998766	0.9990312	0.9992134
	E_0	0.6875186	0.6859956	0.6852707	0.6848437	0.6845643	0.6843678	0.6842214
	Δ_0	3.9440	3.8392	3.7918	3.7649	3.7475	3.7353	3.7263
0.35	E_{Δ_0}	0.995422	0.9977022	0.9985651	0.9989969	0.9992485	0.9994101	0.9995211
	E_0	0.6898739	0.6875347	0.6864131	0.6857525	0.6853177	0.6850133	0.6847826
	Δ_0	3.8528	3.7830	3.7512	3.7331	3.7214	3.7131	3.7071
0.45	E_{Δ_0}	0.9973557	0.9986743	0.9991726	0.9994217	0.9995669	0.99966	0.999724
	E_0	0.6915816	0.6886421	0.6872319	0.6864032	0.6858561	0.6854721	0.6851841
	Δ_0	3.7896	3.7437	3.7227	3.7107	3.7030	3.6975	3.6935

Table 3.5 Maximum and minimum efficiencies of SE and efficiency of PTE at Δ_0 for selected α and $\gamma_0 = 10$

α	n	10	15	20	25	30	35	40
	$E^{\max}$	2.514359	2.59271	2.632221	2.656027	2.671938	2.683321	2.691869
	$E^{\min}$	0.6241089	0.6194624	0.6172487	0.615954	0.6151046	0.6145044	0.6140579
0.05	E_{Δ_0}	0.9723608	0.9858522	0.9910921	0.9937447	0.9953011	0.9963047	0.9969959
	E_0	0.7048151	0.7109485	0.7141065	0.7160225	0.7173123	0.7182322	0.7189266
	Δ_0	3.6558	3.4149	3.3085	3.2486	3.2101	3.1834	3.1637
0.15	E_{Δ_0}	0.9826912	0.9912156	0.9944892	0.9961379	0.9971023	0.9977231	0.9981501
	E_0	0.7154632	0.718267	0.7196676	0.7205036	0.7210576	0.7214511	0.721747
	Δ_0	3.4046	3.2636	3.2003	3.1644	3.1413	3.1252	3.1133
0.25	E_{Δ_0}	0.9890645	0.9944773	0.9965425	0.9975795	0.9981852	0.9985746	0.9988423
	E_0	0.7214757	0.7222815	0.7226722	0.7229054	0.723061	0.7231675	0.723247
	Δ_0	3.2790	3.1867	3.1450	3.1212	3.1058	3.0951	3.0872
0.35	E_{Δ_0}	0.9932963	0.9966255	0.9978902	0.9985241	0.9988939	0.9991315	0.9992948
	E_0	0.7255875	0.7249795	0.7246794	0.7245068	0.7243857	0.7243049	0.7242389
	Δ_0	3.1988	3.1372	3.1092	3.0931	3.0828	3.0755	3.0702
0.45	E_{Δ_0}	0.996118	0.9980503	0.9987822	0.9991485	0.999362	0.9994992	0.9995934
	E_0	0.7285835	0.7269267	0.7261226	0.7256484	0.7253316	0.7251092	0.7249462
	Δ_0	3.1430	3.1025	3.0840	3.0734	3.0666	3.0618	3.0582

h. The derivative of the first term w.r.t h approaches $-\infty$ as $h \to 0^+$ and that of the second term approaches 0 as $h \to 0^+$. Thus,

$$\frac{\partial M_5(\hat{\theta}_n(h))}{\partial h} = \frac{\sigma_\varepsilon^2 K_1}{(K_1+h)^3}(h\theta^2 - \sigma_\varepsilon^2). \tag{3.4.13}$$

Then, we have the following theorem.

Theorem 3.4.3. *A sufficient condition for $M_5(\hat{\theta}_n(h)) \le M_1(\tilde{\theta}_n)$ is that there exists an h such that $0 < h < h^*$ where*

$$h^* = \frac{\sigma_\varepsilon^2}{\theta^2}.$$

Similarly, we differentiate $M_5(\hat{\theta}_n(d))$ w.r.t d giving

$$\frac{\partial M_5(\hat{\theta}_n(d))}{\partial d} = \frac{2\sigma_\varepsilon^2(K_1+d)}{K_1(K_1+1)^2} - \frac{2(1-d)\theta^2}{(K_1+1)^2}$$

so that

$$\left.\frac{M_5(\hat{\theta}_n(d))}{\partial d}\right|_{d=1} = \frac{2\sigma_\varepsilon^2}{K_1(K_1+1)^2} > 0.$$

Therefore, we have the following theorem.

Theorem 3.4.4. *There exists a $d \in (0,1)$ such that $M_5(\hat{\theta}_n(d)) \le M_1(\tilde{\theta}_n)$ and the optimum d_{opt} is given by*

$$d_{\text{opt}} = \frac{K_1(\theta^2 - \sigma_\varepsilon^2)}{K_1\theta^2 + \sigma_\varepsilon^2}.$$

Now, substituting θ^2 and σ_ε^2 by their unbiased estimators $\tilde{\theta}_n^2 - \frac{\tilde{\sigma}_\varepsilon^2}{K_1}$ and $\tilde{\sigma}_\varepsilon^2$, we get

$$\hat{d}_{\text{opt}} = 1 - \frac{\tilde{\sigma}_\varepsilon^2}{\tilde{\theta}_n}\left(1 + \frac{1}{K_1}\right) \tag{3.4.14}$$

and $\hat{d}_{\text{opt}}$ is called the minimum MSE estimator of d.

Comparison of $\hat{\theta}_n(h)$ and $\hat{\theta}_n(d)$: First we plot the graph of $\hat{\theta}_n(h)$ and $\hat{\theta}_n(d)$ (see Figure 3.1). Note that $\hat{\theta}_n(h) = \frac{K_1}{K_1+h}\tilde{\theta}_n$ and $\hat{\theta}_n(d) = \frac{K_1+d}{K_1+1}\tilde{\theta}_n$, so that $\hat{\theta}_n(0) = \tilde{\theta}_n$ and $\hat{\theta}_n(1) = \frac{K_1}{K_1+1}\tilde{\theta}_n$ and $\hat{\theta}_n(0) = \frac{K_1}{K_1+1}\tilde{\theta}_n$ and $\hat{\theta}_n(1) = \tilde{\theta}_n$. Thus, we can draw the graph as in Figure 3.1.

Consider the MSE-difference is given by

$$M_5(\hat{\theta}_n(h)) - M_5(\hat{\theta}_n(d)) = \frac{\sigma_\varepsilon^2}{K_1(K_1+1)^2(K_1+h)^2}\Big\{K_1^2(K_1+1)^2$$

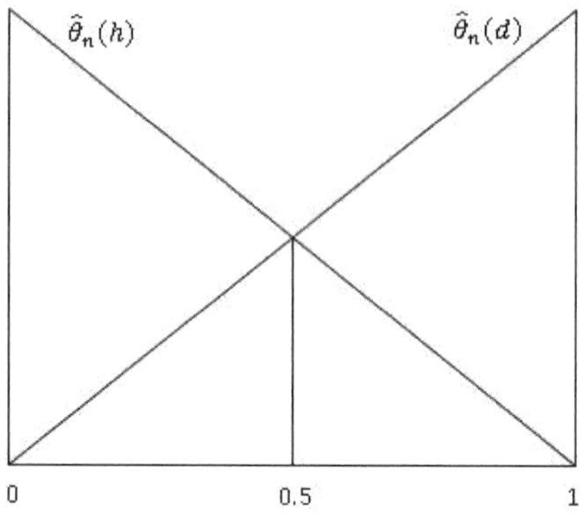

Figure 3.1 Plot of $\hat{\theta}_n(h)$ vs $\hat{\theta}_n(d)$

$$-(K_1+h)^2(K_1+d)^2+\left[(K_1+1)^2h^2 - (K_1+h)^2(1-d)^2\right]\frac{K_1\theta^2}{\sigma_\varepsilon^2}\Bigg\}. \tag{3.4.15}$$

Thus, the MSE-difference is non-negative, i.e., $\hat{\theta}_n(d)$ is superior to $\hat{\theta}_n(h)$ whenever $0<\theta^2<\theta_3^2$, where

$$\theta_3^2 = \frac{(K_1+h)^2(K_1+d)^2 - K_1^2(K_1+1)^2}{K_1(K_1+1)^2h^2 - K_1(K_1+h)^2(1-d)^2}\sigma_\varepsilon^2. \tag{3.4.16}$$

3.5 Various Estimates of Variance

Next, we consider the estimators of σ_ε^2 given below:
(i) the unrestricted estimator of σ_ε^2 is S_u^2, and
(ii) the restricted estimator of σ_ε^2, S_R^2 is defined by

$$(m+1)S_R^2 = mS_u^2 + K_1(\tilde{\theta}_n - \theta_0)^2. \tag{3.5.1}$$

Further, the best invariant estimators of σ_ε^2 are given by

$$\text{(iii)}\quad \tilde{\sigma}_\varepsilon^2 = \frac{mS_u^2}{m+2}, \qquad \text{(iv)}\quad \hat{\sigma}_\varepsilon^2 = \frac{(m+1)S_R^2}{n+3}. \tag{3.5.2}$$

Let c_α be the α-level critical value of the F-distribution with $(1,m)$ d.f. Then we define three more **preliminary test** estimators of σ_ε^2

$$\text{(v)}\qquad S^2_{PT[1]} = \Psi_1(\mathcal{L}_n)m\,S_u^2, \tag{3.5.3}$$

$$\text{(vi)} \qquad S^2_{PT[2]} = \Psi_2(\mathcal{L}_n) m S_u^2, \tag{3.5.4}$$

and

$$\text{(vii)} \qquad S^2_{[s]} = \Psi_s(\mathcal{L}_n) m S_u^2, \tag{3.5.5}$$

where

$$\Psi_1(\mathcal{L}_n) = \frac{1}{m} I(\mathcal{L}_n \geq c_\alpha) + \frac{(1 + \frac{1}{m}\mathcal{L}_n)}{m+1} I(\mathcal{L}_n < c_\alpha), \tag{3.5.6}$$

$$\Psi_2(\mathcal{L}_n) = \frac{1}{m+2} I(\mathcal{L}_n \geq c_\alpha) + \frac{(1 + \frac{1}{m}\mathcal{L}_n)}{m+3} I(\mathcal{L}_n < c_\alpha), \tag{3.5.7}$$

and

$$\Psi_s(\mathcal{L}_n) = \frac{1}{m+2} I\Big(\mathcal{L}_n \geq \frac{m}{m+2}\Big) + \frac{(1 + \frac{1}{m}\mathcal{L}_n)}{m+3} I\Big(\mathcal{L}_n < \frac{m}{m+2}\Big), \tag{3.5.8}$$

respectively.

3.5.1 Bias and MSE Expressions of the Variance Estimators

In the following, we give some expressions for bias and MSE of the scale estimators.

Theorem 3.5.1. *If* $Y \sim M_t^{(n)}(\theta, \sigma^2 \mathbf{V}_n, \gamma_o)$, *then the bias expressions of* S_U^2, S_R^2, $\tilde{\sigma}_\varepsilon^2$, $\hat{\sigma}_\varepsilon^2$, $S^2_{PT[1]}$, $S^2_{PT[2]}$, *and* $S^2_{[s]}$ *are given by*

$$
\begin{aligned}
&\text{(i)} \quad b_1(S_u^2) = 0, \qquad \text{(ii)} \quad b_2(\tilde{\sigma}_\varepsilon^2) = -\frac{2\sigma_\varepsilon^2}{m+2}, \\
&\text{(iii)} \quad b_3(S_R^2) = \frac{\sigma_\varepsilon^2 \Delta^2}{m+1}, \qquad \text{(iv)} \quad b_4(\hat{\sigma}_\varepsilon^2) = \frac{\sigma_\varepsilon^2}{m+3}(\Delta^2 - 2), \\
&\text{(v)} \quad b_5(S^2_{PT[1]}) = -\frac{\sigma_\varepsilon^2}{m+1} \Big\{ G^{(1)}_{1,m+2}(\ell_\alpha; \Delta^2) - G^{(1)}_{3,m}(\ell_\alpha; \Delta^2) \\
&\qquad\qquad -\Delta^2 G^{(2)}_{5,m}(\ell_\alpha; \Delta^2) \Big\}, \\
&\text{(vi)} \quad b_6(S^2_{PT[2]}) = -\frac{\sigma_\varepsilon^2}{m+2} - \frac{\sigma_\varepsilon^2}{(m+2)(m+3)} \Big[m(m+2) G^{(1)}_{1,m+2}(\ell_\alpha; \Delta^2) \\
&\qquad\qquad + G^{(1)}_{3,m}(\ell_\alpha; \Delta^2) - \Delta^2 G^{(2)}_{5,m}(\ell_\alpha; \Delta^2) \Big],
\end{aligned}
\tag{3.5.9}
$$

and

(vii)
$$b_7(S^2_{[s]}) = -\frac{\sigma^2_\varepsilon}{m+2} - \frac{\sigma^2_\varepsilon}{(m+2)(m+3)}\Big[m(m+2)G^{(1)}_{1,m+2}(\ell_\alpha;\Delta^2) + G^{(1)}_{3,m}(\ell_\alpha;\Delta^2) - \Delta^2 G^{(2)}_{5,m}(\ell_\alpha;\Delta^2)\Big].$$

Proof: For (v), we consider

$$\begin{aligned} b_5(S^2_{PT[1]}) &= E[S^2_{PT[1]} - \sigma^2_\varepsilon] \\ &= E[S^2_u - (S^2_u - S^2_R)I(\mathcal{L}_n < c_\alpha)] - \sigma^2_\varepsilon \\ &= -E[(S^2_u - S^2_R)I(\mathcal{L}_n < c_\alpha)]. \end{aligned} \tag{3.5.10}$$

Now

$$(m+1)S^2_R = mS^2_u + K_1(\tilde{\theta}_n - \theta_0)^2.$$

Then

$$S^2_u - \frac{m}{m+1}S^2_u - \frac{K_1(\tilde{\theta}_n - \theta_0)^2}{m+1} = \frac{1}{m+1}\left[S^2_u - K_1(\tilde{\theta}_n - \theta_0)^2\right] \tag{3.5.11}$$

so that

$$E[S^2_u I(\mathcal{L}_n < c_\alpha)] = \sigma^2_\varepsilon G^{(1)}_{1,m+2}(\ell_n;\Delta^2), \tag{3.5.12}$$

and

$$\begin{aligned} E[K_1(\tilde{\theta}_n - \theta_0)^2 I(\mathcal{L}_n < c_\alpha)] &= \sigma^2_\varepsilon G^{(1)}_{3,m}(\ell_n;\Delta^2) \\ &\quad + K_1(\theta - \theta_0)^2 G^{(2)}_{5,m}(\ell_\alpha;\Delta^2), \end{aligned} \tag{3.5.13}$$

so that

$$\begin{aligned} b_5(S^2_{PT[1]}) &= -\frac{\sigma^2_\varepsilon}{m+1}\Big\{G^{(1)}_{1,m+2}(\ell_\alpha;\Delta^2) - G^{(1)}_{3,m}(\ell_n;\Delta^2) \\ &\quad -\Delta^2 G^{(2)}_{5,m}(\ell_\alpha;\Delta^2)\Big\} \end{aligned} \tag{3.5.14}$$

by (3.2.5)–(12.3.8) and $\ell_\alpha = \frac{c_\alpha}{m+c_\alpha}$.

For (vi) we have

$$\begin{aligned} b_6(S^2_{PT[2]}) &= E(S^2_{PT[2]} - \sigma^2_\varepsilon) \\ &= E\left[\frac{mS^2_u}{m+2} - \left(\frac{mS^2_u}{m+2} - \frac{S^2_R}{m+3}\right)I(\mathcal{L}_n < c_\alpha) - \sigma^2_\varepsilon\right] \\ &= -\frac{\sigma^2_\varepsilon}{m+2} - E\Bigg(\frac{mS^2_u}{m+2} - \frac{mS^2_u}{(m+2)(m+3)} \\ &\quad - \frac{K_1(\tilde{\theta}_n - \theta_0)^2}{(m+2)(m+3)}\Bigg)I(\mathcal{L}_n < c_\alpha) \\ &= -\frac{\sigma^2_\varepsilon}{m+2} - \frac{\sigma^2_\varepsilon}{m+3}\Big[mG^{(1)}_{1,m+2}(\ell_\alpha;\Delta^2) \end{aligned}$$

$$
\begin{aligned}
& -\frac{1}{m+2}\Big\{G_{3,m}^{(1)}(\ell_\alpha;\Delta^2)+\Delta^2 G_{5,m}^{(2)}(\ell_\alpha;\Delta^2)\Big\}\Big] \\
= \quad & -\frac{\sigma_\varepsilon}{m+2}-\frac{\sigma_\varepsilon^2}{(m+2)(m+3)}\Big[m(m+2)G_{1,m+2}^{(1)}(\ell_\alpha;\Delta^2) \\
& +G_{3,m}^{(1)}(\ell_\alpha;\Delta^2)-\ \Delta^2 G_{5,m}^{(2)}(\ell_\alpha;\Delta^2)\Big].
\end{aligned}
$$

Similarly,

$$
\begin{aligned}
b_7(S_{[s]}^2) \quad = \quad & -\frac{\sigma_\varepsilon^2}{m+2}-\frac{\sigma_\varepsilon^2}{m+3}\Big[mG_{1,m+2}^{(1)}(\ell_\alpha^*;\Delta^2) \\
& -\frac{1}{m+2}\Big\{G_{3,m}^{(1)}(\ell_\alpha^*;\Delta^2)+\Delta^2 G_{5,m}^{(2)}(\ell_\alpha^*;\Delta^2)\Big\}\Big] \\
= \quad & -\frac{\sigma_\varepsilon^2}{m+2}-\frac{\sigma_\varepsilon^2}{(m+2)(m+3)} \\
& \times\Big[m(m+2)G_{1,m+2}^{(1)}(\ell_\alpha^*;\Delta^2)+G^{(1)}(\ell_\alpha^*;\Delta^2)-\Delta^2 G_{5,m}^{(2)}(\ell_\alpha^*;\Delta^2)\Big]
\end{aligned}
$$

with $\ell_\alpha^* = \frac{m}{m+3}$.

Theorem 3.5.2. *If $\mathbf{Y} \sim M_t^{(n)}(\theta, \sigma^2\mathbf{V}_n, \gamma_o)$, then the MSE expressions of S_u^2, S_R^2, $\tilde{\sigma}_\varepsilon^2$, $\hat{\sigma}_\varepsilon^2$, $S_{PT[1]}^2$, $S_{PT[2]}^2$, and $S_{PT[s]}^2$ are given by*

(i) $$M_1(S_u^2) = \frac{2}{m}\,\sigma_\varepsilon^4$$

(ii) $$M_2(S_R^2) = \left\{\left(\frac{m+3}{m+1}\right)\left[\frac{\kappa^{(2)}}{\left(\kappa^{(1)}\right)^2}\right]-1\right\}\sigma_\varepsilon^4+\frac{\Delta^2(4+\Delta^2)\sigma_\varepsilon^4}{(m+1)^2}$$

(iii) $$M_3(\tilde{\sigma}_\varepsilon^2) = \left(\frac{m}{m+2}\right)\left[\frac{\kappa^{(2)}}{\left(\kappa^{(1)}\right)^2}\right]\sigma_\varepsilon^4-\frac{(m-2)\sigma_\varepsilon^4}{(m+2)}$$

(iv) $$M_4(\hat{\sigma}_\varepsilon^2) = \left\{\left(\frac{m+1}{m+3}\right)\left[\frac{\kappa^{(2)}}{\left(\kappa^{(1)}\right)^2}\right]-1\right\}\sigma_\varepsilon^4+\frac{\sigma_\varepsilon^4(\Delta^4-4\Delta^2+4)}{(m+3)^2}$$

(v) $$
\begin{aligned}
M_5(S_{PT[1]}) = {} & M_1(S_u^2)-\frac{\sigma_\varepsilon^4(m+2)(2m+1)}{(m+1)^2}G_{1,m+4}^{(2)}(\ell_\alpha;\Delta^2) \\
& +\frac{\sigma_\varepsilon^4}{(m+1)^2}\Big[\Big\{3G_{5,m}^{(2)}(\ell_\alpha;\Delta^2)+mG_{3,m+2}^{(2)}(\ell_\alpha;\Delta^2) \\
& +2(m+1)\Big[G_{1,m+2}^{(1)}(\ell_\alpha;\Delta^2)-G_{3,m}^{(1)}(\ell_\alpha;\Delta^2)\Big]\Big\} \\
& +\Delta^2\Big\{6G_{7,m}^{(1)}(\ell_\alpha;\Delta^2)+mG_{5,m+2}^{(1)}(\ell_\alpha;\Delta^2) \\
& -2(m+1)G_{5,m}^{(2)}(\ell_\alpha;\Delta^2)\Big\}+\Delta^2 G_{9,m}^{(2)}(\ell_\alpha;\Delta^2)\Big].
\end{aligned}
$$

$$
\begin{aligned}
M_6\left(S^2_{PT[2]}\right) &= M_3(\tilde{\sigma}^2_{\varepsilon}) - \frac{m(2m+5)}{(m+2)(m+3)}\sigma^4_{\varepsilon}G^{(2)}_{1,m+4}(\ell_\alpha;\Delta^2) \\
&+\frac{3\sigma^4_{\varepsilon}}{m^2(m+3)^2}G^{(2)}_{5,m}(\ell_\alpha;\Delta^2) \\
&+\frac{2\sigma^4_{\varepsilon}}{(m+3)^2}\left\{G^{(2)}_{3,m+2}(\ell_\alpha;\Delta^2)-(m+3)G^{(2)}_{3,m}(\ell_\alpha;\Delta^2)\right\} \\
&+\frac{\Delta^2\sigma^4_{\varepsilon}}{(m+2)(m+3)^2}\left\{2G^{(2)}_{5,m+2}(\ell_\alpha;\Delta^2)+6(m+2)G^{(1)}_{7,m}(\ell_\alpha;\Delta^2)\right. \\
&\left.+\,2(m+3)G^{(1)}_{5,m}(\ell_\alpha;\Delta^2)\right\}+\frac{\Delta^4\sigma^4_{\varepsilon}}{m^2(m+3)^2}G^{(2)}_{9,m}(\ell_\alpha;\Delta^2)
\end{aligned}
$$

$$
\begin{aligned}
M_7(S^2_{[s]}) &= M_3(\tilde{\sigma}^2_{\varepsilon}) - \frac{m(2m+5)}{(m+2)(m+3)}\sigma^4_{\varepsilon}G^{(2)}_{1,m+4}(\ell^*_\alpha;\Delta^2) \\
&+\frac{3\sigma^4_{\varepsilon}}{m^2(m+3)^2}G^{(2)}_{5,m}(\ell^*_\alpha;\Delta^2) \\
&+\frac{2\sigma^4_{\varepsilon}}{(m+3)^2}\left[G^{(2)}_{3,m+2}(\ell^*_\alpha;\Delta^2)-(m+3)G^{(2)}_{3,m}(\ell^*_\alpha;\Delta^2)\right] \\
&+\frac{\Delta^2\sigma^4_{\varepsilon}}{(m+2)(m+3)^2}\left\{2G^{(2)}_{5,m+2}(\ell^*_\alpha;\Delta^2)+6(m+2)G^{(1)}_{7,m}(\ell^*_\alpha;\Delta^2)\right. \\
&\left.+\,2(m+3)G^{(1)}_{5,m}(\ell^*_\alpha;\Delta^2)\right\}+\frac{\Delta^4\sigma^4_{\varepsilon}}{m^2(m+3)^2}G^{(2)}_{9,m}(\ell^*_\alpha;\Delta^2),
\end{aligned}
$$

where $\kappa^{(i)}$ *is given by* (2.6.4) *and* $\ell^*_\alpha = \frac{m}{m+3}$.

Proof: Using (3.2.4), we have

$$
\begin{aligned}
\text{(i)}\quad M_1(S^2_u) &= E(S^2_u-\sigma^2_{\varepsilon})^2 \\
&= \text{Var}(S^2_u) = 2m\sigma^4\left(\kappa^{(1)}\right)^2 \\
&= \frac{2}{m}\,\sigma^4_{\varepsilon}.
\end{aligned}
$$

From (3.3.4), we have

(ii) $\text{Var}(S^2_R)$

$$
\begin{aligned}
&= \text{Var}_t\{E_N[S^2_R|t]\}+E_t\{\text{Var}_N[S^2_R|t]\} \\
&= \text{Var}_t\left\{\frac{\sigma^2t^{-1}}{(m+1)}E_N[\chi^2_{m+1}(\Delta^2_t)]\right\}+E_t\left\{\frac{\sigma^4t^{-2}}{(m+1)^2}\text{Var}_N[\chi^2_{m+1}(\Delta^2_t)]\right\} \\
&= \frac{\sigma^4}{(m+1)^2}\text{Var}_t\left[(m+1)t^{-1}+\frac{K_1(\theta-\theta_0)^2}{\sigma^2}\right]+E_t\left[\frac{\sigma^4t^{-2}}{(m+1)^2}2((m+1)+2\Delta^2_t)\right] \\
&= \frac{\sigma^4}{(m+1)^2}\{(m+1)^2\text{Var}_t\left(t^{-1}\right)\}+\frac{2\sigma^4}{(m+1)}E_t\left(t^{-2}\right)+\frac{4\sigma^2K_1(\theta-\theta_0)^2}{(m+1)^2}E_t\left(t^{-1}\right)
\end{aligned}
$$

$$= \left(\frac{m+3}{m+1}\right)\left[\frac{\kappa^{(2)}}{\left(\kappa^{(1)}\right)^2}\right]\sigma_\varepsilon^4 + \frac{4\Delta^2\sigma_\varepsilon^4}{(m+1)^2} - \sigma_\varepsilon^4.$$

Hence,

$$\begin{aligned} M_2(S_R^2) &= \left(\frac{m+3}{m+1}\right)\left[\frac{\kappa^{(2)}}{\left(\kappa^{(1)}\right)^2}\right]\sigma_\varepsilon^4 - \sigma_\varepsilon^4 + \frac{\Delta^2(4+\Delta^2)\sigma_\varepsilon^4}{(m+1)^2} \\ &= \left\{\left(\frac{m+3}{m+1}\right)\left[\frac{\kappa^{(2)}}{\left(\kappa^{(1)}\right)^2}\right] - 1\right\}\sigma_\varepsilon^4 + \frac{\Delta^2(4+\Delta^2)\sigma_\varepsilon^4}{(m+1)^2}. \end{aligned}$$

(iii)

$$\begin{aligned} \text{Var}[\tilde{\sigma}_\varepsilon^2] &= \text{Var}_t\left\{E_N\left(\frac{mS_u^2}{m+2}\Big|t\right)\right\} + E_t\left\{\text{Var}_N\left[\frac{mS_u^2}{m+2}\Big|t\right]\right\} \\ &= \text{Var}_t\left(\frac{m\sigma^2 t^{-1}}{m+2}\right) + E_t\left\{\left(\frac{\sigma^2}{m+2}\right)^2 t^{-2}(2m)\right\} \\ &= \frac{m^2}{(m+2)^2}\left\{\sigma^4 E_t\left(t^{-2}\right) - \sigma_\varepsilon^4\right\} + \frac{2m\sigma^4}{(m+2)^2}E_t\left(t^{-2}\right) \\ &= \left(\frac{m}{m+2}\right)^2\left\{\sigma^4\kappa^{(2)} - \sigma_\varepsilon^4\right\} + \frac{2m\sigma^4\kappa^{(2)}}{(m+2)^2} \\ &= \left(\frac{m}{m+2}\right)\left[\frac{\kappa^{(2)}}{\left(\kappa^{(1)}\right)^2}\right]\sigma_\varepsilon^4 - \frac{m^2\sigma_\varepsilon^4}{(m+2)^2}. \end{aligned}$$

$$\begin{aligned} M_3(\tilde{\sigma}_\varepsilon^2) &= \left(\frac{m}{m+2}\right)\left[\frac{\kappa^{(2)}}{\left(\kappa^{(1)}\right)^2}\right]\sigma_\varepsilon^4 - \frac{m^2\sigma_\varepsilon^4}{(m+2)^2} + \frac{4\sigma_\varepsilon^4}{(m+2)^2} \\ &= \left(\frac{m}{m+2}\right)\left[\frac{\kappa^{(2)}}{\left(\kappa^{(1)}\right)^2}\right]\sigma_\varepsilon^4 - \frac{(m-2)\sigma_\varepsilon^4}{(m+2)}. \end{aligned}$$

(iv)

$$\begin{aligned} \text{Var}\left[\frac{m+1}{m+3}\,S_R^2\right] &= \left(\frac{m+1}{m+3}\right)^2 \text{Var}(S_R^2) \\ &= \frac{m+1}{m+3}\left[\frac{\kappa^{(2)}}{\left(\kappa^{(1)}\right)^2}\right]\sigma_\varepsilon^4 \\ &\quad -\sigma_\varepsilon^4\left(1 - \frac{4\Delta^2}{(m+3)^2}\right). \end{aligned}$$

$$\begin{aligned} M_4(\hat{\sigma}_\varepsilon^2) &= \left(\frac{m+1}{m+3}\right)\left[\frac{\kappa^{(2)}}{\left(\kappa^{(1)}\right)^2}\right]\sigma_\varepsilon^4 - \sigma_\varepsilon^4\left(1 - \frac{4\Delta^2}{(m+3)^2}\right) + \frac{\sigma_\varepsilon^4(\Delta^2-2)^2}{(m+3)^2} \\ &= \left(\frac{m+1}{m+3}\right)\left[\frac{\kappa^{(2)}}{\left(\kappa^{(1)}\right)^2}\right]\sigma_\varepsilon^4 - \sigma_\varepsilon^4 + \frac{\sigma_\varepsilon^4(\Delta^4 - 4\Delta^2 + 4)}{(m+3)^2} \\ &= \left\{\left(\frac{m+1}{m+3}\right)\left[\frac{\kappa^{(2)}}{\left(\kappa^{(1)}\right)^2}\right] - 1\right\}\sigma_\varepsilon^4 + \frac{\sigma_\varepsilon^4(\Delta^4 - 4\Delta^2 + 4)}{(m+3)^2}. \end{aligned}$$

$$\begin{aligned}
\text{(v)} \qquad M_5(S^2_{PT[1]}) &= E[S^2_{PT[1]} - \sigma^2_{\varepsilon}]^2 \\
&= E[S^2_u - \sigma^2_{\varepsilon}]^2 - \frac{2}{(m+1)} \\
&\quad \times E\Big[(S^2_u - \sigma^2_{\varepsilon})\{S^2_u - K_1(\tilde{\theta}_n - \theta_0)^2\}I(\mathcal{L}_n < c_\alpha)\}\Big] \\
&\quad + \frac{1}{(m+1)^2} E[(S^2_u - K_1(\tilde{\theta}_n - \theta_o)^2)I(\mathcal{L}_n < c_\alpha)]^2 \\
&+ M_1(S^2_u) - \frac{2m+1}{(m+1)^2} E[(S^4_u I(\mathcal{L}_n < c_\alpha)] \\
&\quad + \frac{1}{(m+1)^2} E[K_1^2(\tilde{\theta}_n - \theta_0)^4 I(\mathcal{L}_n < c_\alpha)] \\
&\quad + \frac{m}{(m+1)^2} E[S^2_u K_1(\tilde{\theta}_n - \theta_0)^2 I(\mathcal{L}_n < c_\alpha)] \\
&\quad + \frac{2\sigma^2_{\varepsilon}}{m+1} E[S^2_u I(\mathcal{L}_n < c_\alpha)] \\
&\quad - \frac{2\sigma^2_{\varepsilon}}{m+1} E[K_1(\tilde{\theta}_n - \theta_0)^2 I(\mathcal{L}_n < c_\alpha)].
\end{aligned}$$

Now,

$$\begin{aligned}
E[S^2_u I(\mathcal{L}_n < c_\alpha)] &= E_t\left\{\frac{\sigma^2 t^{-1}}{m} E_N\left[\chi^2_m I\left(\frac{\chi^2_1(\Delta^2_t)}{\chi^2_m} < \frac{1}{m}\, c_\alpha\right)\right]\right\} \\
&= E_t\left[\sigma^2 t^{-1} G_{1,m+2}(\ell_\alpha; \Delta^2_t)\right] = \sigma^2_{\varepsilon} G^{(1)}_{1,m+2}(\ell_\alpha; \Delta^2),
\end{aligned}$$

where $\ell_\alpha = \frac{c_\alpha}{m+c_\alpha}$.

$$\begin{aligned}
E[S^4_U I(\mathcal{L}_n < c_\alpha)] &= E_t\left[\frac{\sigma^2 t^{-1}}{m^2} E_N\left\{\chi^4_m I\left(\frac{\chi^2_1(\Delta^2_t)}{\chi^2_m} < \frac{1}{m} c_\alpha\right)\right]\right\} \\
&= E_t\left[\frac{\sigma^4 t^{-2}}{m^2}\{m(m+2)G_{1,m+4}(\ell_\alpha; \Delta^2_t)\}\right] \\
&= \frac{m+2}{m}\sigma^4_{\varepsilon} G^{(2)}_{1,m+4}(\ell_\alpha; \Delta^2) \quad \text{by (3.9).}
\end{aligned}$$

Next, we have

$$\begin{aligned}
E\big[K_1(\tilde{\theta}_n - \theta_0)^2 I(\mathcal{L}_n < c_\alpha)\big] &= E_t\left\{\sigma^2 t^{-1} E\left[\chi^2_1(\Delta^2_t) I\left(\frac{\chi^2_1(\Delta^2_t)}{\chi^2_m} < \frac{1}{m} C_\alpha\right)\right]\right\} \\
&= E_t\{\sigma^2 t^{-1}\{G^{(1)}_{3,m}(\ell_\alpha; \Delta^2_t) + \Delta^2_t G^{(2)}_{5,m}(\ell_\alpha; \Delta^2)\}\big] \\
&= \sigma^2_{\varepsilon} G^{(1)}_{3,m}(\ell_\alpha; \Delta^2) + \sigma^2_{\varepsilon}\Delta^2 G^{(2)}_{5,m}(\ell_\alpha; \Delta^2),
\end{aligned}$$

and

$$\begin{aligned}
&E\big[K_1^2(\tilde{\theta}_n - \theta_0)^2 I(\mathcal{L}_n < c_\alpha)\big] \\
&= E_t\left\{\sigma^4 t^{-2} E_N\left[\left[\chi^2_1(\Delta^2_t)\right]^2 I\left(\frac{\chi^2_1(\Delta^2_t)}{\chi^2_m} < \frac{1}{m} c_\alpha\right)\right]\right\} \\
&= E_t\left\{\sigma^4 t^{-2}\left[3G^{(2)}_{5,m}(\ell_\alpha; \Delta^2_t) + 6\Delta^2_t G^{(1)}_{7,m}(\ell_\alpha; \Delta^2_t) + \Delta^4_t G^{(2)}_{9,m}(\ell_\alpha; \Delta^2_t)\right]\right\} \\
&= 3\sigma^4_{\varepsilon} G^{(2)}_{5,m}(\ell_\alpha; \Delta^2) + 6\sigma^4_{\varepsilon}\Delta^2 G^{(1)}_{7,m}(\ell_\alpha; \Delta^2) + \sigma^4_{\varepsilon}\Delta^4 G^{(2)}_{9,m}(\ell_\alpha; \Delta^2).
\end{aligned}$$

Further,

$$E\big[S_u^2K_1^2(\tilde{\theta}_n-\theta_0)^2I(\mathcal{L}_n<c_\alpha)\big] = E_t\left\{\frac{\sigma^4t^{-2}}{m}E\left[\chi_m^2\chi_1^2(\Delta_t^2)I\left(\frac{\chi_1^2(\Delta_t^2)}{\chi_m^2}<\frac{1}{m}c_\alpha\right)\right]\right\}$$
$$= E_t\left[\sigma^4t^{-2}\left\{G_{3,m+2}^{(2)}(\ell_\alpha;\Delta_t^2)+\Delta_t^2G_{5,m+2}^{(1)}(\ell_\alpha;\Delta_t^2)\right\}\right]$$
$$= \sigma_\varepsilon^2G_{3,m+2}^{(2)}(\ell_\alpha;\Delta^2)+\sigma_\varepsilon^4\Delta^2G_{5,m+2}^{(1)}(\ell_\alpha;\Delta^2)$$

$$M_6(S_{PT[2]}^2)$$
$$= E\left[(S_{PT[2]}^2)^2\right]-2\sigma_\varepsilon^2E[S_{PT[2]}^2]+\sigma_\varepsilon^4$$
$$= E_t\left\{\frac{t^{-2}}{(m+2)^2}E_N\left(\frac{mS_u^2}{t^{-1}}\right)^2|t\right\}$$
$$+\ \frac{1}{m^2(m+3)^2}E_t\left\{t^{-2}E_N\left[\left(\frac{m}{m+2}-\mathcal{L}_n\right)^2\left(\frac{mS_u^2}{t^{-1}}\right)^2I(\mathcal{L}_n<c_\alpha)|t\right]\right\}$$
$$-\ \frac{2}{m(m+2)(m+3)}E_t\left\{t^{-2}E_N\left[\left(\frac{mS_u^2}{t^{-1}}\right)^2\left(\frac{m}{m+2}-\mathcal{L}_n\right)I(\mathcal{L}_n<c_\alpha)|t\right]\right\}$$
$$+\ \frac{2}{m(m+3)^2}E_t\left\{t^{-2}E_N\left[\left(\frac{mS_u^2}{t^{-1}}\right)\left(\frac{m}{m+2}-\mathcal{L}_n)I(\mathcal{L}_n<c_\alpha)|t\right]\right\}-\sigma_\varepsilon^4$$
$$= \frac{m\sigma^4}{(m+2)}E_t\left(t^{-2}\right)+\frac{1}{m^2(m+3)^2}E_t\left\{t^{-2}E_N\left[\left(\frac{m}{m+2}\right)^2-\frac{2m}{m+2}\mathcal{L}_n+\mathcal{L}_n^2\right]\right.$$
$$\left.\times\left(\frac{mS_u^2}{t^{-1}}\right)^2I(\mathcal{L}_n<c_\alpha)|t\right\}-\frac{2}{m(m+2)(m+3)}E_t\left\{t^{-2}\right.$$
$$\left.\times E_N\left[\left(\frac{m}{m+2}\right)\left(\frac{mS_u^2}{t^{-1}}\right)^2I(\mathcal{L}_n<c_\alpha)-\left(\frac{mS_u^2}{t^{-1}}\right)^2\mathcal{L}_nI(\mathcal{L}_n<c_\alpha)|t\right\}\right.$$
$$+\ \frac{2}{m(m+3)^2}E_t\left\{t^{-2}E_N\left[\left(\frac{m}{m+2}\right)\left(\frac{mS_u^2}{t^{-1}}\right)I(\mathcal{L}_n<c_\alpha)\right.\right.$$
$$\left.\left.-\left(\frac{mS_u^2}{t^{-1}}\right)\mathcal{L}_nI(\mathcal{L}_n<c_\alpha)\right]\right\}-\sigma_\varepsilon^4$$

$$= \frac{m\sigma^4}{m+2}E_t\left(t^{-2}\right)$$
$$+\ \frac{\sigma^4}{m^2(m+3)^2}E_t\left[t^{-2}E_N\left\{\left(\frac{m}{m+2}\right)^2\chi_m^4I(F_{1,m}(\Delta_t^2)<c_\alpha)\right\}\right]$$
$$-\ \frac{2\sigma^4}{(m+2)^2(m+3)^2}E_t\left[t^{-2}E_N\left\{\chi_m^4F_{1,m}(\Delta_t^2)I(F_{1,m}(\Delta_t^2)<c_\alpha)\right\}\right]$$
$$+\ \frac{\sigma^4}{m^2(m+3)^2}E_t\left[t^{-2}E_N\left\{\chi_m^4F_{1,m}(\Delta_t^2)^2I(F_{1,m}(\Delta_t^2)<c_\alpha)\right\}\right]$$

$$
\begin{aligned}
&- \frac{2\sigma^4}{(m+2)^2(m+3)} E_t\left[t^{-2} E_N\left\{\chi_m^4 I(F_{1,m}(\Delta_t^2) < c_\alpha)\right\}\right] \\
&- \frac{2\sigma^4}{(m+2)(m+3)} E_t\left[t^{-2} E_N\left\{\chi_m^4 (F_{1,m}(\Delta_t^2) I(F_{1,m}(\Delta_t^2 < c_\alpha)\right\}\right] \\
&+ \frac{2\sigma^4}{(m+2)(m+3)^2} E_t\left[t^{-2} E_N\left\{\chi_m^4 I(F_{1,m} < c_\alpha)\right\}\right] \\
&- \frac{2\sigma^4}{(m+2)(m+3)^2} E_t\left[t^{-2} E_N\left\{\chi_m^2 F_{1,m}(\Delta_t^2) I(F_{1,m}(\Delta_t^2) < c_\alpha)\right\}\right] - \sigma_\varepsilon^4.
\end{aligned}
$$

Simplification leads to $M_6(S^2_{PT[2]})$. Similarly, $M_7(S^2_{[s]})$ may be obtained by replacing c_α with $\frac{m}{m+2}$.

3.5.2 Analysis of the Estimators of the Variance Parameter

In sections 3.2 and 3.5 we have defined seven estimators of σ_ε^2. The bias and MSE expressions of these estimators are given in section 3.5. In this section, we present the analysis of the MSE expressions.

First, we note that the MSE expression for S_u^2 is constant while the restricted estimate, S_R^2 depends on the departure parameter, Δ^2. Under H_0, i.e., for $\Delta^2 = 0$,

$$
M_2(S_R^2) = \left\{\left(\frac{m+3}{m+1}\right)\left[\frac{\kappa^{(2)}}{\left(\kappa^{(1)}\right)^2}\right] - 1\right\}\sigma_\varepsilon^4
$$

so that $M_2(S_R^2) < M_1(S_u^2)$ provided

$$
\frac{\kappa^{(2)}}{\left(\kappa^{(1)}\right)^2} < \frac{(m+1)(2m+1)}{m+3}. \tag{3.5.15}
$$

The MSE's are equal when Δ^2 equals

$$
\Delta_*^2 = -2 + 2\sqrt{1 + \frac{(m+1)\kappa^{(2)}}{2m\left(\kappa^{(1)}\right)}}. \tag{3.5.16}
$$

Hence, the range of Δ^2 for which S_R^2 dominates S_u^2 is given by $[0, \Delta_*^2]$; otherwise S_u^2 dominates S_R^2. Note that the MSE of S_R^2 is unbounded as $\Delta^2 \to \infty$.

Similarly, under H_0,

$$
M_4(\hat{\sigma}_\varepsilon^2) = \left\{\left(\frac{m+3}{m+1}\right)\left[\frac{\kappa^{(2)}}{\left(\kappa^{(1)}\right)^2}\right] - 1 + \left(\frac{2}{m+3}\right)^2\right\}\sigma_\varepsilon^4
$$

so that $M_4(\hat{\sigma}_\varepsilon^2) < M_3(\tilde{\sigma}_\varepsilon^2)$, provided

$$
\frac{\kappa^{(2)}}{\left(\kappa^{(1)}\right)^2} < \frac{\left[(m+2)(m+3)^2 + 4m + 8 - (m-2)(m+3)^2\right](m+1)}{(m+3)^2\left[(m+3)(m+2) - m(m+1)\right]}. \tag{3.5.17}
$$

Hence, the range of Δ^2 for which $\hat{\sigma}_\varepsilon^2$ dominates $\tilde{\sigma}_\varepsilon^2$ is given by $[0, \Delta^2_{**}]$, where Δ^2_{**} is defined by the solution of the equation

$$\Delta^2(\Delta^2+4) = \frac{2(2m+\kappa^{(2)})(m+3)}{(m+2)\left(\kappa^{(1)}\right)} \tag{3.5.18}$$

that is,

$$\Delta^2_{**} = -2+2\sqrt{1+\frac{(m+3)(2m+\kappa^{(2)})}{2(m+2)\left(\kappa^{(1)}\right)^2}}, \tag{3.5.19}$$

otherwise, $\tilde{\sigma}_\varepsilon^2$ dominates $\hat{\sigma}_\varepsilon^2$.

Now, we show the uniform dominance of $S^2_{[s]}$ over $\tilde{\sigma}_\varepsilon^2$ under the quadratic loss function $\frac{1}{\sigma_\varepsilon^4}(\sigma_x^2 - \sigma_\varepsilon^2)^2$. For this, we consider the risk of $S^2_{[s]}$ with respect to the quadratic loss-function.

Then, we have

$$\begin{aligned}\frac{1}{\sigma_\varepsilon^4}E[mS_u^2\psi_s(\mathcal{L}_n)-\sigma_\varepsilon^2]^2 \;&=\; E_{\mathcal{L}_n}\left\{\psi_s^2(\mathcal{L}_n)E\left[\left(\frac{t^{-1}}{\sigma_\varepsilon^2}\right)^2\left(\frac{mS_u^2}{t^{-1}}\right)^2\middle|\mathcal{L}_n\right]\right.\\ &\left.\quad -2\psi_s(\mathcal{L}_n)E\left[\left(\frac{t^{-1}}{\sigma_\varepsilon^2}\right)\left(\frac{mS_u^2}{t^{-1}}\right)\middle|\mathcal{L}_n\right]+2\right\}.\end{aligned} \tag{3.5.20}$$

Now, consider the term inside the curly bracket of (3.5.20). For fixed Δ^2 and for each $\mathcal{L}_n$, this is a quadratic form in $\psi_s(\mathcal{L}_n)$ with the minimum at

$$\psi_s^*(\mathcal{L}_n) = \frac{E\left[\left(\frac{t^{-1}}{\sigma_\varepsilon^2}\right)\left(\frac{mS_u^2}{t^{-1}}\right)\middle|\mathcal{L}_n\right]}{E\left[\left(\frac{t^{-1}}{\sigma_\varepsilon^2}\right)^2\left(\frac{mS_u^2}{t^{-1}}\right)^2\middle|\mathcal{L}_n\right]}. \tag{3.5.21}$$

which is a function of $\mathcal{L}_n$ and Δ^2 (see Figure 3.2).

The optimum $\psi_0(\mathcal{L}_n)$ is given by

$$\psi_0(\mathcal{L}_n) = \max_{\Delta^2}\psi_s^*(\mathcal{L}_n) = \frac{(1+\frac{1}{m}\mathcal{L}_n)\left(\kappa^{(1)}\right)^2}{(m+3)\kappa^{(2)}}. \tag{3.5.22}$$

If $\mathcal{L}_n < \frac{m}{m+2}$, then $\frac{1+\frac{1}{m}\mathcal{L}_n}{m+3} < \frac{1}{m+2}$, which implies also that

$$\psi_s^*(\mathcal{L}_n) < \psi_0(\mathcal{L}_n) \le \frac{1}{m+2}$$

for all Δ^2, that is, $\psi_0(\mathcal{L}_n)$ is closer to the minimizing value than $\frac{1}{m+2}$. So it is obvious that for each Δ^2 and $\mathcal{L}_n$

$$\frac{1}{\sigma_\varepsilon^4}E\left\{[\psi_s(\mathcal{L}_n)mS_u^2-\sigma_\varepsilon^2]^2\middle|\mathcal{L}_n\right\} \le \frac{1}{\sigma_\varepsilon^4}E\left\{\left[\frac{mS_u^2}{m+2}-\sigma_\varepsilon^2\right]^2\middle|\mathcal{L}_n\right\} \tag{3.5.23}$$

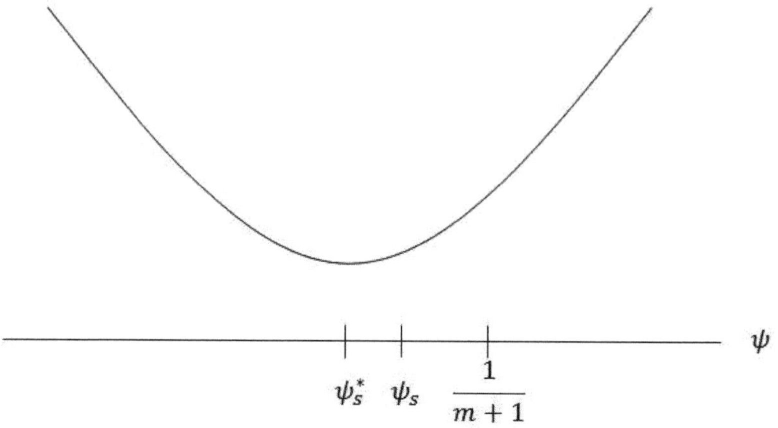

Figure 3.2 Quadratic relation in (3.5.20)

so that $m\psi_s(\mathcal{L}_n)S_u^2$ dominates $\frac{mS_u^2}{m+2} = \tilde{\sigma}_\varepsilon^2$ uniformly in $\Delta^2 \in (0, \infty)$.

Similarly, we consider the $S_{PT[1]}^2$ with the mean square error $M_5(S_{PT[1]}^2)$, which is optimum at the critical value 1 for all $(1, m)$ under H_0. Then,

$$S_{PT[1]}^2 = S_u^2 I\ (\mathcal{L}_n \geq 1) + S_R^2 I\ (\mathcal{L}_n < 1).$$

Using Stein's method, we have optimum ψ-function as

$$\psi_{10}(\mathcal{L}_n) = \frac{1 + \frac{1}{m}\mathcal{L}_n}{m+1} < \frac{1}{m} \quad \text{for } \mathcal{L}_n \leq 1$$

for all Δ^2. This means that $\psi_{10}(\mathcal{L}_n)$ is closer to the minimum value than $1/m$. Hence,

$$\begin{aligned} E\Big[\{\psi_1(\mathcal{L}_n)\chi_m^2 - 1\}^2\Big|\mathcal{L}_n\Big] &\leq E\Big[\Big(\frac{\chi_m^2}{m+2} - 1\Big)^2\Big|\mathcal{L}_n\Big] \\ &\leq \frac{1}{(m+2)^2}E[\{\psi_1(\mathcal{L}_n)\chi_m^2 - (m+2)\}^2\Big|\mathcal{L}_n] \\ &\leq \frac{1}{m^2}E[\{\psi_1(\mathcal{L}_n)\chi_m^2 - m\}^2\Big|\mathcal{L}_n]. \end{aligned} \quad (3.5.24)$$

Thus, the estimator $m\psi_S(\mathcal{L}_n)S_u^2$ dominates the PTE(1) of σ_ε^2 with critical value 1.

Further, $m\psi_S(\mathcal{L}_n)S_u^2 \leq m\psi_2(\mathcal{L}_n)S_U^2$ and equality holds when the critical value is $(m/m+2)$. Thus, Stein-type estimator $m\psi_S(\mathcal{L}_n)S_u^2$ is superior to S_u^2 as well as PTE(1) and PTE(2) uniformly in Δ^2.

To have a better visualization of the performance of different scale estimators comparatively, their MSE expressions are displayed for $n = 30$, different levels of significance ($\alpha = 0.01$ and $\alpha = 0.05$), and degrees of freedom ($\gamma_o = 5, 10, 30, 100$) in Figure 3.2. The left panel is for $\alpha = 0.01$ and the right panel is for $\alpha = 0.05$. The interpretation is clear from the figure.

For particular considerations and more extension, interested readers may refer to Khan and Saleh (1995, 1998, 2003) and Arashi and Tabatabaey (2009).

3.6 Problems

1. Prove that if $Z \sim \mathcal{N}(\Delta, 1)$, then

$$E[\text{sign}Z] = 2\Phi(\Delta) - 1.$$

2. Show that the distribution of S_u^2 in Theorem 3.2.1 is beta type II (inverted beta).
3. Prove that S_u^2 is unbiased for σ_ε^2.
4. Prove (iii)-(iv) of Theorem 3.4.1.
5. Verify (ii)-(vb) of Theorem 3.4.2.
6. Complete the proof of Theorem 3.4.3.
7. Prove Theorem 3.5.1.
8. Prove Theorem 3.5.2 (i) and (ii).

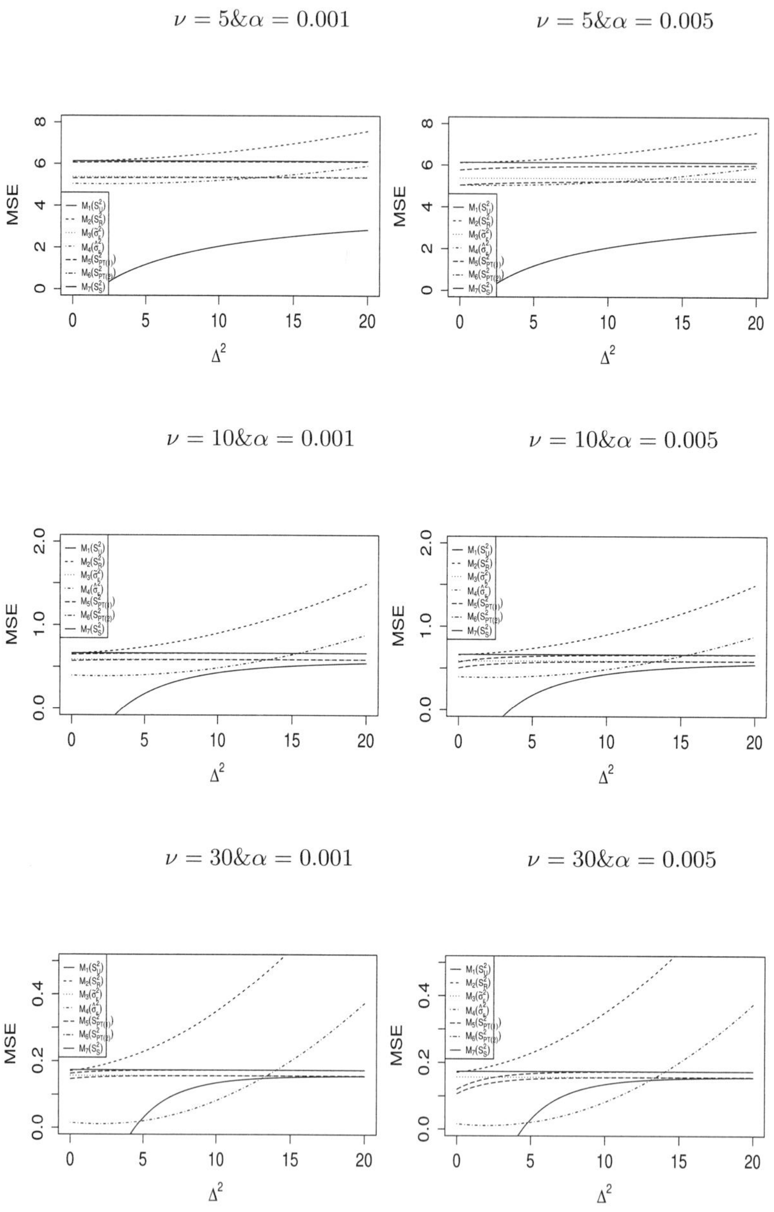

Figure 3.3 Plots of MSE functions (increasing d.f. from up to down).

CHAPTER 4

SIMPLE REGRESSION MODEL

Outline

In this chapter, we discuss some of the basic results from a simple regression model under the assumption that the error-vector is distributed according to the multivariate t-distribution. For normal errors see Saleh (2006) and for t-errors see Khan and Saleh (1997, 2008).

Statistical Inference for Models with Multivariate t-Distributed Errors, First Edition.

A. K. Md. Ehsanes Saleh, M. Arashi, S.M.M. Tabatabaey.

4.1 Introduction

Consider a simple linear model

$$\boldsymbol{Y} = \theta\mathbf{1} + \beta\boldsymbol{x} + \boldsymbol{\varepsilon} = \boldsymbol{A}\boldsymbol{\eta} + \boldsymbol{\varepsilon}, \quad \boldsymbol{A} = [\mathbf{1}, \boldsymbol{x}], \quad \boldsymbol{\eta} = (\theta, \beta)', \tag{4.1.1}$$

where $\boldsymbol{Y} = (Y_1, Y_2, \ldots, Y_n)'$ is the response vector and $\boldsymbol{x} = (x_1, x_2, \ldots, x_n)'$ is a fixed vector of known constants, while $\boldsymbol{\varepsilon} = (\varepsilon_1, \varepsilon_2, \ldots, \varepsilon_n)'$ is the error vector distributed according to the low belonging to the class of multivariate t-distributions, say, $M_t^{(n)}(\mathbf{0}, \sigma^2\mathbf{V}_n, \gamma_o)$.

As in Chapter 3, the covariance matrix is formulated as

$$\begin{aligned} E(\boldsymbol{\varepsilon}'\boldsymbol{\varepsilon}) &= \int_0^\infty W_o(t) Cov\left\{\mathcal{N}_n(\mathbf{0}, t^{-1}\sigma^2\mathbf{V}_n)\right\} dt \\ &= \sigma^2\kappa^{(1)}\mathbf{V}_n = \sigma_{\boldsymbol{\varepsilon}}^2\mathbf{V}_n. \end{aligned} \tag{4.1.2}$$

4.2 Estimation and Testing of $\boldsymbol{\eta}$

In this section, we consider LSE of $\boldsymbol{\eta}$ and test of hypothesis, $H_0 : \boldsymbol{\eta} = \boldsymbol{\eta}_o$ vs $H_A : \boldsymbol{\eta} \neq \boldsymbol{\eta}_o$.

4.2.1 Estimation of $\boldsymbol{\eta}$

For the LSE of $\boldsymbol{\eta}$, we minimize

$$\min_{\boldsymbol{\eta}} (\boldsymbol{Y} - \boldsymbol{A}\boldsymbol{\eta})'\mathbf{V}_n^{-1}(\boldsymbol{Y} - \boldsymbol{A}\boldsymbol{\eta})$$

to obtain the LSE of $\boldsymbol{\eta}$ as

$$\begin{aligned} \tilde{\boldsymbol{\eta}} &= \left(\boldsymbol{A}'\mathbf{V}_n^{-1}\boldsymbol{A}\right)^{-1}\left(\boldsymbol{A}'\mathbf{V}_n^{-1}\mathbf{Y}\right) \\ &= \boldsymbol{K}^{-1}\left(\boldsymbol{A}'\mathbf{V}_n^{-1}\mathbf{Y}\right) \\ &= \begin{pmatrix} \tilde{\theta}_n \\ \tilde{\beta}_n \end{pmatrix}, \end{aligned} \tag{4.2.1}$$

where

$$\begin{aligned} \boldsymbol{K} &= \boldsymbol{A}'\mathbf{V}_n^{-1}\boldsymbol{A} \\ &= \begin{pmatrix} K_1 & K_3 \\ K_3 & K_2 \end{pmatrix}^{-1} \begin{bmatrix} \mathbf{1}'\mathbf{V}_n^{-1}\mathbf{Y} \\ \boldsymbol{x}'\mathbf{V}_n^{-1}\mathbf{Y} \end{bmatrix}, \end{aligned} \tag{4.2.2}$$

with $K_1 = \mathbf{1}'\mathbf{V}_n^{-1}\mathbf{1}$, $K_2 = \boldsymbol{x}'\mathbf{V}_n^{-1}\boldsymbol{x}$, and $K_3 = \mathbf{1}'\mathbf{V}_n^{-1}\boldsymbol{x} = \boldsymbol{x}'\mathbf{V}_n^{-1}\mathbf{1}$.

Theorem 4.2.1. *Assume in the simple linear model* (4.1.1), $\boldsymbol{Y} \sim M_t^{(n)}(\boldsymbol{\eta}, \sigma^2\mathbf{V}_n, \gamma_o)$; *then we have* $\tilde{\boldsymbol{\eta}} \sim M_t^{(2)}(\boldsymbol{\eta}, \sigma^2\boldsymbol{K}^{-1}, \gamma_o)$.

Proof: Under the assumption $\boldsymbol{Y}|t \sim \mathcal{N}_n(\boldsymbol{\eta}, \sigma^2 t^{-1}\mathbf{V}_n)$, the exact distribution of $\tilde{\boldsymbol{\eta}}$ follows $\mathcal{N}_2(\boldsymbol{\eta}, \sigma^2 t^{-1}\boldsymbol{K}^{-1})$, where

$$\begin{aligned}\boldsymbol{K}^{-1} = (\boldsymbol{A}'\mathbf{V}_n^{-1}\boldsymbol{A})^{-1} &= \begin{pmatrix} K_1 & K_3 \\ K_3 & K_2 \end{pmatrix}^{-1} \\ &= \frac{1}{K_1K_2 - K_3^2}\begin{pmatrix} K_2 & -K_3 \\ -K_3 & K_1 \end{pmatrix}.\end{aligned}$$

Thus, we get

$$f_{\boldsymbol{Y}}(\boldsymbol{y}) = \int_0^\infty W_o(t)\mathcal{N}_2\left(\boldsymbol{\eta}, \sigma^2 t^{-1}\boldsymbol{K}^{-1}\right) dt,$$

which completes the proof.

Also the unbiased estimator of σ_ε^2 is S_u^2 given by

$$S_u^2 = m^{-1}(\boldsymbol{Y} - \boldsymbol{A}\tilde{\boldsymbol{\eta}})'\mathbf{V}_n^{-1}(\boldsymbol{Y} - \boldsymbol{A}\tilde{\boldsymbol{\eta}})\,; \qquad m = n - 2. \tag{4.2.3}$$

4.2.2 Test of Intercept Parameter

At this step, first we propose a test statistic of the parameter $\boldsymbol{\eta}$, and then we focus on the problem of estimation of the intercept parameter in a more precise setup.

Theorem 4.2.2. *For the test of the null hypothesis* $H_0 : \boldsymbol{\eta} = \boldsymbol{\eta}_o$ *vs* $H_A : \boldsymbol{\eta} \neq \boldsymbol{\eta}_o$, *consider parameter space* Ω *and subspace* ω *as defined below*

$$\begin{aligned}\Omega_1 &= \{(\boldsymbol{\eta}, \sigma, \mathbf{V}_n) : \boldsymbol{\eta} \in \mathbb{R}^2, \sigma \in \mathbb{R}^+, \mathbf{V}_n > 0\}, \textit{and}, \\ \Omega_0 &= \{(\boldsymbol{\eta}, \sigma, \mathbf{V}_n) : \boldsymbol{\eta} = \boldsymbol{\eta}_o = (\theta_o, \beta_o)', \boldsymbol{\eta}_o \in \mathbb{R}^2, \sigma \in \mathbb{R}^+, \mathbf{V}_n > 0\}.\end{aligned}$$

Then the LR criterion for testing the hypothesis $H_o : \boldsymbol{\eta} = \boldsymbol{\eta}_o$ *is given by*

$$\mathcal{L}_n^{**} = S_u^{-2}\left[\frac{1}{2}(\tilde{\boldsymbol{\eta}} - \boldsymbol{\eta}_o)'\boldsymbol{K}(\tilde{\boldsymbol{\eta}} - \boldsymbol{\eta}_o)\right]$$

and it has the following generalized noncentral F *distribution with the pdf*

$$g_{2,m}^{**}(\mathcal{L}_n) = \sum_{r\geq 0} \frac{\left(\frac{2}{m}\right)^{\frac{1}{2}(2+2r)} \mathcal{L}_n^{\frac{1}{2}(2r)} K_r^{(0)}(\Delta_{**}^2)}{r!\, B\left(\frac{2r+2}{2}, \frac{m}{2}\right)\left(1 + \frac{2}{m}\mathcal{L}_n\right)^{\frac{1}{2}(2+2r+m)}}$$

where $\Delta^2_{**} = \xi_{**}/\sigma^2_{\varepsilon}$ *for* $\xi_{**} = (\boldsymbol{\eta} - \boldsymbol{\eta}_o)'\boldsymbol{K}(\boldsymbol{\eta} - \boldsymbol{\eta}_o)$, *and* $K_r^{(0)}(\Delta^2_{**})$ *is given by equation* (2.6.11).

Proof: For the test of the null hypothesis $H_o : \boldsymbol{\eta} = \boldsymbol{\eta}_o$ versus $H_A : \boldsymbol{\eta} \neq \boldsymbol{\eta}_o$, let

$$\begin{aligned}\tilde{\sigma}^2_{\varepsilon} &= \frac{1}{m}(\boldsymbol{Y} - \boldsymbol{A}\tilde{\boldsymbol{\eta}})'\mathbf{V}_n^{-1}(\boldsymbol{Y} - \boldsymbol{A}\tilde{\boldsymbol{\eta}}),\\ \hat{\sigma}^2_{\varepsilon} &= \frac{1}{m+1}(\boldsymbol{Y} - \boldsymbol{A}\boldsymbol{\eta}_o)'\mathbf{V}_n^{-1}(\boldsymbol{Y} - \boldsymbol{A}\boldsymbol{\eta}_o).\end{aligned}$$

Further, let L_i, $i = 0, 1$, be the largest value that the likelihood function takes in region Ω_i, $i = 0, 1$. Then, we have

$$\begin{aligned}\Lambda &= \frac{L_0}{L_1}\\ &= a\left[\frac{(\boldsymbol{Y} - \boldsymbol{A}\tilde{\boldsymbol{\eta}})'\mathbf{V}_n^{-1}(\boldsymbol{Y} - \boldsymbol{A}\tilde{\boldsymbol{\eta}})}{(\boldsymbol{Y} - \boldsymbol{A}\boldsymbol{\eta}_o)'\mathbf{V}_n^{-1}(\boldsymbol{Y} - \boldsymbol{A}\boldsymbol{\eta}_o)}\right]^{\frac{n}{2}}\\ &= a\left(\frac{mS_u^2}{mS_u^2 + (\boldsymbol{\eta} - \boldsymbol{\eta}_o)'\boldsymbol{K}(\boldsymbol{\eta} - \boldsymbol{\eta}_o)}\right)^{\frac{n}{2}}\\ &= a\left(\frac{1}{1 + \frac{1}{m}\mathcal{L}_n^{**}}\right)^{\frac{n}{2}},\end{aligned}$$

where

$$a = \left(1 + \frac{1}{m}\right)^{\frac{n}{2}}\left(\frac{m + \gamma_o}{m + 1 + \gamma_o}\right)^{\frac{1}{2}(n+\gamma_o)}.$$

Hence, $\mathcal{L}_n^{**}$ is the LR test for testing the underlying null hypothesis. For its non-null distribution, we note that under normality, $\mathcal{L}_n^{**}$ follows the noncentral F-distribution with $(1, m)$ d.f. and noncentrality parameter $(\Delta^2_{**})_t = \frac{t(\boldsymbol{\eta}-\boldsymbol{\eta}_o)'\boldsymbol{K}(\boldsymbol{\eta}-\boldsymbol{\eta}_o)}{\sigma^2}$. Then, integrating over t w.r.t. the inverse gamma distribution W_o, the proof is completed.

Accordingly, we have

Corollary 4.2.2.1. *Under* H_o, *the pdf of* $\mathcal{L}_n^{**}$ *is given by*

$$g^{**}_{2,m}(\mathcal{L}_n) = \frac{\left(\frac{2}{m}\right)}{B\left(1, \frac{m}{2}\right)\left(1 + \frac{2}{m}\mathcal{L}_n\right)^{\frac{1}{2}(m+2)}},$$

which is the central F-distribution with $(2, m)$ d.f.

Corollary 4.2.2.2. *The power function at* γ*-level of significance of* $\mathcal{L}_n^{**}$, *say, generalized noncentral F cumulative distribution function of the statistic,* $\mathcal{L}_n^{**}$ *is given by* $1 - \mathcal{G}_{2,m}(l_\gamma; \Delta^2_{**})$ *where*

$$\mathcal{G}_{2,m}(l_\gamma; \Delta^2_{**}) = \sum_{r\geq 0}\frac{1}{r!}K_r^{(0)}(\Delta^2_{**})I_x\left[\frac{1}{2}(2 + 2r), \frac{m}{2}\right],$$

where $I_x(.,.)$ *is the incomplete Beta function,* $x = \frac{l_\gamma}{m+l_\gamma}$ *and* $l_\gamma = F_{1,m}(\gamma)$.

Straightforward consequences of Theorem 4.2.2 provide the test statistics for individuals $H_o : \theta = \theta_0$ and $H_o : \beta = \beta_o$. In order to test the null hypothesis $H_o : \beta = \beta_o$ against an alternative $H_A : \beta \neq \beta_o$, one uses the test statistic $\mathcal{L}_n^*$, defined by

$$\mathcal{L}_n = \frac{(\tilde{\beta}_n - \beta_o)^2 K_4}{S_u^2}; \qquad K_4 = \left(\frac{K_1}{K_1K_2 - K_3^2}\right)^{-1}.$$

Then, the exact distribution of $\mathcal{L}_n$ under H_o has the central F-distribution with $(1, m)$ d.f. Similarly, for the test of $H_o : \theta = \theta_o$ against $H_A : \theta \neq \theta_o$ one uses the test statistic

$$\mathcal{L}_n^* = \frac{(\tilde{\theta}_n - \theta_o)^2 K_5}{S_u^2}; \qquad K_5 = \left(\frac{K_2}{K_1K_2 - K_3^2}\right)^{-1}. \tag{4.2.4}$$

The exact distribution of $\mathcal{L}_n^*$ under H_o is central F-distribution with $(1, m)$ d.f. Note that, based on the virtue of (4.2.4), one can directly conclude the following result.

Lemma 4.2.1. *The LRC $\mathcal{L}_n$ for testing the hypothesis $H_o : \theta = \theta_o$ has the following distribution with the pdf*

$$g_{1,m}^*(\mathcal{L}_n) = \sum_{r\geq 0} \frac{\left(\frac{1}{m}\right)^{\frac{1}{2}(1+2r)} \mathcal{L}_n^{\frac{1}{2}(2r-1)} K_r^(0)(\Delta_*^2)}{r!\; B\left(\frac{2r+1}{2}, \frac{m}{2}\right)\left(1 + \frac{1}{m}\mathcal{L}_n\right)^{\frac{1}{2}(1+2r+m)}},$$

where $\Delta_*^2 = \xi_*/\sigma_\varepsilon^2$ *for* $\xi_* = K_5(\theta - \theta_o)^2$.

Now, we turn our attention to the estimation of **intercept and slope parameters** θ and β when it is suspected that the slope parameter β may be β_o. As a special case, it covers the two-sample problem of estimating one mean when it is suspected that the two means may be equal. Also, one-sample estimation of mean is obtained by letting $\boldsymbol{x} = \mathbf{0}$ and prior information $\theta = \theta_o$

4.2.3 Estimators of β and θ

In addition to $\tilde{\beta}_n$ and $\tilde{\theta}_n$, we present a few more estimators of β and θ. First, we consider the estimators of β. (i) The unrestricted estimator (UE) is $\tilde{\beta}_n$ and (ii) the restricted estimator (RE) is β_o. Thus, the preliminary test estimator (PTE) of β may be written as

$$\text{(iii)} \quad \hat{\beta}_n^{PT} = \tilde{\beta}_n - (\tilde{\beta}_n - \beta_o) I(\mathcal{L}_n < F_{1,m}(\alpha)), \tag{4.2.5}$$

and the Stein-type estimator of β is given by

$$\text{(iv)} \quad \hat{\beta}_n^S = \tilde{\beta}_n - c\frac{S_u(\tilde{\beta}_n - \beta_o)}{\sqrt{K_4}|\tilde{\beta}_n - \beta_o|}. \tag{4.2.6}$$

First of all, note that we have

$$\begin{aligned}\text{(i)}\quad \tilde{\theta}_n &= K_1^{-1}\mathbf{1}'\mathbf{V}_n^{-1}\mathbf{Y} - K_1^{-1}K_3\tilde{\beta}_n \\ &= K_1^{*}\mathbf{Y} - K_2^{*}\tilde{\beta}_n, \qquad K_1^{*} = K_1^{-1}\mathbf{1}'\mathbf{V}_n^{-1}, K_2^{*} = K_1^{-1}K_3. \quad (4.2.7)\end{aligned}$$

Replacing $\mathbf{V}_n$ by $\boldsymbol{I}_n$ in (4.2.7) results in $\tilde{\theta}_n = \bar{Y} - \bar{x}\tilde{\beta}_n$, as in Saleh (2006, p. 56). If we suspect β to be β_o, then the restricted estimator (RE) of θ is given by

$$\text{(ii)}\quad \hat{\theta}_n = K_1^{*}\mathbf{Y} - K_2^{*}\beta_o. \tag{4.2.8}$$

Now, following Saleh (2006), we define the estimators given below:

Preliminary test estimator (PTE) of θ is given by

$$\begin{aligned}\text{(iii)}\quad \hat{\theta}_n^{PT} &= \hat{\theta}_n\, I(\mathcal{L}_n < F_{1,m}(\alpha)) + \tilde{\theta}_n\, I(\mathcal{L}_n \geq F_{1,m}(\alpha)) \\ &= \tilde{\theta}_n + (\tilde{\beta}_n - \beta_o)K_2^{*} I(\mathcal{L}_n < F_{1,m}(\alpha)),\end{aligned} \tag{4.2.9}$$

where $F_{1,m}(\alpha)$ is the α-level upper critical value of a central F-distribution with $(1, m)$ d.f. and $I(A)$ is the indicator function of the set A.

Shrinkage-type estimators (SEs) of θ are given by

$$\begin{aligned}\text{(iv)}\quad \hat{\theta}_n^{S} &= \tilde{\theta}_n + \frac{c}{\sqrt{K_4}}\, S_u K_2^{*} \frac{(\tilde{\beta}_n - \beta_o)}{|(\tilde{\beta}_n - \beta_o)|}, \qquad c > 0 \\ \text{(va)}\quad \hat{\theta}_n(h) &= \frac{K_5}{K_5 + h}\tilde{\theta}_n, \quad h > 0 \\ \text{(vb)}\quad \hat{\theta}_n(d) &= \frac{K_5 + d}{K_5 + 1}\tilde{\theta}_n, \quad 0 < d < 1.\end{aligned} \tag{4.2.10}$$

4.3 Properties of Intercept Parameter

In this section, we derive the exact bias and MSE expressions for the proposed estimators of the intercept parameter.

4.3.1 Bias Expressions of the Estimators

The biases of $\tilde{\theta}_n$ and $\hat{\theta}_n$ are obvious and given by

$$b_1(\tilde{\theta}_n) = 0, \quad b_2(\hat{\theta}_n) = K_2^{*}(\beta - \beta_o). \tag{4.3.1}$$

For the PTE, let $Z = \frac{(\tilde{\beta}_n - \beta_o)}{\sqrt{t^{-1}\sigma^2 K_4}}$; then $Z|t \sim \mathcal{N}(\Delta_t, 1)$, where $\Delta_t = \frac{(\beta - \beta_o)}{\sqrt{t^{-1}\sigma^2 K_4}}$. Further, conditioning on t (under normality), $\frac{mS_u^2}{t^{-1}\sigma^2}|t \sim \chi_m^2$ is independent of $Z|t$. Thus we obtain

$$b_3(\hat{\theta}_n^{PT}) = E\left[\tilde{\theta}_n - \theta + (\tilde{\beta}_n - \beta_o)K_2^{*} I\left(\mathcal{L}_n < F_{1,m}(\alpha)\right)\right]$$

$$
\begin{aligned}
&= K_2^* E_t \left\{ E_N \left[\sqrt{t^{-1}\sigma^2 K_4} Z I \left(\frac{t^{-1}\sigma^2 Z^2}{S_u^2} < F_{1,m}(\alpha) \right) |t \right] \right\} \\
&= K_2^* \sqrt{K_4} E_t \left\{ \sqrt{t^{-1}\sigma^2} Z I \left(\frac{m\chi_1^2(\Delta_t)}{\chi_m^2} < F_{1,m}(\alpha) \right) |t \right\} \\
&= K_2^* \sqrt{K_4} \sigma_{\varepsilon} \Delta G_{3,m}^{(2)} \left(\frac{1}{3} F_{1,m}(\alpha); \Delta^2 \right), \qquad (4.3.2)
\end{aligned}
$$

where $\Delta^2 = \xi/\sigma_{\varepsilon}^2$ for $\xi = K_4(\beta - \beta_o)^2$.

Finally, for the bias expression of SE, we have

$$
\begin{aligned}
b_4(\hat{\theta}_n^S) &= E\left[\tilde{\theta}_n + c(\tilde{\beta}_n - \beta_o) K_2^* \left| \mathcal{L}_n^{\frac{1}{2}} \right|^{-1} - \theta \right] \\
&= K_2^* E \left[c(\tilde{\beta}_n - \beta_o) \frac{S_u}{(\tilde{\beta}_n - \beta_o)\sqrt{K_4}} \right] \\
&= c K_2^* K_4^{-\frac{1}{2}} E_t \left\{ E \left[Z \left| \frac{S_u}{Z} \right| |t \right] \right\}. \qquad (4.3.3)
\end{aligned}
$$

Inasmuch as $Z|t \sim \mathcal{N}(\Delta_t, 1)$ is independent of $S_u^2|t$, the expression in (4.3.3) simplifies to

$$
\begin{aligned}
b_4(\hat{\theta}_n^S) &= c K_2^* K_4^{-\frac{1}{2}} \int_0^\infty W_o(t) E \left[Z \left| \frac{S_u}{Z} \right| |t \right] dt \\
&= c K_2^* K_4^{-\frac{1}{2}} \int_0^\infty W_o(t) E \left[\sqrt{\frac{m S_u^2}{t^{-1}\sigma^2}} \sqrt{\frac{t^{-1}\sigma^2}{m}} |t \right] E \left[\frac{Z}{|Z|} |t \right] dt \\
&= c K_2^* K_4^{-\frac{1}{2}} \int_0^\infty W_o(t) \sqrt{\frac{t^{-1}\sigma^2}{m}} E \left[\frac{Z}{|Z|} |t \right] dt \\
&= c K_2^* K_4^{-\frac{1}{2}} \frac{\Gamma\left(\frac{m+1}{2}\right)}{\Gamma\left(\frac{m}{2}\right)} \sqrt{\frac{\sigma^2}{2m}} \int_0^\infty t^{-\frac{1}{2}} \left[1 - 2\Phi(-\Delta_t)\right] W_o(t) dt. \qquad (4.3.4)
\end{aligned}
$$

4.3.2 MSE Expressions of the Estimators

Using Theorem 4.2.1, we get

$$
M_1(\tilde{\theta}_n) = \sigma_{\varepsilon}^2 K_2 (K_1 K_2 - K_3^2)^{-1}. \qquad (4.3.5)
$$

For the restricted estimator, applying Theorem 4.2.1, we have

$$\begin{aligned}
M_2(\hat{\theta}_n) &= E\left[(\tilde{\theta}_n-\theta)+K_2^*(\tilde{\beta}_n-\beta_o)\right]^2 \\
&= M_1(\tilde{\theta}_n)+K_2^{*2}E(\tilde{\beta}_n-\beta_o)^2+2K_2^*E\left[(\tilde{\theta}_n-\theta)(\tilde{\beta}_n-\beta_o)\right] \\
&= \sigma_\varepsilon^2K_2(K_1K_2-K_3^2)^{-1}+K_2^{*2}\left[\frac{K_1\sigma_\varepsilon^2}{K_1K_2-K_3^2}+(\beta-\beta_o)^2\right] \\
&\quad -2K_2^*\frac{K_3\sigma_\varepsilon^2}{K_1K_2-K_3^2} \\
&= \left(K_1^{-1}+\Delta^2K_4{}^{-1}\right)\sigma_\varepsilon^2. \qquad (4.3.6)
\end{aligned}$$

For the MSE of PTE, using equation (3.2.9b) of Saleh (2006), we can obtain

$$\begin{aligned}
M_3(\hat{\theta}_n^{PT}) &= E\left[(\tilde{\theta}_n-\theta)+K_2^*(\tilde{\beta}_n-\beta_o)I(\mathcal{L}_n<F_{1,m}(\alpha))\right]^2 \\
&= M_1(\tilde{\theta}_n) \\
&\quad +K_2^{*2}K_4{}^{-1}E_t\left\{E_N\left[(t^{-1}\sigma_\varepsilon^2)Z^2I\left(\frac{Z^2}{\chi_m^2/m}<F_{1,m}(\alpha)\right)|t\right]\right\} \\
&\quad -2K_2^{*2}K_4{}^{-1}E_t\left\{E_N\left[(t^{-1}\sigma_\varepsilon^2)Z^2I\left(\frac{Z^2}{\chi_m^2/m}<F_{1,m}(\alpha)\right)|t\right]\right\} \\
&\quad +2K_2^{*2}K_4{}^{-1}\sigma_\varepsilon\Delta E_t\left\{E\left[\sqrt{t^{-1}\sigma_\varepsilon^2}ZI\left(\frac{Z^2}{\chi_m^2/m}<F_{1,m}(\alpha)\right)|t\right]\right\} \\
&= \sigma_\varepsilon^2K_2(K_1K_2-K_3^2)^{-1} \\
&\quad +2\sigma_\varepsilon^2\Delta^2K_2^{*2}K_4{}^{-1}\left[G_{3,m}^{(2)}\left(\frac{1}{3}F_{1,m}(\alpha);\Delta^2\right)\right] \\
&\quad -\sigma_\varepsilon^2K_2^{*2}K_4{}^{-1}\left\{G_{3,m}^{(1)}\left(\frac{1}{3}F_{1,m}(\alpha);\Delta^2\right)\right. \\
&\quad \left.+\Delta^2G_{5,m}^{(2)}\left(\frac{1}{5}F_{1,m}(\alpha);\Delta^2\right)\right\}. \qquad (4.3.7)
\end{aligned}$$

Finally, for the shrinkage estimator, by making use of equation (3.3.3), we have

$$\begin{aligned}
M_4(\hat{\theta}_n^S) &= E\left[(\tilde{\theta}_n-\theta)+c\frac{S_u}{\sqrt{K_4}}\frac{(\tilde{\beta}_n-\beta_o)}{|\tilde{\beta}_n-\beta_o|}\right]^2 \\
&= E(\tilde{\theta}_n-\theta)^2+\frac{c^2\sigma_\varepsilon^2}{K_4}+\frac{2c}{\sqrt{K_4}}E(S_u)E\left[\frac{(\tilde{\theta}_n-\theta)(\tilde{\beta}_n-\beta_o)}{|\tilde{\beta}_n-\beta_o|}\right] \\
&= \frac{\sigma_\varepsilon^2}{K_5}+\frac{c^2\sigma_\varepsilon^2}{K_4}+\frac{2cc_n\sigma_\varepsilon}{\sqrt{K_4}}E\left\{E_N\left[(\tilde{\theta}_n-\theta)\Big|(\tilde{\beta}_n-\beta_o)\right]\frac{(\tilde{\beta}_n-\beta_o)}{|\tilde{\beta}_n-\beta_o|}\right\} \\
&= \frac{\sigma_\varepsilon^2}{K_5}+\frac{c^2\sigma_\varepsilon^2}{K_4}+\frac{2cc_n\sigma_\varepsilon}{\sqrt{K_4}}E\left\{-\frac{K_3}{K_1}\left[(\tilde{\beta}_n-\beta)-(\beta-\beta_o)\right]\frac{(\tilde{\beta}_n-\beta_o)}{|\tilde{\beta}_n-\beta_o|}\right\}
\end{aligned}$$

$$\begin{aligned}
&= \frac{\sigma_\varepsilon^2}{K_5} + \frac{c^2\sigma_\varepsilon^2}{K_4} - \frac{2cc_n\sigma_\varepsilon}{\sqrt{K_4}}\frac{K_3}{K_1}E\left[\frac{(\tilde{\beta}_n-\beta_o)^2}{|\tilde{\beta}_n-\beta_o|}\right] \\
&\quad + \frac{2cc_n\sigma_\varepsilon}{\sqrt{K_4}}\frac{K_3(\beta-\beta_o)}{K_1}E\left[\frac{(\tilde{\beta}_n-\beta_o)}{|\tilde{\beta}_n-\beta_o|}\right] \\
&= \frac{\sigma_\varepsilon^2}{K_5} + \frac{c^2\sigma_\varepsilon^2}{K_4} - \frac{2cc_n\sigma_\varepsilon^2}{\sqrt{K_4}}\frac{K_3}{K_1}E_t\left\{E_N\left[\,|Z| - \Delta_t \mathrm{sign}(Z)\right]|t\right\} \\
&= \frac{\sigma_\varepsilon^2}{K_5} + \frac{c^2\sigma_\varepsilon^2}{K_4} - \frac{2cc_n\sigma_\varepsilon^2}{\sqrt{K_4}}\frac{K_3}{K_1}E_t\left\{\sqrt{\frac{2}{\pi}}e^{-\frac{\Delta_t^2}{2}}\right\} \\
&= \frac{\sigma_\varepsilon^2}{K_5} + \frac{c^2\sigma_\varepsilon^2}{K_4} - \frac{2cc_n\sigma_\varepsilon^2}{\sqrt{K_4}}\frac{K_3}{K_1}\sqrt{\frac{2}{\pi}}\left(1+\frac{\Delta^2}{\gamma_o-2}\right)^{-\frac{\gamma_o}{2}} \\
&= \frac{\sigma_\varepsilon^2}{K_5} + \frac{\sigma_\varepsilon^2}{K_4}\left[c^2 - 2cc_n\frac{K_3}{K_1}\sqrt{\frac{2}{\pi}}\left(1+\frac{\Delta^2}{\gamma_o-2}\right)^{-\frac{\gamma_o}{2}}\right].
\end{aligned}$$

Minimize $M_4(\hat{\theta}_n^S)$ w.r.t. c to get

$$c_{\text{opt}} = \sqrt{\frac{2}{\pi}}c_n\frac{K_3}{K_1}\left(1+\frac{\Delta^2}{\gamma_o-2}\right)^{-\frac{\gamma_o}{2}}.$$

It is obvious that c_{opt} depends on Δ^2, but we want it to be free of Δ^2. Thus, we choose

$$c^* = \frac{K_3}{K_1}\sqrt{\frac{2}{\pi}}c_n.$$

Substituting c^* in the expression of $M_4(\hat{\theta}_n^S)$ with c in it, we obtain

$$\begin{aligned}
M_4(\hat{\theta}_n^S) &= \frac{\sigma_\varepsilon^2}{K_5} + \frac{\sigma_\varepsilon^2}{K_4}\left[\frac{2}{\pi}\frac{K_3^2}{K_1^2}c_n^2 - \frac{2}{\pi}\frac{K_3^2}{K_1}c_n^2\left(1+\frac{\Delta^2}{\gamma_o-2}\right)^{-\frac{\gamma_o}{2}}\right] \\
&= \frac{\sigma_\varepsilon^2}{K_5} - \frac{2}{\pi}\frac{\sigma_\varepsilon^2}{K_4}\frac{K_3^2}{K_1^2}c_n\left[2\left(1+\frac{\Delta^2}{\gamma_o-2}\right)^{-\frac{\gamma_o}{2}} - 1\right] \\
&= \frac{\sigma_\varepsilon^2 K_2}{(K_1K_2 - K_3^2)}\left[1 - \frac{2}{\pi}\frac{K_3^2}{K_1K_2}c_n^2\left\{2\left(1+\frac{\Delta^2}{\gamma_o-2}\right)^{-\frac{\gamma_o}{2}} - 1\right\}\right].
\end{aligned}$$

The derivation of $M_5(\hat{\theta}_n(h))$ and $M_5(\hat{\theta}_n(d))$ are obvious.

4.4 Comparison

In this section, we compare the proposed estimators with respect to their MSE functions. The mean-square relative efficiency (MRE) of $\hat{\theta}_n$ compared to $\tilde{\theta}_n$ may be written as

$$\begin{aligned}
\text{MRE}(\hat{\theta}_n;\tilde{\theta}_n) &= \frac{M_1(\tilde{\theta}_n)}{M_2(\hat{\theta}_n)} \\
&= \frac{(K_1K_2 - K_3^2)^{-1}\sigma_\varepsilon^2}{(K_1^{-1} + \Delta^2 K_4^{-1})\sigma_\varepsilon^2} \\
&= \frac{K_1K_4K_2}{(K_4 + \Delta^2 K_1)(K_1K_2 - K_3^2)} \\
&= \frac{K_2}{K_4 + \Delta^2 K_1}. \qquad (4.4.1)
\end{aligned}$$

The efficiency is a decreasing function of Δ^2. Under $H_o : \beta = \beta_o$, it has the maximum

$$\text{MRE}(\hat{\theta}_n;\tilde{\theta}_n) = \frac{K_2}{K_4}. \qquad (4.4.2)$$

In general to compare $\hat{\theta}_n$ and $\tilde{\theta}_n$ using (4.4.1), $\text{MRE}(\hat{\theta}_n;\tilde{\theta}_n) > 1$ whenever $\Delta^2 < (\frac{K_3}{K_1})^2$.

The mean-square relative efficiency of $\hat{\theta}_n^{PT}$ compared to $\tilde{\theta}_n$ is given by

$$\text{MRE}(\hat{\theta}_n^{PT};\tilde{\theta}_n) = [1 + g(\Delta^2)]^{-1}, \qquad (4.4.3)$$

where

$$\begin{aligned}
g(\Delta^2) = \quad & -\frac{K_2^{*2}K_1}{K_2}\left\{G_{3,m}^{(1)}\left(\frac{1}{3}F_{1,m}(\alpha);\Delta^2\right)\right. \\
&\left.+\Delta^2\left(G_{5,m}^{(2)}\left(\frac{1}{5}F_{1,m}(\alpha);\Delta^2\right) - 2G_{3,m}^{(2)}\left(\frac{1}{3}F_{1,m}(\alpha);\Delta^2\right)\right)\right\}.
\end{aligned} \qquad (4.4.4)$$

Under H_o, it has the maximum value

$$\text{MRE}(\hat{\theta}_n^{PT};\tilde{\theta}_n) = \left\{1 - \frac{K_2^{*2}K_1}{K_2}G_{3,m}^{(1)}\left(\frac{1}{3}F_{1,m}(\alpha);0\right)\right\}^{-1}. \qquad (4.4.5)$$

In general, $\text{MRE}(\hat{\theta}_n^{PT};\tilde{\theta}_n) \gtreqless 1$ according to

$$\Delta^2 \lesseqgtr \frac{G_{3,m}^{(1)}\left(\frac{1}{3}F_{1,m}(\alpha);\Delta^2\right)}{2G_{3,m}^{(2)}\left(\frac{1}{3}F_{1,m}(\alpha);\Delta^2\right) - G_{5,m}^{(2)}\left(\frac{1}{5}F_{1,m}(\alpha);\Delta^2\right)}. \qquad (4.4.6)$$

The mean-square relative efficiency of $\hat{\theta}_n^S$ compared to $\tilde{\theta}_n$ is given by

$$\text{MRE}(\hat{\theta}_n^S;\tilde{\theta}_n) = \left[1 - \frac{2}{\pi}\frac{K_3^2}{K_1K_2}c_n^2\left\{2\left(1 + \frac{\Delta^2}{\gamma_o - 2}\right)^{-\frac{\gamma_o}{2}} - 1\right\}\right]^{-1}$$

$$= \quad [1 + h(\Delta^2)]^{-1}, \tag{4.4.7}$$

where

$$h(\Delta^2) \quad = \quad \frac{2}{\pi}\frac{K_3^2}{K_1K_2}c_n^2\left\{1 - 2\left(1 + \frac{\Delta^2}{\gamma_o - 2}\right)^{-\frac{\gamma_o}{2}}\right\}, \tag{4.4.8}$$

is an increasing function with respect to Δ^2. Under H_o, $\text{MRE}(\hat{\theta}_n^S; \tilde{\theta}_n)$ simplifies to

$$\text{MRE}(\hat{\theta}_n^S; \tilde{\theta}_n) = \left[1 - \frac{2}{\pi}\frac{K_3^2}{K_1K_2}c_n^2\right]^{-1} > 1. \tag{4.4.9}$$

As $\Delta^2 \to \infty$, $\text{MRE}(\hat{\theta}_n^S; \tilde{\theta}_n) \to 1$ and it crosses the 1-line at $\Delta^2 = (\gamma_o - 2)\left(2^{\frac{2}{\gamma_o}} - 1\right)$ $(\gamma_o \geq 3)$.

Thus, at $\Delta^2 = 0$, $\text{MRE}(\hat{\theta}_n^S; \tilde{\theta}_n)$ has maximum value $[1 - \frac{2}{\pi}\frac{K_3^2}{K_1K_2}c_n^2]^{-1} \geq 1$ then drops to a minimum value $[1 + \frac{2}{\pi}\frac{K_3^2}{K_1K_2}c_n^2]^{-1}$ as $\Delta^2 \to \infty$. The loss of efficiency is $1 - [1 + \frac{2}{\pi}\frac{K_3^2}{K_1K_2}c_n^2]^{-1}$ while the gain in efficiency is $[1 - \frac{2}{\pi}\frac{K_3^2}{K_1K_2}c_n^2]^{-1}$. Further, $\hat{\theta}_n^S$ performs better than $\tilde{\theta}_n$ in the interval $0 \leq \Delta^2 \leq (\gamma_o - 2)\left(2^{\frac{2}{\gamma_o}} - 1\right)$ and outside of this interval $\tilde{\theta}_n$ performs better.

4.4.1 Optimum Level of Significance of $\hat{\theta}_n^{PT}$

Following Section 3.2.4 of Saleh (2006), denote the relative efficiency of $\hat{\theta}_n^{PT}$ compared to $\tilde{\theta}_n$ by $\text{MRE}(\alpha, \Delta^2)$. Its maximum value occurs at $\Delta^2 = 0$ as given in (4.4.5), i.e., $\max_{\Delta^2} \text{MRE}(\alpha, \Delta^2) = \text{MRE}(\alpha, 0)$. Subsequently, in order to obtain preliminary test estimator with a minimum guaranteed efficiency E_0, for example, we adopt the following procedure: If $\Delta^2 \leq 1$, we always choose $\tilde{\theta}_n$. However, in general, Δ^2 is unknown, so there is no way to choose an estimator that is uniformly best. For this reason, we select an estimator with minimum guaranteed efficiency, such as E_0, and look for a suitable α from the set $A_0 = \left\{\alpha | \text{MRE}(\alpha, \Delta^2) \geq E_0\right\}$. The estimator chosen maximizes $\text{MRE}(\alpha, \Delta^2)$ over all $\alpha \in A_0$ and Δ^2. Thus, we solve the following equation for the optimum α^*:

$$\min_{\Delta^2} \text{MRE}(\alpha, \Delta^2) = E(\alpha, \Delta_0^2(\alpha)) = E_0. \tag{4.4.10}$$

The solution α^* obtained this way gives the PTE with minimum guaranteed efficiency E_0.

4.5 Numerical Illustration

According to Arashi, Tabatabaey and Soleimani (2012), assume that the known matrix $\mathbf{V}_6$ $(n = 6)$ is given by

$$\mathbf{V}_6 = \begin{bmatrix} 2.57 & 0.85 & 1.56 & 1.79 & 1.33 & 0.42 \\ 0.85 & 37.00 & 3.34 & 13.47 & 7.59 & 0.52 \\ 1.56 & 3.34 & 8.44 & 5.77 & 2.00 & 0.50 \\ 1.79 & 13.47 & 5.77 & 34.01 & 10.50 & 1.77 \\ 1.33 & 7.59 & 2.00 & 10.50 & 23.01 & 3.43 \\ 0.42 & 0.52 & 0.50 & 1.77 & 3.43 & 4.59 \end{bmatrix}.$$

Further, assume that $\boldsymbol{x}' = (2\ 6\ 1\ 8\ 3\ 4)$.

The graphs of PTE and SE biases versus Δ are displayed in Figures 4.1– 4.3. As it can be realized, when both the level of significance α and degrees of freedom ν increase, the bias of PTE decreases. The bias of SE performs the same as ν increases. Similar conclusions can be made for the MSE graphs in Figures 4.4 – 4.7.

For the MRE graphs in Figures 4.8 – 4.11, it can be concluded that the efficiency of $\hat{\theta}_n$ relative to $\tilde{\theta}_n$ is a decreasing function. MRE$(\hat{\theta}_n^{PT}; \tilde{\theta}_n)$ is a decreasing function relative to Δ and also for small level of significance α, the unrestricted estimator performs better than the PTE. This scenario has a little bit change for the d.f. γ_o; its behavior can be verified from Figure 4.10. Finally, the shrinkage estimator performs better than the unrestricted estimator as ν increases.

To conclude this section, Table 4.1 gives selected values of $\xi = \frac{K_2^{*2} K_1}{K_2}$ and $\alpha = 0.05(0.05)0.35$ for the procedure of choosing the level α^* of significance.

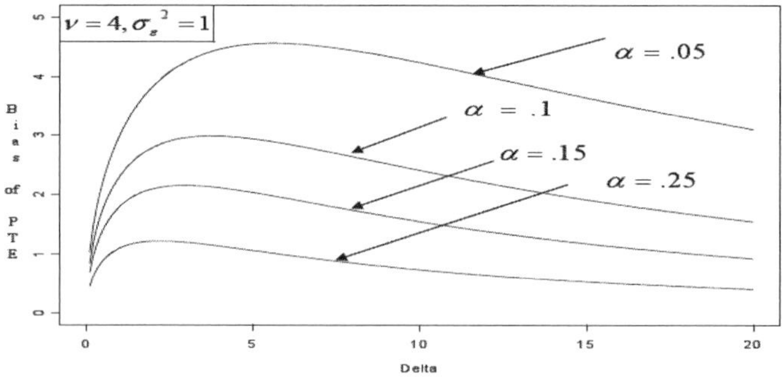

Figure 4.1 Graph of bias function for PTE

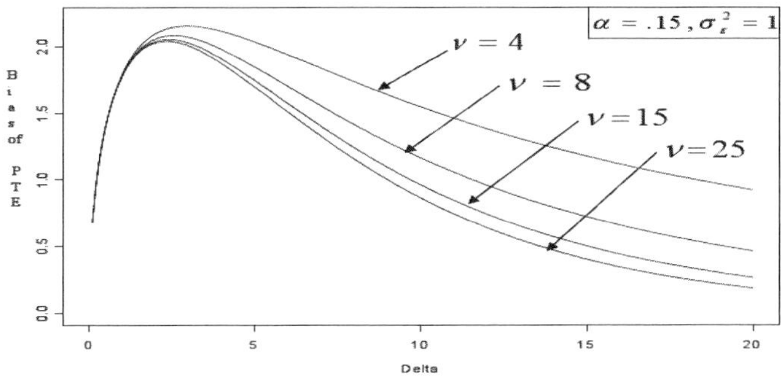

Figure 4.2 Graph of bias function for PTE

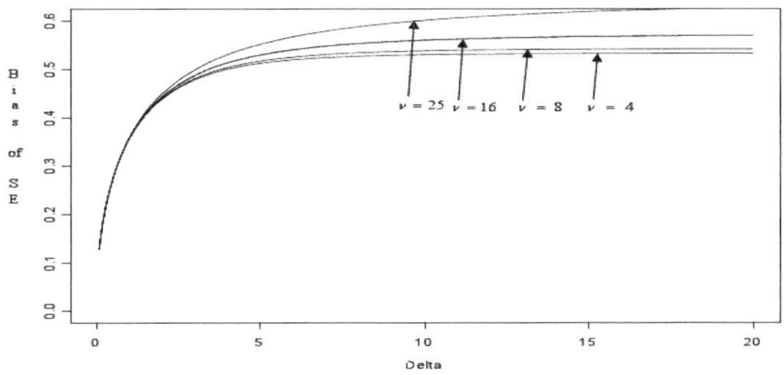

Figure 4.3 Graph of bias function for SE

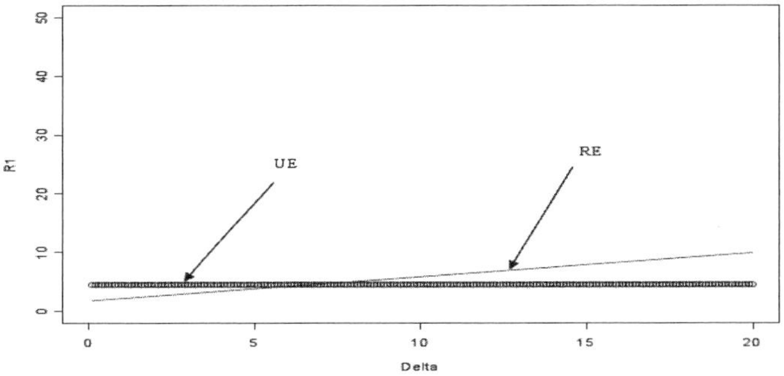

Figure 4.4 Graph of risk function for UE and RE

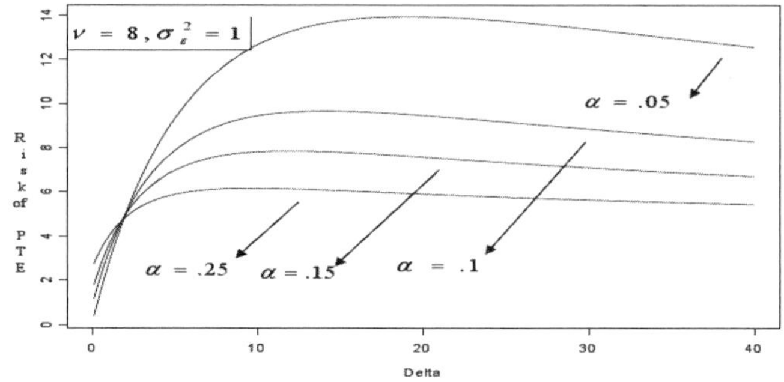

Figure 4.5 Graph of risk function for PTE

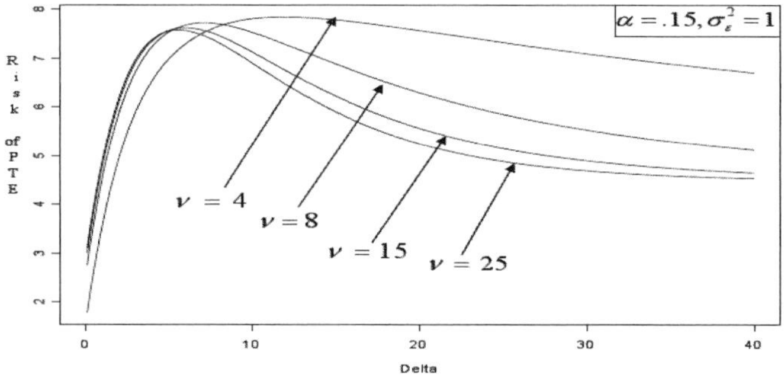

Figure 4.6 Graph of risk function for PTE

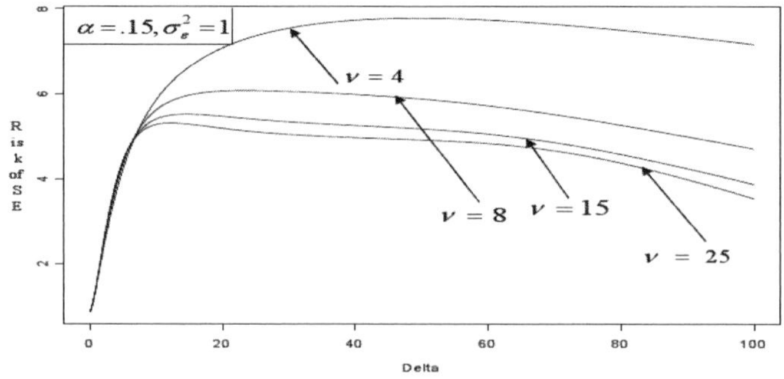

Figure 4.7 Graph of risk function for SE

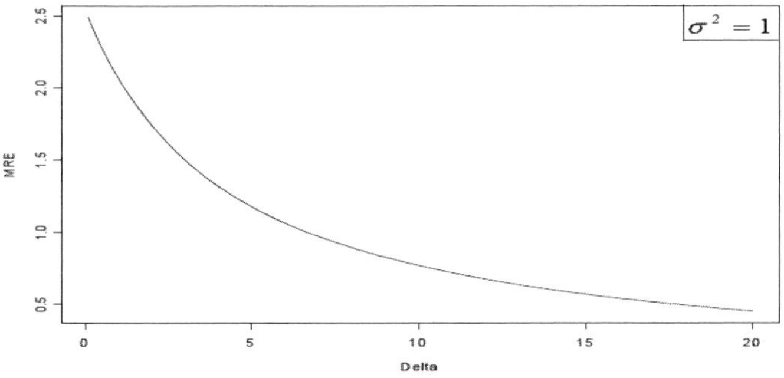

Figure 4.8 Graph of MRE (RE vs UE)

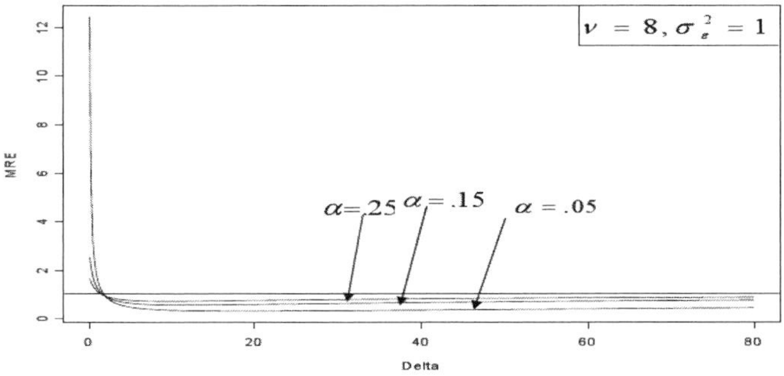

Figure 4.9 Graph of MRE (PTE vs UE)

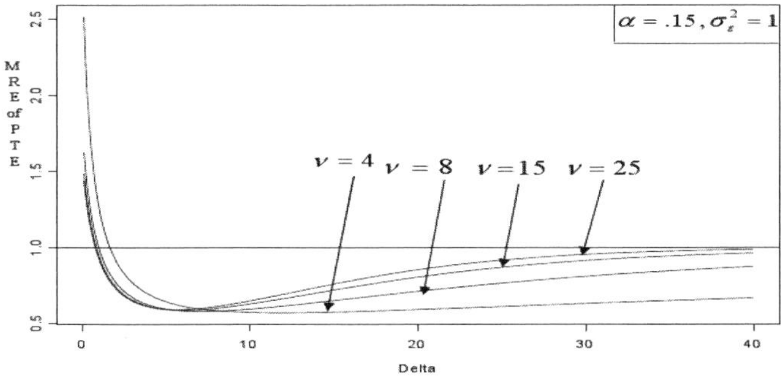

Figure 4.10 Graph of MRE (PTE vs UE)

Table 4.1 Maximum and minimum guaranteed efficiencies for n=6

α	ξ	0.1	0.2	0.3	0.4	0.5	0.6	0.7	0.8	0.9
0.05	E_{max}	1.12	1.27	1.47	1.75	2.16	2.82	4.06	7.23	32.83
	E_{min}	0.75	0.60	0.50	0.43	0.38	0.34	0.30	0.27	0.25
	Δ^2_{min}	12.10	12.10	12.10	12.10	12.10	12.10	12.10	12.10	12.10
0.1	E_{max}	1.09	1.21	1.35	1.54	1.78	2.11	2.60	3.38	4.81
	E_{min}	0.84	0.73	0.64	0.57	0.52	0.47	0.44	0.40	0.38
	Δ^2_{min}	8.70	8.70	8.70	8.70	8.70	8.70	8.70	8.70	8.70
0.15	E_{max}	1.07	1.16	1.27	1.40	1.56	1.75	2.01	2.35	2.83
	E_{min}	0.89	0.80	0.73	0.67	0.62	0.58	0.54	0.50	0.40
	Δ^2_{min}	7.10	7.10	7.10	7.10	7.10	7.10	7.10	7.10	7.10
0.20	E_{max}	1.06	1.13	1.21	1.30	1.41	1.53	1.69	1.87	2.10
	E_{min}	0.92	0.85	0.79	0.74	0.70	0.66	0.62	0.59	0.56
	Δ^2_{min}	6.20	6.20	6.20	6.20	6.20	6.20	6.20	6.20	6.20
0.25	E_{max}	1.04	1.10	1.16	1.23	1.30	1.39	1.49	1.60	1.73
	E_{min}	0.94	0.88	0.84	0.80	0.76	0.72	0.69	0.66	0.64
	Δ^2_{min}	5.60	5.60	5.60	5.60	5.60	5.60	5.60	5.60	5.60
0.30	E_{max}	1.03	1.08	1.12	1.17	1.23	1.29	1.35	1.42	1.51
	E_{min}	0.95	0.91	0.87	0.84	0.81	0.78	0.75	0.73	0.70
	Δ^2_{min}	5.20	5.20	5.20	5.20	5.20	5.20	5.20	5.20	5.20

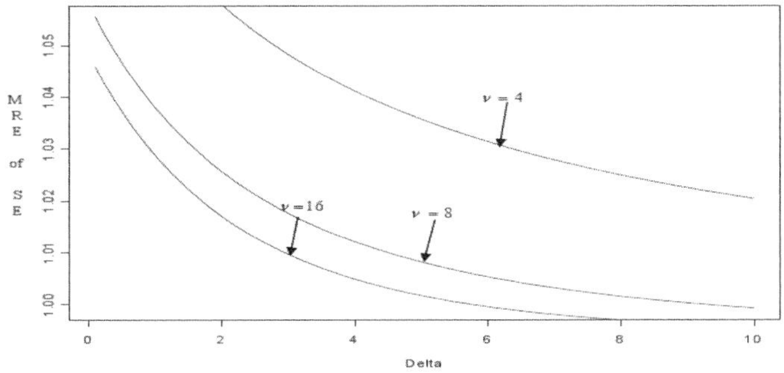

Figure 4.11 Graph of MRE (SE vs UE)

4.6 Problems

1. Refer to (4.1.1) as $\boldsymbol{Y} = \theta\mathbf{1} + \beta\boldsymbol{x} + \boldsymbol{\varepsilon} = \boldsymbol{A\eta} + \varepsilon$, $\boldsymbol{A} = [\mathbf{1}, \boldsymbol{x}]$, $\boldsymbol{\eta} = (\theta, \beta)'$, where $\boldsymbol{Y} = (Y_1, \ldots, Y_n)'$, $\boldsymbol{x} = (x_1, \ldots, x_n)'$, and $\boldsymbol{\varepsilon} = (\varepsilon_1, \ldots, \varepsilon_n)' \sim \mathcal{N}_n(\mathbf{0}, \sigma^2\boldsymbol{I}_n)$. Find the distribution of $\tilde{\boldsymbol{\eta}} = (\tilde{\theta}_n, \tilde{\beta}_n)'$.

2. For the joint test of the hypothesis $H_0 : \boldsymbol{\eta} = (\theta, \beta)' = (\theta_o, \beta_o)' = \boldsymbol{\eta}_o$ against the alternative $H_A : \boldsymbol{\eta} \neq \boldsymbol{\eta}_o$, find the LR test.

3. Find the power function of the test under the assumptions of problems 1 and 2.

4. Refer to section 4.2.3 and consider the restricted estimator $\hat{\theta}_n$ of the intercept parameter θ when β is suspected to be β_o. Find the bias and MSE of $\hat{\theta}_n$ under normality assumption.

5. Consider the preliminary test estimator $\hat{\theta}_n^{PT}$ of θ when β is suspected to be β_o. Find the bias and MSE of $\hat{\theta}_n^{PT}$ under $\mathcal{N}_n(\mathbf{0}, \sigma^2\boldsymbol{I}_n)$.

6. Consider the shrinkage-type estimator $\hat{\theta}_n^S$ of the intercept parameter θ when β is suspected to be β_o. Find the bias and MSE of $\hat{\theta}_n$ under $\mathcal{N}_n(\mathbf{0}, \sigma^2\boldsymbol{I}_n)$.

CHAPTER 5

ANOVA

Outline

In this chapter, we will discuss the ANOVA model in detail, beginning with the estimation of the model parameters and test of hypothesis. In addition, we shall provide some improved estimators of the ANOVA parameters with their dominance properties. Also included are various estimators of the scale parameter in the model along with their comparisons.

Statistical Inference for Models with Multivariate t-Distributed Errors, First Edition.

A. K. Md. Ehsanes Saleh, M. Arashi, S.M.M. Tabatabaey.

5.1 Model Specification

The analysis of variance (ANOVA) is a set of statistical techniques for studying the variability from different sources and comparing them to understand the relative importance of each the sources. It is also used to make inferences about the populations through the test of significance, including the very important comparison of two or more means of the population concerned. ANOVA is one of the most important models among many models belonging to the class of general linear models. This model may be written compactly as

$$\boldsymbol{Y} = \boldsymbol{B\theta} + \boldsymbol{E}, \tag{5.1.1}$$

where

$$\boldsymbol{Y} = (y_{11}, \cdots, y_{1n_1}, \cdots, y_{p1}, \cdots, y_{pn_p})', \tag{5.1.2}$$

in which y_{ij} is the value of measurement,

$$\boldsymbol{B} = \text{ Diag}(\mathbf{1}_{n_1}, ...\mathbf{1}_{n_p}) \tag{5.1.3}$$

is the block diagonal vectors, where $\mathbf{1}_{n_\alpha} = (1, \cdots, 1)'$ is a n_α-tuple of 1's,

$$\boldsymbol{\theta} = (\theta_1, \cdots, \theta_p)', \tag{5.1.4}$$

is the vector of treatments, and

$$\boldsymbol{E} = (\epsilon_1, \cdots, \epsilon_p)' \tag{5.1.5}$$

is the n-vector of random errors. Here, $n = n_1 + n_2 + \cdots + n_p$ is the total size of the samples collected from the p population.

The distribution of $\boldsymbol{E}$ is given by the pdf

$$g(\boldsymbol{E}) = \frac{\gamma_o^{\frac{\gamma_o}{2}}}{\pi^{\frac{n}{2}}\sigma^n\Gamma\left(\frac{\gamma_o}{2}\right)}\left(1 + \frac{1}{\gamma_o\sigma^2}\boldsymbol{E}'\boldsymbol{E}\right)^{-\frac{n+\gamma_o}{2}}.$$

In other words, y_{ij} represent the jth observation $(j = 1, \cdots, n_i)$ on the ith treatment $(i = 1, \cdots, p)$. In this chapter, we extend the classical normal theory of analysis of variance to the analysis of variance based on multivariate t-error distribution, say $M_t^{(n)}(\mathbf{0}, \sigma^2\mathbf{V}_n, \gamma_o)$, where $\mathbf{V}_n$ is a known p.d. symmetric matrix of size $n \times n$. Note that $\mathbf{V}_n$ is a block diagonal matrix

$$\mathbf{V}_n = \text{Diag}(\mathbf{V}_{n_1}, ...\mathbf{V}_{n_p}), \tag{5.1.6}$$

where each $\mathbf{V}_{n_\alpha}$ is a known p.d. symmetric matrix of size $n_\alpha \times n_\alpha$.

5.2 Proposed Estimators and Testing

In this section, we present the least square/maximum likelihood (LS/ML) estimates of $\boldsymbol{\theta}$ and σ^2. We suspect that $\boldsymbol{\theta}$ may satisfy the constraint $\boldsymbol{\theta} = \theta_0\mathbf{1}_p$, where θ_0 is unknown. In addition to the unrestricted estimator of $\boldsymbol{\theta}$, we exhibit the restricted estimator (RE), preliminary test estimator (PTE), Stein-type shrinkage estimator (SSE), and positive-rule shrinkage estimator (PRSE). Also, we obtain the LR criterion for testing the uncertain restriction, $\boldsymbol{\theta} = \theta_0 \mathbf{1}_p$.

The unrestricted LSE of $\boldsymbol{\theta}$ may be obtained by minimizing

$$\min_{\boldsymbol{\theta}} (\boldsymbol{Y} - \boldsymbol{B\theta})' \mathbf{V}_n^{-1} (\boldsymbol{Y} - \boldsymbol{B\theta}). \tag{5.2.1}$$

Then, the LSE of $\boldsymbol{\theta}$ is given by

$$\tilde{\boldsymbol{\theta}}_n = (\boldsymbol{B}'\mathbf{V}_n^{-1}\boldsymbol{B})^{-1}(\boldsymbol{B}\mathbf{V}_n^{-1}\boldsymbol{Y}), \tag{5.2.2}$$

which may be written as

$$\tilde{\boldsymbol{\theta}}_n = \boldsymbol{K}^{-1}\boldsymbol{Y}^*, \tag{5.2.3}$$

where

$$\boldsymbol{K} = \boldsymbol{B}'\mathbf{V}_n^{-1}\boldsymbol{B} = \text{Diag}\,(k_1, \cdots, k_p)\,, \quad k_\alpha = \mathbf{1}'_{n_\alpha} {\mathbf{V}_n^{(\alpha)}}^{-1} \mathbf{1}_{n_\alpha} \tag{5.2.4}$$

$$\boldsymbol{Y}^* = \boldsymbol{B}'\mathbf{V}_n^{-1}\boldsymbol{Y} = (y_1, \cdots, y_p)\,, \qquad y_\alpha = \mathbf{1}'_{n_\alpha} {\mathbf{V}_n^{(\alpha)}}^{-1} \boldsymbol{Y}_\alpha. \tag{5.2.5}$$

Also, the LSE of σ^2 is given by

$$\boldsymbol{S}_u = \frac{1}{n-p}(\boldsymbol{Y} - \boldsymbol{B}\tilde{\boldsymbol{\theta}}_n)'\mathbf{V}_n^{-1}(\boldsymbol{Y} - \boldsymbol{B}\tilde{\boldsymbol{\theta}}_n) = \frac{\widetilde{\boldsymbol{E}}'\mathbf{V}_n^{-1}\widetilde{\boldsymbol{E}}}{n-p}, \text{ where } \widetilde{\boldsymbol{E}} = \boldsymbol{Y} - \boldsymbol{B}\tilde{\boldsymbol{\theta}}_n. \tag{5.2.6}$$

Generally, the main object of ANOVA problems is the test of the null hypothesis $H_0 : \boldsymbol{\theta} = \theta_0 \mathbf{1}_p$ against the alternative $H_A : \boldsymbol{\theta} \neq \theta_0 \mathbf{1}_p$. Before revealing the underlying test, we propose the RE of the vector-treatment $\boldsymbol{\theta}$ under conditions in which the treatments are equal.

We now consider the restricted estimators of the parameters under the hypothesis $H_0 : \boldsymbol{\theta} = \theta_0 \mathbf{1}_p$. Then the RE of $\boldsymbol{\theta}$ is given by

$$\hat{\boldsymbol{\theta}}_n = \frac{1}{n}\,\mathbf{1}_p\mathbf{1}'_p\boldsymbol{K}\tilde{\boldsymbol{\theta}}_n. \tag{5.2.7}$$

The corresponding RE of σ^2 is given by

$$\boldsymbol{S}_u^R = \frac{1}{m}\,\hat{\boldsymbol{E}}'\mathbf{V}_n^{-1}\hat{\boldsymbol{E}} \text{ where } \hat{\boldsymbol{E}} = \boldsymbol{Y} - \boldsymbol{B}\hat{\boldsymbol{\theta}}_n, \quad m = n - p. \tag{5.2.8}$$

We also have

$$Cov(\tilde{\boldsymbol{\theta}}_n) = \sigma_{\boldsymbol{\varepsilon}}^2\boldsymbol{K}^{-1}, \quad Bias(\tilde{\boldsymbol{\theta}}_n) = \mathbf{0}$$
$$Cov(\hat{\boldsymbol{\theta}}_n) = \frac{\sigma_{\boldsymbol{\varepsilon}}^2}{n}\mathbf{1}_n\mathbf{1}'_n, \quad Bias(\hat{\boldsymbol{\theta}}_n) = \boldsymbol{\delta} = \boldsymbol{H\theta}, \quad \boldsymbol{H} = \boldsymbol{I}_p - \frac{1}{n}\mathbf{1}_p\mathbf{1}'_p\boldsymbol{K}.$$

Under $\boldsymbol{\theta} = \theta_0\mathbf{1}_p$, $E(\hat{\boldsymbol{\theta}}_n) = \theta_0\mathbf{1}_p$.

Now, we propose the likelihood ratio criterion (LRC) for testing the hypothesis $H_0 : \boldsymbol{\theta} = \theta_0 \mathbf{1}_p$ and its non-null distribution in the following theorem.

Theorem 5.2.1. *Let*

$$\begin{aligned}\Omega_1 &= \{(\boldsymbol{\theta}, \sigma, \mathbf{V}_n) : \boldsymbol{\theta} \in \mathbb{R}^p, \sigma \in \mathbb{R}^+, \mathbf{V}_n > 0\}, \textit{and} \\ \Omega_0 &= \{(\boldsymbol{\theta}, \sigma, \mathbf{V}_n) : \boldsymbol{\theta} = \theta_0 \mathbf{1}_p, \theta_0 \in \mathbb{R}, \sigma \in \mathbb{R}^+, \mathbf{V}_n > 0\}.\end{aligned}$$

Then, the LRC for testing the underlying hypothesis is given by

$$\mathcal{L}_n = \frac{\tilde{\boldsymbol{\theta}}_n' \boldsymbol{H}' \boldsymbol{K} \boldsymbol{H} \tilde{\boldsymbol{\theta}}_n}{(p-1)\boldsymbol{S}_u}, \quad \boldsymbol{H} = \boldsymbol{I}_p - \frac{1}{n} \mathbf{1}_p \mathbf{1}_p' \boldsymbol{K}, \tag{5.2.9}$$

and it has the following generalized noncentral F distribution:

$$g^*_{p,m}(\mathcal{L}_n) = \sum_{r \geq 0} \frac{\left(\frac{p-1}{m}\right)^{\frac{1}{2}(p+2r-1)} \mathcal{L}_n^{\frac{1}{2}(p+2r-3)} K_r^{(0)}(\Delta^2)}{r!\, B\left(\frac{p+2r-1}{2}, \frac{m}{2}\right) \left(1 + \frac{p-1}{m} \mathcal{L}_n\right)^{\frac{1}{2}(p+m+2r-1)}}, \tag{5.2.10}$$

where $\Delta^2 = \xi / \sigma_{\boldsymbol{\varepsilon}}^2$ *for* $\xi = \boldsymbol{\theta}' \boldsymbol{H}' \boldsymbol{K} \boldsymbol{H} \boldsymbol{\theta}$.

Proof: Because $\boldsymbol{Y} | \boldsymbol{\theta}, \sigma^2 \sim M_t^{(n)}(\boldsymbol{B}\boldsymbol{\theta}, \sigma^2 \mathbf{V}_n, \gamma_o)$, according to the regions Ω_i, $i = 0, 1$, the LR is given by

$$\begin{aligned}\Lambda &= \frac{L_0}{L_1} \\ &= a \left\{ \frac{\widetilde{\boldsymbol{E}}' \mathbf{V}_n^{-1} \widetilde{\boldsymbol{E}} / (n-1)}{\hat{\boldsymbol{E}}' \mathbf{V}_n^{-1} \hat{\boldsymbol{E}} / m} \right\}^{\frac{n}{2}} \\ &= a \left\{ \frac{m \widetilde{\boldsymbol{E}}' \mathbf{V}_n^{-1} \widetilde{\boldsymbol{E}}}{(n-1) \left[\widetilde{\boldsymbol{E}}' \mathbf{V}_n^{-1} \widetilde{\boldsymbol{E}} + \tilde{\boldsymbol{\theta}}_n' \boldsymbol{H}' \boldsymbol{K} \boldsymbol{H} \tilde{\boldsymbol{\theta}}_n \right]} \right\}^{\frac{n}{2}} \\ &= a \left(\frac{1}{1 + \frac{p-1}{m} \mathcal{L}_n} \right)^{\frac{n}{2}},\end{aligned}$$

where

$$a = \left(\frac{m + \gamma_o}{m + 1 + \gamma_o} \right)^{\frac{1}{2}(n + \gamma_o)}.$$

Hence, $\mathcal{L}_n$ is the LR test for testing the null hypothesis $H_0 : \boldsymbol{\theta} = \theta_0 \mathbf{1}_p$.

For the non-null distribution of $\mathcal{L}_n$, suppose $f_{(\boldsymbol{\theta}, s_e^2)}(.,.)$ denotes the joint distribution of $(\boldsymbol{\theta}, s_e^2)$, where s_e^2 is an unbiased estimator of σ^2 given by $\frac{n-p}{m} \boldsymbol{S}_u$. Then, according to (2.6.3), we can write

$$f_{(\boldsymbol{\theta}, s_e^2)}(\boldsymbol{x}, z) = \int_0^\infty W_o(t) f^*_{(\boldsymbol{\theta}, s_e^2)}(\boldsymbol{x}, z)\, dt, \tag{5.2.11}$$

where $f^*_{(\boldsymbol{\theta},s_e^2)}(.,.)$ denotes the joint distribution of $(\boldsymbol{\theta}, s_e^2)$ under the (normality) assumption $\boldsymbol{E} \sim \mathcal{N}_n(\mathbf{0}, \sigma^2 t^{-1}\mathbf{V}_n)$. In this case, because $\boldsymbol{\theta}$ is independent of s_e^2, we may have

$$f^*_{(\boldsymbol{\theta},s_e^2)}(\boldsymbol{x}, z) = \ddot{f}_{\boldsymbol{\theta}}(\boldsymbol{x}) h_{s_e^2}(z), \tag{5.2.12}$$

where $\ddot{f}_{\boldsymbol{\theta}}(.)$ is the pdf of $\mathcal{N}_p(\boldsymbol{\theta}, \sigma^2 t^{-1}\boldsymbol{K}^{-1})$ and $h_{s_e^2}(.)$ is the density of s_e^2.

Because of this special representation using (5.2.11) and (5.2.12), we are able to derive the distribution of $\mathcal{L}_n$ as follows:

$$g^*_{p,m}(\mathcal{L}_n) = \int_0^\infty W_o(t)\mathcal{G}_{p,m}(\mathcal{L}_n, \Delta^2)\, dt, \tag{5.2.13}$$

where $\mathcal{G}_{p,m}(\mathcal{L}_n, \Delta^2)$ denotes the distribution of $\mathcal{L}_n$ under the (normality) assumption $\boldsymbol{E} \sim \mathcal{N}_n(\mathbf{0}, \sigma^2 t^{-1}\mathbf{V}_n)$.

Define $\boldsymbol{z}_1 = \mathbf{V}_n^{-1/2}(\boldsymbol{Y} - \widetilde{\boldsymbol{Y}})$ for $\widetilde{\boldsymbol{Y}} = \boldsymbol{B}\tilde{\boldsymbol{\theta}}_n$. Under normality, it follows that $t^{\frac{1}{2}}\sigma^{-1}(\boldsymbol{I}_n - \boldsymbol{\mathcal{A}})^{-1/2}\boldsymbol{z}_1 \sim \mathcal{N}_n(\mathbf{0}, \boldsymbol{I}_n)$, where $\boldsymbol{\mathcal{A}} = \mathbf{V}_n^{-1/2}\boldsymbol{B}\boldsymbol{K}^{-1}\boldsymbol{B}'\mathbf{V}_n^{-1/2}$ is a symmetric idempotent matrix and so is $(\boldsymbol{I}_n - \boldsymbol{\mathcal{A}})$. It follows that

$$rank(\boldsymbol{I}_n - \boldsymbol{\mathcal{A}}) = tr(\boldsymbol{I}_n - \boldsymbol{\mathcal{A}}) = n - p.$$

Therefore, we obtain

$$s_e^2 = \frac{\boldsymbol{z}_1'(\boldsymbol{I}_n - \boldsymbol{\mathcal{A}})^{-1/2}(\boldsymbol{I}_n - \boldsymbol{\mathcal{A}})(\boldsymbol{I}_n - \boldsymbol{\mathcal{A}})^{-1/2}\boldsymbol{z}_1}{n-p} \sim \sigma^2 t^{-1}\chi^2_{n-p}. \tag{5.2.14}$$

Also, we have $\tilde{\boldsymbol{\theta}}_n \sim \mathcal{N}_p(\boldsymbol{\theta}, \sigma^2 t^{-1}\boldsymbol{K}^{-1})$.

Now, let $\boldsymbol{\Gamma} = (\boldsymbol{\Gamma}_1, \boldsymbol{\Gamma}_2)$ be an orthogonal matrix such that for a $p \times (p-1)$ matrix $\boldsymbol{\Gamma}_1$ and $p \times 1$ vector $\boldsymbol{\Gamma}_2$, we have $\boldsymbol{\Gamma}_2'\boldsymbol{\Gamma}_1 = 0$ and $\boldsymbol{\Gamma}_1\boldsymbol{\Gamma}_1' + \boldsymbol{\Gamma}_2\boldsymbol{\Gamma}_2' = \boldsymbol{I}_p$. Then, for the symmetric idempotent matrix $\boldsymbol{\mathcal{H}} = \boldsymbol{K}^{-1/2}\boldsymbol{H}'\boldsymbol{K}\boldsymbol{H}\boldsymbol{K}^{-1/2}$ of rank $(p-1)$, we have $\boldsymbol{\Gamma}\boldsymbol{\mathcal{H}}\boldsymbol{\Gamma}' = \begin{pmatrix} \boldsymbol{I}_{p-1} & \mathbf{0} \\ \mathbf{0} & \mathbf{0} \end{pmatrix}$. Define $\boldsymbol{z}_2 = t^{1/2}\sigma^{-1}\boldsymbol{\Gamma}\boldsymbol{K}^{-1/2}\tilde{\boldsymbol{\theta}}_n$. Then, under normality of errors, $\boldsymbol{z}_2 \sim \mathcal{N}_p(\sigma^{-1}t^{1/2}\boldsymbol{\Gamma}\boldsymbol{K}^{-1/2}\boldsymbol{\theta}, \boldsymbol{I}_p)$. Partitioning $\boldsymbol{z}_2 = (\dot{\boldsymbol{z}}_1', \dot{\boldsymbol{z}}_2')'$ where $\dot{\boldsymbol{z}}_1$ is an $(p-1)$-vector, we get

$$\tilde{\boldsymbol{\theta}}_n'\boldsymbol{H}'\boldsymbol{K}\boldsymbol{H}\tilde{\boldsymbol{\theta}}_n = \sigma^2 t^{-1}\dot{\boldsymbol{z}}_1'\dot{\boldsymbol{z}}_1 \sim \sigma^2 t^{-1}\chi^2_{p-1}\left(\frac{\xi}{\sigma^2 t^{-1}}\right). \tag{5.2.15}$$

Consequently, using (5.2.13) and (5.2.15), we obtain $\mathcal{L}_n|t \sim F_{p-1,m}\left(\frac{\xi}{\sigma^2 t^{-1}}\right)$ with the following pdf:

$$\mathcal{G}_{p,m}(\mathcal{L}_n) = \sum_{r\geq 0} \frac{\left(\frac{p-1}{m}\right)^{\frac{1}{2}(p+2r-1)} \mathcal{L}_n^{\frac{1}{2}(p+2r-3)} \left(\frac{\xi}{2\sigma^2 t^{-1}}\right)^r e^{\frac{-\xi t}{2\sigma^2}}}{r!\, \boldsymbol{B}\left(\frac{p+2r-1}{2}, \frac{m}{2}\right)\left(1 + \frac{p-1}{m}\mathcal{L}_n\right)^{\frac{1}{2}(p+m+2r-1)}}.$$

Integrating out over the inverse-gamma distribution, we get the desired result.

Then, we propose the following theorem.

Theorem 5.2.2. *Assuming* $\boldsymbol{Y}|\boldsymbol{\theta},\sigma^2 \sim M_t^{(n)}(\boldsymbol{B\theta},\sigma^2\mathbf{V}_n,\gamma_o)$ *in the ANOVA model* (5.1.1), *we have*

$$\begin{aligned}\tilde{\boldsymbol{\theta}}_n &\sim M_t^{(p)}\left(\boldsymbol{\theta},\sigma^2\boldsymbol{K}^{-1},\gamma_o\right),\\ \tilde{\boldsymbol{\theta}}_n-\hat{\boldsymbol{\theta}}_n &\sim M_t^{(p)}\left(\boldsymbol{H\theta},\sigma^2\boldsymbol{HK}^{-1},\gamma_o\right).\end{aligned}$$

Proof: From (5.2.5) and (5.2.7) for $\dot{\boldsymbol{\theta}}_n=\tilde{\boldsymbol{\theta}}_n-\hat{\boldsymbol{\theta}}_n$ we get

$$\begin{aligned}Cov(\tilde{\boldsymbol{\theta}}_n) &= \sigma_{\boldsymbol{\varepsilon}}^2\boldsymbol{K}^{-1},\\ Cov(\dot{\boldsymbol{\theta}}_n) &= \sigma_{\boldsymbol{\varepsilon}}^2\boldsymbol{HK}^{-1}.\end{aligned}$$

Then conditioning on t (under normality), using Theorem 5.1.3.2 of Saleh (2006) we have

$$\begin{aligned}\tilde{\boldsymbol{\theta}}_n &\sim \mathcal{N}_p\left(\boldsymbol{\theta},\sigma^2t^{-1}\boldsymbol{K}^{-1}\right),\\ \dot{\boldsymbol{\theta}}_n &\sim \mathcal{N}_p\left(\boldsymbol{H\theta},\sigma^2t^{-1}\boldsymbol{HK}^{-1}\right).\end{aligned}$$

Therefore

$$\begin{aligned}f_{\tilde{\boldsymbol{\theta}}_n}(\boldsymbol{x}) &= \int_0^\infty W_o(t)\mathcal{N}_p\left(\boldsymbol{\theta},\sigma^2t^{-1}\boldsymbol{K}^{-1}\right)\,dt,\\ f_{\dot{\boldsymbol{\theta}}_n}(\boldsymbol{x}) &= \int_0^\infty W_o(t)\mathcal{N}_p\left(\boldsymbol{H\theta},\sigma^2t^{-1}\boldsymbol{HK}^{-1}\right)\,dt.\end{aligned}$$

Now, we propose three more estimators following Saleh (2006) and consider the preliminary test estimator (PTE), Stein-type shrinkage estimators (SE), and positive-rule estimators (PRSE). Accordingly, the PT estimator of $\boldsymbol{\theta}$ is given by

$$\begin{aligned}\hat{\boldsymbol{\theta}}_n^{PT} &= \tilde{\boldsymbol{\theta}}_n-(\tilde{\boldsymbol{\theta}}_n-\hat{\boldsymbol{\theta}}_n)I\left(\mathcal{L}_n<F_{p-1,m}(\gamma)\right)\\ &= \tilde{\boldsymbol{\theta}}_n-I\left(\mathcal{L}_n<F_{p-1,m}(\gamma)\right)\boldsymbol{H}\tilde{\boldsymbol{\theta}}_n,\end{aligned}\tag{5.2.16}$$

where $I(A)$ is the indicator function of the set A and $F_{p-1,m}(\gamma)$ is the upper γ-quantile of the central F distribution with $p-1$ and m degrees of freedom.

The PTE has the disadvantage that it depends on $\gamma(0<\gamma<1)$, and the level of significance, and the choice of estimators remains between the two values $\tilde{\boldsymbol{\theta}}_n$ and $\hat{\boldsymbol{\theta}}_n$. To eliminate this difficulty, we define the S estimator

$$\begin{aligned}\hat{\boldsymbol{\theta}}_n^{S} &= \tilde{\boldsymbol{\theta}}_n-c\mathcal{L}_n^{-1}(\tilde{\boldsymbol{\theta}}_n-\hat{\boldsymbol{\theta}}_n),\\ &= \tilde{\boldsymbol{\theta}}_n-c\mathcal{L}_n^{-1}\boldsymbol{H}\tilde{\boldsymbol{\theta}}_n\end{aligned}\tag{5.2.17}$$

where

$$c=\frac{(p-3)m}{(p-1)(m+2)},\quad p>3.\tag{5.2.18}$$

The constant c is defined based on the degrees of freedom $(p-1, m)$. Consider that the latter estimator is not a convex combination, so it may exceed the value $\hat{\boldsymbol{\theta}}_n$. Thus, in order to shrink toward the value $\hat{\boldsymbol{\theta}}_n$, we utilize a convex combination of $\hat{\boldsymbol{\theta}}_n$ and $\hat{\boldsymbol{\theta}}_n^S$, via the preliminary test procedure to obtain the PRS estimator of $\boldsymbol{\theta}$ for $p \geq 4$, given by

$$\begin{aligned} \hat{\boldsymbol{\theta}}_n^{S+} &= \hat{\boldsymbol{\theta}}_n + \left(1 - c\mathcal{L}_n^{-1}\right) I(\mathcal{L}_n > c)(\tilde{\boldsymbol{\theta}}_n - \hat{\boldsymbol{\theta}}_n) \\ &= \hat{\boldsymbol{\theta}}_n^S - \left(1 - c\mathcal{L}_n^{-1}\right) I(\mathcal{L}_n \leq c)\boldsymbol{H}\tilde{\boldsymbol{\theta}}_n. \end{aligned} \quad (5.2.19)$$

5.3 Bias, MSE, and Risk Expressions

Suppose $\boldsymbol{\theta}_n^*$ is an estimator of $\boldsymbol{\theta}$. In this section, we derive bias, MSE matrix, and weighted quadratic risk, respectively expressing by $\boldsymbol{b}(\boldsymbol{\theta}_n^*)$, $\boldsymbol{M}(\boldsymbol{\theta}_n^*)$ and $R(\boldsymbol{\theta}_n^*; \boldsymbol{W})$ for each estimator stated in section 5.2. We will proceed by the following definitions:

$$\begin{aligned} \boldsymbol{b}(\boldsymbol{\theta}_n^*) &= E(\boldsymbol{\theta}_n^* - \boldsymbol{\theta}_0), \\ \boldsymbol{M}(\boldsymbol{\theta}_n^*) &= E(\boldsymbol{\theta}_n^* - \boldsymbol{\theta}_0)(\boldsymbol{\theta}_n^* - \boldsymbol{\theta}_0)', \\ R(\boldsymbol{\theta}^*; \boldsymbol{W}) &= E(\boldsymbol{\theta}_n^* - \boldsymbol{\theta}_0)'\boldsymbol{W}(\boldsymbol{\theta}_n^* - \boldsymbol{\theta}_0), \end{aligned}$$

where $\boldsymbol{W}$ is a known p.d. weight matrix.

Using Theorem 5.2.2, we have

$$\begin{aligned} \boldsymbol{b}(\tilde{\boldsymbol{\theta}}_n) &= \boldsymbol{0}, \\ \boldsymbol{M}(\tilde{\boldsymbol{\theta}}_n) &= \sigma_\varepsilon^2 \boldsymbol{K}^{-1}, \\ R(\tilde{\boldsymbol{\theta}}_n, \boldsymbol{W}) &= \sigma_\varepsilon^2 tr(\boldsymbol{W}\boldsymbol{K}^{-1}), \end{aligned} \quad (5.3.1)$$

$$\begin{aligned} \boldsymbol{b}(\hat{\boldsymbol{\theta}}_n) &= -\boldsymbol{H}\boldsymbol{\theta} = -\boldsymbol{\delta}, \\ \boldsymbol{M}(\hat{\boldsymbol{\theta}}_n) &= \frac{\sigma_\varepsilon^2}{n} \boldsymbol{1}_p \boldsymbol{1}_p', \\ R(\hat{\boldsymbol{\theta}}_n, \boldsymbol{W}) &= \frac{\sigma_\varepsilon^2}{n} \boldsymbol{1}_p' \boldsymbol{W} \boldsymbol{1}_p + \boldsymbol{\delta}'\boldsymbol{W}\boldsymbol{\delta}, \end{aligned} \quad (5.3.2)$$

and

$$\begin{aligned} \boldsymbol{b}(\hat{\boldsymbol{\theta}}_n^{PT}) &= -\boldsymbol{\delta} G_{p+1,m}^{(2)}(l_\gamma; \Delta^2), \\ \boldsymbol{M}(\hat{\boldsymbol{\theta}}_n^{PT}) &= \sigma_\varepsilon^2 \boldsymbol{K}^{-1} - \sigma_\varepsilon^2 (\boldsymbol{H}\boldsymbol{K}^{-1}\boldsymbol{H}') G_{p+1,m}^{(1)}(l_\gamma; \Delta^2) + \boldsymbol{\delta}\boldsymbol{\delta}' \\ &\quad \times \{2G_{p+1,m}^{(2)}(l_\gamma; \Delta^2) - G_{p+3,m}^{(2)}(l_\gamma; \Delta^2)\}, \\ R(\hat{\boldsymbol{\theta}}_n^{PT}, \boldsymbol{W}) &= \sigma_\varepsilon^2 tr(\boldsymbol{W}\boldsymbol{K}^{-1}) - \sigma_\varepsilon^2 tr(\boldsymbol{W}\boldsymbol{H}\boldsymbol{K}^{-1}\boldsymbol{H}') G_{p+1,m}^{(1)}(l_\gamma; \Delta^2) \\ &\quad + \boldsymbol{\delta}'\boldsymbol{W}\boldsymbol{\delta}\{2G_{p+1,m}^{(2)}(l_\gamma; \Delta^2) - G_{p+3,m}^{(2)}(l_\gamma; \Delta^2)\}, \end{aligned} \quad (5.3.3)$$

where $G^{(2-j)}_{\nu_1,\nu_2}(.;\Delta^2)$, $j=1,2$ is given by

$$G^{(2-h)}_{p,m}(l_\gamma;\Delta^2)=\sum_{r\geq 0}K^{(h)}_r(\Delta^2)I_{l_\gamma}\left[\frac{1}{2}(p+2r),\frac{m}{2}\right],\quad h=1,2, \tag{5.3.4}$$

where $K^h_r(\Delta^2)$ is given by (2.6.16), $l_\gamma=\frac{(p-1)F_{p-1,m}(\gamma)}{m+(p-1)F_{p-1,m}(\gamma)}$.

$$\begin{aligned}
\boldsymbol{b}(\hat{\boldsymbol{\theta}}^S_n) &= -c(p-1)\boldsymbol{\delta}E^{(2)}[\chi^{-2}_{p+1}(\Delta^2)]\\
\boldsymbol{M}(\hat{\boldsymbol{\theta}}^S_n) &= \sigma^2_{\boldsymbol{\varepsilon}}\boldsymbol{K}^{-1}-c(p-1)\sigma^2_{\boldsymbol{\varepsilon}}\boldsymbol{H}\boldsymbol{K}^{-1}\boldsymbol{H}'\{(p-3)E^{(1)}[\chi^{-4}_{p+1}(\Delta^2)]\\
&\quad +2\Delta^2E^{(1)}[\chi^{-4}_{p+3}(\Delta^2)]\}+c(p^2-1)\boldsymbol{\delta}\boldsymbol{\delta}'E^{(2)}[\chi^{-4}_{p+3}(\Delta^2)]\\
R(\hat{\boldsymbol{\theta}}^S_n;\boldsymbol{W}) &= \sigma^2_{\boldsymbol{\varepsilon}}tr(\boldsymbol{W}\boldsymbol{K}^{-1})-c(p-1)\sigma^2_{\boldsymbol{\varepsilon}}tr(\boldsymbol{W}\boldsymbol{H}\boldsymbol{K}^{-1}\boldsymbol{H}')\\
&\quad \times\{(p-3)E^{(1)}[\chi^{-4}_{p+1}(\Delta^2)]+2\Delta^2E^{(1)}[\chi^{-4}_{p+3}(\Delta^2)]\}\\
&\quad +c(p^2-1)\boldsymbol{\delta}'\boldsymbol{W}\boldsymbol{\delta}E^{(2)}[\chi^{-4}_{p+3}(\Delta^2)],
\end{aligned} \tag{5.3.5}$$

where

$$E^{(2-h)}[\chi^{-2}_{p+s}(\Delta^2)]=\sum_{r\geq 0}K^{(h)}_r(\Delta^2)(p+s-2+2r)^{-1}, \tag{5.3.6}$$

$$E^{(2-h)}[\chi^{-4}_{p+s}(\Delta^2)]=\sum_{r\geq 0}K^{(h)}_r(\Delta^2)(q+s-2+2r)^{-1}(p+s-4+2r)^{-1}. \tag{5.3.7}$$

$$\begin{aligned}
\boldsymbol{b}(\hat{\boldsymbol{\theta}}^{S+}_n) &= c_1\boldsymbol{\delta}\{E^{(2)}[F^{-1}_{p+1,m}(\Delta^2)]-E^{(2)}[F^{-1}_{p+1,m}(\Delta^2)I(F^{-1}_{p+1,m}(\Delta^2)<c_1)]\}\\
&\quad +\boldsymbol{\delta}H^{(2)}_{p+1,m}(c_1;\Delta^2)
\end{aligned}$$

$$\begin{aligned}
\boldsymbol{M}(\hat{\boldsymbol{\theta}}^{S+}_n) &= \boldsymbol{M}(\hat{\boldsymbol{\theta}}^S_n)\\
&\quad -\sigma^2_{\boldsymbol{\varepsilon}}\boldsymbol{H}\boldsymbol{K}^{-1}\boldsymbol{H}'E^{(1)}[(1-c_1F^{-1}_{p+1,m}(\Delta^2))^2I(F_{p+1,m}(\Delta^2)<c_1)]\\
&\quad +\boldsymbol{\delta}\boldsymbol{\delta}'\{2E^{(2)}[(1-c_1F^{-1}_{p+1,m}(\Delta^2))I(F_{p+1,m}(\Delta^2)<c_1)]\\
&\quad -E^{(2)}[(1-c_2F^{-1}_{p+3,m}(\Delta^2))^2I(F_{p+3,m}(\Delta^2)<c_2)]\}
\end{aligned}$$

$$\begin{aligned}
R(\hat{\boldsymbol{\theta}}^{S+}_n;\boldsymbol{W}) &= \boldsymbol{R}(\hat{\boldsymbol{\theta}}^S_n;\boldsymbol{W})-\sigma^2_{\boldsymbol{\varepsilon}}tr(\boldsymbol{W}\boldsymbol{H}\boldsymbol{K}^{-1}\boldsymbol{H}')\\
&\quad \times E^{(1)}[(1-c_1F^{-1}_{p+1,m}(\Delta^2))^2I(F_{p+1,m}(\Delta^2)<c_1)]\\
&\quad +\boldsymbol{\delta}'\boldsymbol{W}\boldsymbol{\delta}\{2E^{(2)}[(1-c_1F^{-1}_{p+1,m}(\Delta^2))I(F_{p+1,m}(\Delta^2)<c_1)]\\
&\quad -E^{(2)}[(1-c_2F^{-1}_{p+3,m}(\Delta^2))^2I(F_{p+3,m}(\Delta^2)<c_2)]\},
\end{aligned} \tag{5.3.8}$$

where

$$\begin{aligned} E^{(2-h)}[F_{p+s,m}^{-1}(\Delta^2)I(F_{p+s,m}(\Delta^2) < c_i)] &= \sum_{r\geq 0} K_r^{(h)}(\Delta^2) \\ &\quad \times (p+s)(p+s-2+2r)^{-1} \\ &\quad \times I_{x'}\left[\frac{p+s-2+2r}{2}, \frac{m+2}{2}\right]. \end{aligned} \tag{5.3.9}$$

$$\begin{aligned} E^{(2-h)}[F_{p+s,m}^{-2}(\Delta^2)I(F_{p+s,m}(\Delta^2) < c_i)] = \sum_{r\geq 0} \frac{(q+s)^2}{m} K_r^{(h)}(\Delta^2) \\ \times [(p+s-2+2r)(p+s-4+2r)]^{-1} \\ \times I_{x'}\left[\frac{p+s-4+2r}{2}, \frac{m+4}{2}\right], \end{aligned} \tag{5.3.10}$$

under which for $s = 1$ we use $c_1 = \frac{c(p-1)}{p+1}$ and for $s = 3$ we use $c_2 = \frac{c(p-1)}{p+3}$ and $x' = \frac{c(p-1)}{m+c(p-1)}$.

5.4 Risk Analysis

In this section, we provide the superiority conditions of underlying estimators compared with the others in four subsections. We may say that the estimator $\boldsymbol{\delta}_1$ performs better than the estimator $\boldsymbol{\delta}_2$ in the sense that $\boldsymbol{R}(\boldsymbol{\delta}_1;\boldsymbol{\theta}) \leq \boldsymbol{R}(\boldsymbol{\delta}_2;\boldsymbol{\theta})$. In this case, we use the notation $\boldsymbol{\delta}_1 \succeq \boldsymbol{\delta}_2$.

5.4.1 Comparison of $\hat{\theta}_n$ and $\tilde{\theta}_n$

Using equations (5.3.1) and (5.3.2), we have

$$\begin{aligned} R(\tilde{\boldsymbol{\theta}}_n;\boldsymbol{\theta}) - R(\hat{\boldsymbol{\theta}}_n;\boldsymbol{\theta}) &= \sigma_\varepsilon^2 tr(\boldsymbol{W}\boldsymbol{K}^{-1}) \\ &\quad -[\sigma_\varepsilon^2 tr(\boldsymbol{W}\boldsymbol{K}^{-1}) - \sigma_\varepsilon^2 tr(\boldsymbol{W}\boldsymbol{H}\boldsymbol{K}^{-1}\boldsymbol{H}') + \boldsymbol{\delta}'\boldsymbol{W}\boldsymbol{\delta}] \\ &= \sigma_\varepsilon^2 tr(\boldsymbol{W}\boldsymbol{H}\boldsymbol{K}^{-1}\boldsymbol{H}') - \boldsymbol{\delta}'\boldsymbol{W}\boldsymbol{\delta}. \end{aligned} \tag{5.4.1}$$

The risk difference in (5.4.1) is negative, i.e., ($\tilde{\boldsymbol{\theta}}_n \succeq \hat{\boldsymbol{\theta}}_n$) iff

$$\sigma_\varepsilon^{-2}\boldsymbol{\delta}'\boldsymbol{W}\boldsymbol{\delta} \leq tr(\boldsymbol{W}\boldsymbol{H}\boldsymbol{K}^{-1}\boldsymbol{H}')$$

and $\hat{\boldsymbol{\theta}}_n \succeq \tilde{\boldsymbol{\theta}}_n$ provided

$$\sigma_\varepsilon^{-2}\boldsymbol{\delta}'\boldsymbol{W}\boldsymbol{\delta} \geq tr(\boldsymbol{W}\boldsymbol{H}\boldsymbol{K}^{-1}\boldsymbol{H}').$$

5.4.2 Comparison of $\hat{\boldsymbol{\theta}}_n^{PT}$ and $\tilde{\boldsymbol{\theta}}_n(\hat{\boldsymbol{\theta}}_n)$

From equations (5.3.1) and (5.3.3), we get

$$\begin{aligned}\boldsymbol{R}(\tilde{\boldsymbol{\theta}}_n;\boldsymbol{\theta}) - \boldsymbol{R}(\hat{\boldsymbol{\theta}}_n^{PT};\boldsymbol{\theta}) &= \sigma_{\varepsilon}^2 tr(\boldsymbol{W}\boldsymbol{H}\boldsymbol{K}^{-1}\boldsymbol{H}')G_{p+1,m}^{(1)}(l_\gamma;\Delta^2) - \boldsymbol{\delta}'\boldsymbol{W}\boldsymbol{\delta} \\ &\quad \times\{2G_{p+1,m}^{(2)}(l_\gamma;\Delta^2) - G_{p+3,m}^{(2)}(l_\gamma;\Delta^2)\}. \quad (5.4.2)\end{aligned}$$

Using the Courant theorem (see Theorem A.2.4 of Anderson, 2003), we have

$$\xi\lambda_{min}(\boldsymbol{W}\boldsymbol{K}^{-1}) \;\leq\; \boldsymbol{\delta}'\boldsymbol{W}\boldsymbol{\delta} \;\leq\; \xi\lambda_{max}(\boldsymbol{W}\boldsymbol{K}^{-1})$$

or

$$\sigma_{\varepsilon}^2\Delta^2\lambda_{min}(\boldsymbol{W}\boldsymbol{K}^{-1}) \;\leq\; \boldsymbol{\delta}'\boldsymbol{W}\boldsymbol{\delta} \;\leq\; \sigma_{\varepsilon}^2\Delta^2\lambda_{max}(\boldsymbol{W}\boldsymbol{K}^{-1}), \qquad (5.4.3)$$

where $\lambda_{min}(\boldsymbol{A})$ and $\lambda_{max}(\boldsymbol{A})$ are the minimum and maximum eigenvalues of $\boldsymbol{A}$, respectively.

Therefore, using (5.4.2) and (5.4.3), $\hat{\boldsymbol{\theta}}_n^{PT} \succeq \tilde{\boldsymbol{\theta}}_n$ iff

$$\Delta^2 \;\leq\; \frac{tr(\boldsymbol{W}\boldsymbol{H}\boldsymbol{K}^{-1}\boldsymbol{H}')G_{p+1,m}^{(1)}(l_\gamma;\Delta^2)}{\lambda_{max}(\boldsymbol{W}\boldsymbol{K}^{-1})[2G_{p+1,m}^{(2)}(l_\gamma;\Delta^2) - G_{p+3,m}^{(2)}(l_\gamma;\Delta^2)]},$$

while $\tilde{\boldsymbol{\theta}}_n \succeq \hat{\boldsymbol{\theta}}_n^{PT}$ iff

$$\Delta^2 \;\geq\; \frac{tr(\boldsymbol{W}\boldsymbol{H}\boldsymbol{K}^{-1}\boldsymbol{H}')G_{p+1,m}^{(1)}(l_\gamma;\Delta^2)}{\lambda_{min}(\boldsymbol{W}\boldsymbol{K}^{-1})[2G_{p+1,m}^{(2)}(l_\gamma;\Delta^2) - G_{p+3,m}^{(2)}(l_\gamma;\Delta^2)]}.$$

Consider under H_0, the equation (5.4.2) reduces to

$$R(\tilde{\boldsymbol{\theta}}_n;\boldsymbol{\theta}) - R(\hat{\boldsymbol{\theta}}_n^{PT};\boldsymbol{\theta}) \;=\; \sigma_{\varepsilon}^2 tr(\boldsymbol{W}\boldsymbol{H}\boldsymbol{K}^{-1}\boldsymbol{H}')H_{p+1,m}^{(1)}(l_\gamma;\Delta^2).$$

Thus, $\hat{\boldsymbol{\theta}}_n^{PT} \succeq \tilde{\boldsymbol{\theta}}_n$ under H_0.

From equations (5.3.2) and (5.3.3), we have

$$\begin{aligned}R(\hat{\boldsymbol{\theta}}_n;\boldsymbol{\theta}) - R(\hat{\boldsymbol{\theta}}_n^{PT};\boldsymbol{\theta}) &= -\sigma_{\varepsilon}^2 tr(\boldsymbol{W}\boldsymbol{H}\boldsymbol{K}^{-1}\boldsymbol{H}')[1 - G_{p+1,m}^{(1)}(l_\gamma;\Delta^2)] \\ &\quad +\boldsymbol{\delta}'\boldsymbol{W}\boldsymbol{\delta}\{1 - 2G_{p+1,m}^{(2)}(l_\gamma;\Delta^2) + G_{p+3,m}^{(2)}(l_\gamma;\Delta^2)\}. \quad (5.4.4)\end{aligned}$$

Therefore by (5.4.3) and (5.4.4), we get $\hat{\boldsymbol{\theta}}_n \succeq \hat{\boldsymbol{\theta}}_n^{PT}$ whenever

$$\Delta^2 \;\leq\; \frac{tr(\boldsymbol{W}\boldsymbol{H}\boldsymbol{K}^{-1}\boldsymbol{H}')[1 - G_{p+1,m}^{(1)}(l_\gamma;\Delta^2)]}{\lambda_{max}(\boldsymbol{W}\boldsymbol{K}^{-1})[1 - 2G_{p+1,m}^{(2)}(l_\gamma;\Delta^2) + G_{p+3,m}^{(2)}(l_\gamma;\Delta^2)]},$$

while $\hat{\boldsymbol{\theta}}_n^{PT} \succeq \hat{\boldsymbol{\theta}}_n$ whenever

$$\Delta^2 \;\geq\; \frac{tr(\boldsymbol{W}\boldsymbol{H}\boldsymbol{K}^{-1}\boldsymbol{H}')[1 - G_{p+1,m}^{(1)}(l_\gamma;\Delta^2)]}{\lambda_{min}(\boldsymbol{W}\boldsymbol{K}^{-1})[1 - 2G_{p+1,m}^{(2)}(l_\gamma;\Delta^2) + G_{p+3,m}^{(2)}(l_\gamma;\Delta^2)]}.$$

Also, under H_0 using equation (5.4.4), we have

$$R(\hat{\boldsymbol{\theta}}_n;\boldsymbol{\theta}) - R(\hat{\boldsymbol{\theta}}_n^{PT};\boldsymbol{\theta}) = -\sigma_{\varepsilon}^2 tr(\boldsymbol{W}\boldsymbol{H}\boldsymbol{K}^{-1}\boldsymbol{H}')[1 - G_{p+1,m}^{(1)}(l_{\gamma};\Delta^2)] < 0.$$

Thus, under H_0, $\hat{\boldsymbol{\theta}}_n$ performs better than $\hat{\boldsymbol{\theta}}_n^{PT}$.

5.4.3 Comparison of $\hat{\boldsymbol{\theta}}_n^S$, $\tilde{\boldsymbol{\theta}}_n$, $\hat{\boldsymbol{\theta}}_n$, and $\hat{\boldsymbol{\theta}}_n^{PT}$

Comparing $\hat{\boldsymbol{\theta}}_n^S$ relative to $\tilde{\boldsymbol{\theta}}_n$, the risk difference for $p \geq 4$ is given by

$$\begin{aligned} R(\tilde{\boldsymbol{\theta}}_n;\boldsymbol{W}) - R(\hat{\boldsymbol{\theta}}_n^S;\boldsymbol{W}) &= c(p-1)\sigma_{\varepsilon}^2 tr(\boldsymbol{W}\boldsymbol{H}\boldsymbol{K}^{-1}\boldsymbol{H}') \\ &\quad \times\{(p-3)E^{(1)}[\chi_{p+1}^{-4}(\Delta^2)] + 2\Delta^2 E^{(1)}[\chi_{p+3}^{-4}(\Delta^2)]\} \\ &\quad -c(p^2-1)\boldsymbol{\delta}'\boldsymbol{W}\boldsymbol{\delta}E^{(2)}[\chi_{p+3}^{-4}(\Delta^2)] \\ &= c(p-1)\sigma_{\varepsilon}^2 tr(\boldsymbol{W}\boldsymbol{H}\boldsymbol{K}^{-1}\boldsymbol{H}') \\ &\quad \times\Big\{(p-3)E^{(1)}[\chi_{p+1}^{-1}(\Delta^2)] + 2\Delta^2 E^{(2)}[\chi_{p+3}^{-4}(\Delta^2)] \\ &\quad \times\left[1 - \frac{(p+1)\boldsymbol{\delta}\boldsymbol{W}\boldsymbol{\delta}}{2(\boldsymbol{\delta}'\boldsymbol{K}\boldsymbol{\delta})tr(\boldsymbol{W}\boldsymbol{H}\boldsymbol{K}^{-1}\boldsymbol{H})}\right]\Big\}. \qquad (5.4.5) \end{aligned}$$

The equation (5.4.5) is positive for all $\mathcal{C}$ such that

$$\left\{\mathcal{C}: \frac{tr(\boldsymbol{W}\boldsymbol{H}\boldsymbol{K}^{-1}\boldsymbol{H}')}{\lambda_{max}(\boldsymbol{W}\boldsymbol{K}^{-1})} \geq \frac{p+1}{2}\right\},$$

which asserts $\hat{\boldsymbol{\theta}}_n^S$ uniformly dominates $\tilde{\boldsymbol{\theta}}_n$.

In order to compare $\hat{\boldsymbol{\theta}}_n^S$ with $\hat{\boldsymbol{\theta}}_n$, first note that under H_0, $\Delta^2 = 0$ and, therefore,

$$\begin{aligned} R(\hat{\boldsymbol{\theta}}_n;\boldsymbol{W}) - R(\hat{\boldsymbol{\theta}}_n^S;\boldsymbol{W}) &= \frac{\sigma_{\varepsilon}^2}{n}\mathbf{1}_p'\boldsymbol{W}\mathbf{1}_p - \sigma_{\varepsilon}^2 tr(\boldsymbol{W}\boldsymbol{K}^{-1}) \\ &\quad +c(p-1)(p-3)\sigma_{\varepsilon}^2 tr(\boldsymbol{W}\boldsymbol{H}\boldsymbol{K}^{-1}\boldsymbol{H}')E^{(1)}[\chi_{p+1}^{-4}(0)] \\ &= -\sigma_{\varepsilon}^2 tr(\boldsymbol{W}\boldsymbol{H}\boldsymbol{K}^{-1}\boldsymbol{H}') + c(p-1)(p-3)\sigma_{\varepsilon}^2 \\ &\quad \times tr(\boldsymbol{W}\boldsymbol{H}\boldsymbol{K}^{-1}\boldsymbol{H}')E^{(1)}[\chi_{p+1}^{-4}(0)] \\ &= \sigma_{\varepsilon}^2 tr(\boldsymbol{W}\boldsymbol{H}\boldsymbol{K}^{-1}\boldsymbol{H}') \\ &\quad \times\left\{c(p-1)(p-3)E^{(1)}[\chi_{p+1}^{-4}(0)] - 1\right\}. \qquad (5.4.6) \end{aligned}$$

Hence, $\hat{\boldsymbol{\theta}}_n \succeq \hat{\boldsymbol{\theta}}_n^S$ under H_0 provided $E^{(1)}[\chi_{p+1}^{-4}(0)] \leq \frac{m+2}{m(p-3)^2}$. But as Δ^2 departs from zero, the risk of $\hat{\boldsymbol{\theta}}_n$ becomes unbounded while the risk of $\hat{\boldsymbol{\theta}}_n^S$ remains below $\tilde{\boldsymbol{\theta}}_n$. In this case, $\hat{\boldsymbol{\theta}}_n^S$ performs better than $\hat{\boldsymbol{\theta}}_n$.

Comparing $\hat{\boldsymbol{\theta}}_n^S$ with $\hat{\boldsymbol{\theta}}_n^{PT}$ under H_0 using the risk difference, we have

$$\begin{aligned} R(\hat{\boldsymbol{\theta}}_n^{PT};\boldsymbol{W}) - R(\hat{\boldsymbol{\theta}}_n^S;\boldsymbol{W}) &= c(p-1)(p-3)\sigma_{\boldsymbol{\varepsilon}}^2 tr(\boldsymbol{W}\boldsymbol{H}\boldsymbol{K}^{-1}\boldsymbol{H}')E^{(1)}[\chi_{p+1}^{-4}(0)] \\ &\quad -\sigma_{\boldsymbol{\varepsilon}}^2 tr(\boldsymbol{W}\boldsymbol{H}\boldsymbol{K}^{-1}\boldsymbol{H}')G_{p+1,m}^{(1)}(l_\gamma;0) \\ &= \sigma_{\boldsymbol{\varepsilon}}^2 tr(\boldsymbol{W}\boldsymbol{H}\boldsymbol{K}^{-1}\boldsymbol{H}') \\ &\quad \times\left\{\frac{(p-3)^2 m}{m+2}E^{(1)}[\chi_{p+1}^{-4}(0)] - H_{p+1,m}^{(1)}(l_\gamma;0)\right\}. \end{aligned} \tag{5.4.7}$$

The right-hand side (R.H.S.) of (5.4.7) is nonpositive, that is, $\hat{\boldsymbol{\theta}}_n^{PT} \succeq \hat{\boldsymbol{\theta}}_n^S$, for such γ satisfying

$$m(p-3)^2 E^{(1)}[\chi_{p+1}^{-4}(0)] \leq (m+2)H_{p+1,m}^{(1)}(l_\gamma;0). \tag{5.4.8}$$

However, as Δ^2 moves away from zero, the risk of $\hat{\boldsymbol{\theta}}_n^{PT}$ exceeds the risks of $\tilde{\boldsymbol{\theta}}_n$, while the risk of $\hat{\boldsymbol{\theta}}_n^S$ remains below the risk of $\tilde{\boldsymbol{\theta}}_n$ and both ultimately merge with the risk of $\tilde{\boldsymbol{\theta}}_n$ as $\Delta^2 \to \infty$. Thus, $\hat{\boldsymbol{\theta}}_n^S$ dominates $\hat{\boldsymbol{\theta}}_n^{PT}$ outside a small neighborhood of the origin.

5.4.4 Comparison of $\hat{\boldsymbol{\theta}}_n^S$ and $\hat{\boldsymbol{\theta}}_n^{S+}$

In the comparison of $\hat{\boldsymbol{\theta}}_n^{S+}$ with $\hat{\boldsymbol{\theta}}_n^S$, using the risk difference, we may write

$$\begin{aligned} R(\hat{\boldsymbol{\theta}}_n^S;\boldsymbol{W}) - R(\hat{\boldsymbol{\theta}}_n^{S+};\boldsymbol{W}) &= \sigma_{\boldsymbol{\varepsilon}}^2 tr(\boldsymbol{W}\boldsymbol{H}\boldsymbol{K}^{-1}\boldsymbol{H}') \\ &\quad \times E^{(1)}\left[\left(1-c_1 F_{p+1,m}^{-1}(\Delta^2)\right)^2 I\left(F_{p+1,m}(\Delta^2)<c_1\right)\right] \\ &\quad -2\boldsymbol{\delta}'\boldsymbol{W}\boldsymbol{\delta}E^{(2)}\Bigg[\left(1-c_1 F_{p+1,m}^{-1}(\Delta^2)\right) \\ &\quad \times I\left(F_{p+1,m}(\Delta^2)<c_1\right)\Bigg] + \boldsymbol{\delta}'\boldsymbol{W}\boldsymbol{\delta} \\ &\quad \times E^{(2)}\left[\left(1-c_2 F_{p+3,m}^{-1}(\Delta^2)\right)^2 I\left(F_{p+3,m}(\Delta^2)<c_2\right)\right]. \end{aligned} \tag{5.4.9}$$

The R.H.S. of (5.4.9) is definitely positive since, for $F_{p+1,m}(\Delta^2) < c_1$, we have

$$\left(1-c_1 F_{p+1,m}^{-1}(\Delta^2)\right) < 0,$$

and, therefore,

$$-2\boldsymbol{\delta}'\boldsymbol{W}\boldsymbol{\delta}E^{(2)}\left[\left(1-c_1 F_{p+1,m}^{-1}(\Delta^2)\right) I\left(F_{p+1,m}(\Delta^2)<c_1\right)\right] > 0.$$

Also, the expectation of a positive random variable is positive, so

$$E^{(2)}\left[\left(1-c_1F_{p+1,m}^{-1}(\Delta^2)\right)^2 I\left(F_{p+1,m}(\Delta^2)<c_1\right)\right]>0.$$

Thus, we conclude $\hat{\boldsymbol{\theta}}_n^{S+}$ strictly performs better than $\hat{\boldsymbol{\theta}}_n^{S}$ and uniformly $\hat{\boldsymbol{\theta}}_n^{S+} \succeq \hat{\boldsymbol{\theta}}_n^{S}$. This analysis shows that, if we can demonstrate the superiority of $\hat{\boldsymbol{\theta}}_n^{S}$ over each other estimator, then we will conclude the same result for $\hat{\boldsymbol{\theta}}_n^{S+}$.

Further, we show that the shrinkage factor c of the Stein-type estimator is robust with respect to the treatment parameters and the unknown mixing distributions.

Theorem 5.4.1. *Consider the model* (5.1.1) *where the error-vector distributed as* $M_t^{(n)}(\mathbf{0},\sigma^2\mathbf{V}_n,\gamma_o)$*, where* $\gamma_o>4$*. Then the Stein-type estimator* $\hat{\boldsymbol{\theta}}_n^{S}$ *of the treatment given by*

$$\begin{aligned}\hat{\boldsymbol{\theta}}_n^{S} &= \tilde{\boldsymbol{\theta}}_n - c\boldsymbol{Q}^{-1}\boldsymbol{H}\tilde{\boldsymbol{\theta}}_n\mathcal{L}_n^{-1}\\ &= \tilde{\boldsymbol{\theta}}_n - c^*(ms_e^2)\boldsymbol{Q}^{-1}\boldsymbol{H}\tilde{\boldsymbol{\theta}}_n(\tilde{\boldsymbol{\theta}}_n'\boldsymbol{H}'\boldsymbol{Q}^{-1}\boldsymbol{H}\tilde{\boldsymbol{\theta}}_n)^{-1}\\ &= \tilde{\boldsymbol{\theta}}_n - c^*(ms_e^2)\boldsymbol{Q}^{-1}\mathbf{Z}(\mathbf{Z}'\boldsymbol{Q}^{-1}\mathbf{Z})^{-1},\end{aligned}$$

where $\mathbf{Z}=\boldsymbol{H}\tilde{\boldsymbol{\theta}}_n$*, uniformly dominates the unrestricted estimator* $\tilde{\boldsymbol{\theta}}_n$ *with respect to a quadratic loss-function with the positive-definite matrix* $\boldsymbol{Q}$ *if and only if* c^* *satisfies the condition* $0<c^*\le\frac{2(p-3)}{m+2}$*. The largest reduction of the risk is attained when* $c^*=\frac{p-3}{m+2}$.

Proof: The risk-difference of the SE and the UE is given by

$$\begin{aligned}&E(\hat{\boldsymbol{\theta}}_n^{S}-\boldsymbol{\theta})'\boldsymbol{Q}(\hat{\boldsymbol{\theta}}_n^{S}-\boldsymbol{\theta}) - E(\tilde{\boldsymbol{\theta}}_n-\boldsymbol{\theta})'\boldsymbol{Q}(\tilde{\boldsymbol{\theta}}_n-\boldsymbol{\theta})\\ =\;& (c^*)^2E[(ms_e^2)^2(\mathbf{Z}'\boldsymbol{Q}^{-1}\mathbf{Z})^{-1}] - 2c^*E[(ms_e^2)(\tilde{\boldsymbol{\theta}}_n-\boldsymbol{\theta})'\boldsymbol{H}\tilde{\boldsymbol{\theta}}_n(\mathbf{Z}'\boldsymbol{Q}^{-1}\mathbf{Z})^{-1})]\\ =\;& (c^*)^2m(m+2)E_t[E_N\mathbf{Z}'\boldsymbol{Q}^{-1}\mathbf{Z}^{-1}] - 2c^*m(p-3)E_t[E_N(\mathbf{Z}'\boldsymbol{Q}^{-1}\mathbf{Z})^{-1}]\le 0,\end{aligned}$$

if and only if $0<c*\le\frac{2(p-3)}{m+2}$, since (by (2.6.4))

$$\begin{aligned}\int_0^\infty \frac{1}{\mathbf{Z}'\boldsymbol{Q}^{-1}\mathbf{Z}}t^{-2}W_o(t)dt &= \frac{1}{\mathbf{Z}'\boldsymbol{Q}^{-1}\mathbf{Z}}\kappa^{(2)}\\ &= \frac{1}{\mathbf{Z}'\boldsymbol{Q}^{-1}\mathbf{Z}}\left(\frac{\gamma_o^2}{(\gamma_o-2)(\gamma_o-4)}\right)\\ &> 0, \quad \text{for} \quad \gamma_o>4. \qquad (5.4.10)\end{aligned}$$

Maximum reduction occurs when $c^*=\frac{p-3}{m+2}$. Thus, we choose $c=\frac{(p-3)m}{(p-1)(m+2)}$.

Remarkably, the behavior of all estimators compared with the others under multivariate t is exactly the same as under normal theory, as exhibited in Saleh (2006). This phenomenon shows the domination order of estimators and regarding substantial conditions under normal theory are significantly robust.

5.5 Problems

1. Derive the restricted unbiased estimator of σ^2.

2. Prove Corollaries 4.2.1.1 and 4.2.1.2.

3. Estimate the unknown scalar θ_0 in linear constraint $\boldsymbol{\theta} = \theta_0 \mathbf{1}$.

4. Let $\boldsymbol{X}_{(n)} = (\tilde{\boldsymbol{\theta}}_n - \boldsymbol{\theta})$, $\boldsymbol{Y}_{(n)} = (\tilde{\boldsymbol{\theta}}_n - \hat{\boldsymbol{\theta}}_n)$, and $\boldsymbol{Z}_{(n)} = (\hat{\boldsymbol{\theta}}_n - \theta_0 \mathbf{1})$. Derive the distributions of $\begin{bmatrix} \boldsymbol{X}_{(n)} \\ \boldsymbol{Y}_{(n)} \end{bmatrix}$ and $\begin{bmatrix} \boldsymbol{Z}_{(n)} \\ \boldsymbol{Y}_{(n)} \end{bmatrix}$.
   Hint: Apply Theorem 4.2.2 and Theorem 5.1.3.2 of Saleh (2006).

5. Display graph of the risk functions based on multivariate t-errors.

6. Determine the determinants of $\boldsymbol{M}(\tilde{\boldsymbol{\theta}}_n)$, $\boldsymbol{M}(\hat{\boldsymbol{\theta}}_n)$, $\boldsymbol{M}(\hat{\boldsymbol{\theta}}_n^{PT})$, $\boldsymbol{M}(\hat{\boldsymbol{\theta}}_n^{S})$, and $\boldsymbol{M}(\hat{\boldsymbol{\theta}}_n^{S+})$.

7. Consider a general shrinkage estimator defined as $\hat{\boldsymbol{\theta}}_n(\boldsymbol{C}) = \boldsymbol{C}\tilde{\boldsymbol{\theta}}_n$, where $\boldsymbol{C}$ takes two choices as $\boldsymbol{C} = \left(\boldsymbol{I}_p + k_o \boldsymbol{K}^{-1}\right)^{-1}$, for $k_o \geq 0$ and for $0 < d \leq 1$, $(\boldsymbol{K} + d\boldsymbol{I}_p)(\boldsymbol{K} + \boldsymbol{I}_p)^{-1}$, resulting in a ridge-type estimator and Liu-type estimator, respectively. Pre-multiply each different choice of $\boldsymbol{C}$ in the estimators defined in section 5.2 to get ridge and Liu type versions of them and then derive bias, MSE matrix, and risk expressions.

CHAPTER 6

PARALLELISM MODEL

Outline

In this chapter, we discuss the parallelism model in detail, beginning with the estimation of the model parameters and test of hypothesis. In addition, we provide some improved estimators of the parallelism parameters with their dominance properties.

6.1 Model Specification

Consider several laboratories (say, p of them) dealing with similar experiments in the analysis of bioassay data or shelf-life determination of the pharmaceutical products.

Statistical Inference for Models with Multivariate t-Distributed Errors, First Edition.

A. K. Md. Ehsanes Saleh, M. Arashi, S.M.M. Tabatabaey.

Generally, the statistical analysis of the data is based on simple linear models with independent normal errors, and the basic problem is the estimation of the intercept and slope parameters of these models and how to combine these results from the concerned laboratories to make improved analysis of the common problem.

In combining the results of several linear models, one may tentatively suspect that the slopes of the linear models may be the same while intercepts are different, yielding to the parallelism models for the analysis and combination of the results from several laboratories. The initial studies of such problems have been initiated by Lambert, Saleh, and Sen (1985); Akritus, Saleh, and Sen (1985); and Saleh and Sen (1985), among others, encompassing normal theory and nonparametric methods. Details of the developments are given in Saleh (2006). Also, interested readers may refer to Khan (2003, 2006a), Khan and Saleh (2006), and Arashi, Saleh, and Tabatabaey (2010) for some recent further references.

In this chapter, we consider the parallelism problem in a more general setup involving the dependent errors discussed earlier. It will be shown that the improved estimators considered in this chapter retain the same dominance properties as under the normal theory. Further, it will be illustrated that that the shrinkage factor of the Stein estimators is robust with respect to the regression parameters of the multivariate t-model.

Thus, we consider the parallelism model given by

$$\boldsymbol{Y} = \boldsymbol{B}\boldsymbol{\Theta} + \boldsymbol{X}\boldsymbol{\beta} + \boldsymbol{E}, \tag{6.1.1}$$

where

$$\boldsymbol{Y} = (\boldsymbol{Y}_1', \cdots, \boldsymbol{Y}_p')' \tag{6.1.2}$$

is an n-vector in which each $\boldsymbol{Y}_\alpha = (\boldsymbol{Y}_{\alpha 1}, \cdots, \boldsymbol{Y}_{\alpha n_\alpha})'$ for $\alpha = 1, \cdots, p$ such that $n = n_1 + \cdots + n_p$,

$$\boldsymbol{B} = \text{Diag}(\mathbf{1}_{n_1}, \cdots, \mathbf{1}_{n_p}), \tag{6.1.3}$$

where $\mathbf{1}_{n_\alpha} = (1, \cdots, 1)'$ is a n_α-tuple of 1's,

$$\boldsymbol{\Theta} = (\theta_1, \cdots, \theta_p)', \tag{6.1.4}$$

the intercept vector,

$$\boldsymbol{X} = \text{Diag}(\boldsymbol{X}_1, ...\boldsymbol{X}_p), \tag{6.1.5}$$

where $\boldsymbol{X}_\alpha = (x_{\alpha 1}, \cdots, x_{\alpha n_\alpha})'$,

$$\boldsymbol{\beta} = (\beta_1, \cdots, \beta_p), \tag{6.1.6}$$

the slope vector.
Similarly,

$$\boldsymbol{E} = (\boldsymbol{E}_1', \cdots, \boldsymbol{E}_p')' \tag{6.1.7}$$

is an n-vector of errors in which each $\boldsymbol{E}_\alpha = (\epsilon_{\alpha 1}, \cdots, \epsilon_{\alpha n_\alpha})'$ such that $\boldsymbol{E}$ is distributed according to multivariate t-distribution, i.e., $M_t^{(n)}(\mathbf{0}, \sigma^2\mathbf{V}_n, \gamma_o)$, for

$$\mathbf{V}_n = \text{Diag}(\mathbf{V}_1, ...\mathbf{V}_p), \tag{6.1.8}$$

where each $\mathbf{V}_\alpha$ is a known positive definite symmetric matrix of size $n_\alpha \times n_\alpha$.

6.2 Estimation of the Parameters and Test of Parallelism

For the unrestricted estimators (UEs) of intercept ($\boldsymbol{\Theta}$) and slope ($\boldsymbol{\beta}$) parameters, we minimize $(\boldsymbol{Y} - \boldsymbol{B\Theta} - \boldsymbol{X\beta})'\boldsymbol{V}_n^{-1}(\boldsymbol{Y} - \boldsymbol{B\Theta} - \boldsymbol{X\beta})$ w.r.t. $\boldsymbol{\Theta}$ and $\boldsymbol{\beta}$ to obtain the normal system of equations

$$\begin{cases} \boldsymbol{K}_1\boldsymbol{\Theta} + \boldsymbol{K}_3\boldsymbol{\beta} = \boldsymbol{Y}_1^*, \\ \boldsymbol{K}_3\boldsymbol{\Theta} + \boldsymbol{K}_2\boldsymbol{\beta} = \boldsymbol{Y}_2^*. \end{cases},$$

where

$$\boldsymbol{Y}_1^* = \boldsymbol{B}'\mathbf{V}_n^{-1}\boldsymbol{Y}, \;\; \boldsymbol{Y}_2^* = \boldsymbol{X}'\mathbf{V}_n^{-1}\boldsymbol{Y},$$
$$\boldsymbol{K}_1 = \boldsymbol{B}'\mathbf{V}_n^{-1}\boldsymbol{B}, \;\; \boldsymbol{K}_2 = \boldsymbol{X}'\mathbf{V}_n^{-1}\boldsymbol{X}, \;\; \boldsymbol{K}_3 = \boldsymbol{B}'\mathbf{V}_n^{-1}\boldsymbol{X},$$

and

$$\begin{aligned}
\boldsymbol{Y}_1^* &= \left(y_1^{(1)}, \cdots, y_p^{(1)}\right), \; y_\alpha^{(1)} = \mathbf{1}_{n_\alpha}'\mathbf{V}_\alpha^{-1}\boldsymbol{Y}_\alpha, \\
\boldsymbol{Y}_2^* &= \left(y_1^{(2)}, \cdots, y_p^{(2)}\right), \; y_\alpha^{(2)} = \boldsymbol{X}_\alpha'\mathbf{V}_\alpha^{-1}\boldsymbol{Y}_\alpha, \\
\boldsymbol{K}_1 &= \text{Diag}\left(k_1^{(1)}, \cdots, k_p^{(1)}\right), \; k_\alpha^{(1)} = \mathbf{1}_{n_\alpha}'\mathbf{V}_\alpha^{-1}\mathbf{1}_{n_\alpha}, \\
\boldsymbol{K}_2 &= \text{Diag}\left(k_1^{(2)}, \cdots, k_p^{(2)}\right), \; k_\alpha^{(2)} = \boldsymbol{X}_\alpha'\mathbf{V}_\alpha^{-1}\boldsymbol{X}_\alpha, \\
\boldsymbol{K}_3 &= \text{Diag}\left(k_1^{(3)}, \cdots, k_p^{(3)}\right), \; k_\alpha^{(3)} = \mathbf{1}_{n_\alpha}'\mathbf{V}_\alpha^{-1}\boldsymbol{X}_\alpha.
\end{aligned}$$

Then the LSE estimators of $\boldsymbol{\Theta}$ and $\boldsymbol{\beta}$ are given by

$$\begin{aligned}
\begin{bmatrix} \tilde{\boldsymbol{\Theta}}_n \\ \tilde{\boldsymbol{\beta}}_n \end{bmatrix} &= \begin{bmatrix} \boldsymbol{K}_1 & \boldsymbol{K}_3 \\ \boldsymbol{K}_3 & \boldsymbol{K}_2 \end{bmatrix}^{-1} \begin{bmatrix} \boldsymbol{Y}_1^* \\ \boldsymbol{Y}_2^* \end{bmatrix} \\
&= \begin{bmatrix} \boldsymbol{C}_{22}^{-1} & -\boldsymbol{K}_1^{-1}\boldsymbol{K}_3\boldsymbol{C}_{11}^{-1} \\ -\boldsymbol{K}_2^{-1}\boldsymbol{K}_3\boldsymbol{C}_{22}^{-1} & \boldsymbol{C}_{11}^{-1} \end{bmatrix} \begin{bmatrix} \boldsymbol{Y}_1^* \\ \boldsymbol{Y}_2^* \end{bmatrix},
\end{aligned} \tag{6.2.1}$$

where $\boldsymbol{C}_{11}$ and $\boldsymbol{C}_{22}$ are Schur complements of $\boldsymbol{K}_1$ and $\boldsymbol{K}_2$, respectively, given by

$$\boldsymbol{C}_{11} = \boldsymbol{K}_2 - \boldsymbol{K}_3\boldsymbol{K}_1^{-1}\boldsymbol{K}_3, \quad \boldsymbol{C}_{22} = \boldsymbol{K}_1 - \boldsymbol{K}_3\boldsymbol{K}_2^{-1}\boldsymbol{K}_3$$

and

$$\begin{aligned}
\boldsymbol{C}_{11}^{-1} &= \text{Diag}\left(c_1^{(11)}, \cdots, c_p^{(11)}\right), \\
c_\alpha^{(11)} &= \left[k_\alpha^{(2)} - \frac{\left(k_\alpha^{(3)}\right)^2}{k_\alpha^{(1)}}\right]^{-1} = \frac{k_\alpha^{(1)}}{k_\alpha^{(1)}k_\alpha^{(2)} - \left(k_\alpha^{(3)}\right)^2}, \quad \alpha = 1, \ldots, p, \\
\boldsymbol{C}_{22}^{-1} &= \text{Diag}\left(c_1^{(22)}, \cdots, c_p^{(22)}\right),
\end{aligned}$$

$$c_\alpha^{(22)} = \left[k_\alpha^{(1)} - \frac{\left(k_\alpha^{(3)}\right)^2}{k_\alpha^{(2)}} \right]^{-1} = \frac{k_\alpha^{(2)}}{k_\alpha^{(1)} k_\alpha^{(2)} - \left(k_\alpha^{(3)}\right)^2}, \quad \alpha = 1, \ldots, p.$$

Note that $\boldsymbol{K}_1^{-1}\boldsymbol{K}_3\boldsymbol{C}_{11}^{-1} = \boldsymbol{K}_2^{-1}\boldsymbol{K}_3\boldsymbol{C}_{22}^{-1}$.

Also, the UE of σ^2 is given by

$$\widetilde{\sigma}^2 = \frac{1}{n-1} \widetilde{\boldsymbol{E}}' \mathbf{V}_n^{-1} \widetilde{\boldsymbol{E}}, \text{ where } \widetilde{\boldsymbol{E}} = \boldsymbol{Y} - \boldsymbol{B}\tilde{\boldsymbol{\Theta}}_n - \boldsymbol{X}\tilde{\boldsymbol{\beta}}_n.$$

And

$$s_{\boldsymbol{\varepsilon}}^2 = \frac{n-1}{n-2p} \widetilde{\sigma}^2 \tag{6.2.2}$$

is the unbiased estimator of $\sigma_{\boldsymbol{\varepsilon}}^2$.

Theorem 6.2.1. *Assuming* $\boldsymbol{Y}|\boldsymbol{\Theta}, \boldsymbol{\beta}, \sigma^2 \sim M_t^{(n)}(\boldsymbol{B\Theta} + \boldsymbol{X\beta}, \sigma^2 \mathbf{V}_n, \gamma_o)$ *we have*

$$\begin{bmatrix} \tilde{\boldsymbol{\Theta}}_n \\ \tilde{\boldsymbol{\beta}}_n \end{bmatrix} \sim M_t^{(2p)} \left\{ \begin{bmatrix} \boldsymbol{\Theta} \\ \boldsymbol{\beta} \end{bmatrix}, \sigma^2 \boldsymbol{K}^{-1}, \gamma_o \right\}, \tag{6.2.3}$$

where

$$\boldsymbol{K} = \begin{bmatrix} \boldsymbol{K}_1 & \boldsymbol{K}_3 \\ \boldsymbol{K}_3 & \boldsymbol{K}_2 \end{bmatrix}. \tag{6.2.4}$$

Proof: From (6.2.1), we get

$$Cov \begin{bmatrix} \tilde{\boldsymbol{\Theta}}_n \\ \tilde{\boldsymbol{\beta}}_n \end{bmatrix} = \sigma^2 \begin{bmatrix} \boldsymbol{K}_1 & \boldsymbol{K}_3 \\ \boldsymbol{K}_3 & \boldsymbol{K}_2 \end{bmatrix}^{-1}.$$

Using Theorem 6.1.1 of Saleh (2006), it can be obtained that

$$\begin{bmatrix} \tilde{\boldsymbol{\Theta}}_n \\ \tilde{\boldsymbol{\beta}}_n \end{bmatrix} \sim \mathcal{N}_{2p} \left\{ \begin{bmatrix} \boldsymbol{\Theta} \\ \boldsymbol{\beta} \end{bmatrix}, \sigma^2 \boldsymbol{K}^{-1} \right\}$$

under normality theory. Therefore, we conclude

$$f_{\tilde{\boldsymbol{\Theta}}_n, \tilde{\boldsymbol{\beta}}_n}(\boldsymbol{x}, \boldsymbol{y}) = \int_0^\infty W_o(t) \mathcal{N}_{2p} \left\{ \begin{bmatrix} \boldsymbol{\Theta} \\ \boldsymbol{\beta} \end{bmatrix}, \sigma^2 t^{-1} \boldsymbol{K}^{-1} \right\} dt,$$

which gives the result using (2.6.3).

It is desirable to consider the parallelism hypothesis $H_0 : \boldsymbol{\beta} = \beta_0 \mathbf{1}_p$ holds. Before we consider the test of parallelism, we propose the restricted estimators (REs) of the underlying parameters under the conditions that the slope parameters are equal, that is $\boldsymbol{\beta} = \beta_0 \mathbf{1}_p$, where β_0 is unknown common slope (scalar).

Then, for the restricted estimators of the parameters under the hypothesis $H_0 : \boldsymbol{\beta} = \beta_0 \mathbf{1}_p$, i.e., parallelism, we minimize $\boldsymbol{E}'\mathbf{V}_n^{-1}\boldsymbol{E} + 2\boldsymbol{\lambda}'(\boldsymbol{\beta} - \beta_0 \mathbf{1}_p)$ by the method of Lagrangian multipliers. Accordingly, we have the following system of equations:

$$\begin{cases} \boldsymbol{K}_1\boldsymbol{\Theta} + \boldsymbol{K}_3\boldsymbol{\beta} = \boldsymbol{Y}_1^* \\ \boldsymbol{K}_3\boldsymbol{\Theta} + \boldsymbol{K}_2\boldsymbol{\beta} + \boldsymbol{\lambda} = \boldsymbol{Y}_2^* \\ \boldsymbol{\beta} = \beta_0 \mathbf{1}_p \end{cases}.$$

Then, REs of $\boldsymbol{\Theta}$ and $\boldsymbol{\beta}$ are obtained by solving the following system of equations in terms of $\boldsymbol{\lambda}$ (the Lagrangian parameter) as

$$\begin{bmatrix} \hat{\boldsymbol{\Theta}}_n \\ \hat{\boldsymbol{\beta}}_n \end{bmatrix} = \begin{bmatrix} \boldsymbol{C}_{22}^{-1} & -\boldsymbol{K}_1^{-1}\boldsymbol{K}_3\boldsymbol{C}_{11}^{-1} \\ -\boldsymbol{K}_2^{-1}\boldsymbol{K}_3\boldsymbol{C}_{22}^{-1} & \boldsymbol{C}_{11}^{-1} \end{bmatrix} \begin{bmatrix} \boldsymbol{Y}_1^* \\ \boldsymbol{Y}_2^* - \boldsymbol{\lambda} \end{bmatrix}. \tag{6.2.5}$$

It can be easily seen that $\hat{\boldsymbol{\Theta}}_n = \tilde{\boldsymbol{\Theta}}_n - \boldsymbol{K}_1^{-1}\boldsymbol{K}_3(\tilde{\boldsymbol{\beta}}_n - \hat{\boldsymbol{\beta}}_n)$, where

$$\begin{aligned} \hat{\boldsymbol{\beta}}_n &= \hat{\beta}_{0n}\mathbf{1}_p, \quad \hat{\beta}_{0n} \text{ is the LSE of } \beta_0 \\ &= \frac{\mathbf{1}_p'\boldsymbol{C}_{11}^{-1}\tilde{\boldsymbol{\beta}}_n}{\mathbf{1}_p'\boldsymbol{C}_{11}^{-1}\mathbf{1}_p}\mathbf{1}_p = \hat{\beta}_{0n}\mathbf{1}_p. \end{aligned}$$

Therefore, the REs of $\boldsymbol{\Theta}$ and $\boldsymbol{\beta}$ are given by

$$\hat{\boldsymbol{\Theta}}_n = \tilde{\boldsymbol{\Theta}}_n + \boldsymbol{K}_1^{-1}\boldsymbol{K}_3\boldsymbol{H}\tilde{\boldsymbol{\beta}}_n, \text{ and } \hat{\boldsymbol{\beta}}_n = \frac{\mathbf{1}_p'\boldsymbol{C}_{11}^{-1}\tilde{\boldsymbol{\beta}}_n}{\mathbf{1}_p'\boldsymbol{C}_{11}^{-1}\mathbf{1}_p}\mathbf{1}_p, \tag{6.2.6}$$

where

$$\boldsymbol{H} = \boldsymbol{I}_p - \frac{\mathbf{1}_p\mathbf{1}_p'\boldsymbol{C}_{11}^{-1}}{\mathbf{1}_p'\boldsymbol{C}_{11}^{-1}\mathbf{1}_p}.$$

Accordingly, the RE of σ_ε^2 is given by

$$\hat{\sigma}^2 = \frac{1}{m}\hat{\boldsymbol{E}}'\mathbf{V}_n^{-1}\hat{\boldsymbol{E}}, \text{ where } \hat{\boldsymbol{E}} = \boldsymbol{Y} - \boldsymbol{B}\hat{\boldsymbol{\Theta}}_n - \boldsymbol{X}\hat{\boldsymbol{\beta}}_n \text{ and } m = n - 2p. \tag{6.2.7}$$

6.2.1 Test of Parallelism

Now, we propose the likelihood ratio criterion (LRC) for testing the parallelism hypothesis $H_0 : \boldsymbol{\beta} = \beta_0 \mathbf{1}_p$ and its non-null distribution in the following theorem.

Theorem 6.2.2. *Let*

$$\begin{aligned}\Omega_1 &= \{(\boldsymbol{\Theta},\boldsymbol{\beta},\sigma,\mathbf{V}_n) : \boldsymbol{\Theta},\boldsymbol{\beta} \in \mathbb{R}^p, \sigma \in \mathbb{R}^+, \mathbf{V}_n > 0\},\\ \Omega_0 &= \{(\boldsymbol{\Theta},\boldsymbol{\beta},\sigma,\mathbf{V}_n) : \boldsymbol{\Theta},\boldsymbol{\beta} \in \mathbb{R}^p, \boldsymbol{\beta} = \beta_0\mathbf{1}_p, \sigma \in \mathbb{R}^+, \mathbf{V}_n > 0\}.\end{aligned}$$

Then the LRC for testing parallelism hypothesis is given by

$$\mathcal{L}_n = \frac{\tilde{\boldsymbol{\beta}}_n'\boldsymbol{H}'\boldsymbol{C}_{11}\boldsymbol{H}\tilde{\boldsymbol{\beta}}_n}{(p-1)s_{\varepsilon}^2}, \tag{6.2.8}$$

having the following generalized noncentral F distribution:

$$g_{p,m}^*(\mathcal{L}_n) = \sum_{r\geq 0} \frac{\left(\frac{p-1}{m}\right)^{\frac{1}{2}(p+2r-1)} \mathcal{L}_n^{\frac{1}{2}(p+2r-3)} K_r^{(0)}(\Delta^2)}{r!\, B\left(\frac{p+2r-1}{2}, \frac{m}{2}\right)\left(1+\frac{p-1}{m}\mathcal{L}_n\right)^{\frac{1}{2}(p+m+2r-1)}}, \tag{6.2.9}$$

where $m = n - 2p$*,* $\Delta^2 = \xi/\varepsilon$ *for* $\xi = \boldsymbol{\beta}'\boldsymbol{H}'\boldsymbol{C}_{11}\boldsymbol{H}\boldsymbol{\beta}$.

Proof: As usual, the LR is $\Lambda = L_0/L_1$, where L_i, $i = 0, 1$, is the largest value that the likelihood function takes in region Ω_i, $i = 0, 1$.
Because $\boldsymbol{Y}|\boldsymbol{\Theta},\boldsymbol{\beta},\sigma^2 \sim M_t^{(n)}(\boldsymbol{B\Theta}+\boldsymbol{X\beta},\sigma^2\mathbf{V}_n,\gamma_o)$, we have

$$\begin{aligned}L_0 &= \frac{\Gamma\left(\frac{n+\gamma_o}{2}\right)}{(\pi\gamma_o)^{\frac{n}{2}}\Gamma\left(\frac{\gamma_o}{2}\right)(\hat{\sigma}^2)^{\frac{n}{2}}}\left[1+\frac{\hat{\boldsymbol{E}}'\mathbf{V}_n^{-1}\hat{\boldsymbol{E}}}{\gamma_o\hat{\sigma}^2}\right]^{-\frac{1}{2}(n+\gamma_o)}\\ &= \frac{\Gamma\left(\frac{n+\gamma_o}{2}\right)}{(\pi\gamma_o)^{\frac{n}{2}}\Gamma\left(\frac{\gamma_o}{2}\right)(\hat{\sigma}^2)^{\frac{n}{2}}}\left[1+\frac{m}{\gamma_o}\right]^{-\frac{1}{2}(n+\gamma_o)}\end{aligned}$$

and

$$\begin{aligned}L_1 &= \frac{\Gamma\left(\frac{n+\gamma_o}{2}\right)}{(\pi\gamma_o)^{\frac{n}{2}}\Gamma\left(\frac{\gamma_o}{2}\right)(\tilde{\sigma}^2)^{\frac{n}{2}}}\left[1+\frac{\tilde{\boldsymbol{E}}'\mathbf{V}_n^{-1}\tilde{\boldsymbol{E}}}{\gamma_o\tilde{\sigma}^2}\right]^{-\frac{1}{2}(n+\gamma_o)}\\ &= \frac{\Gamma\left(\frac{n+\gamma_o}{2}\right)}{(\pi\gamma_o)^{\frac{n}{2}}\Gamma\left(\frac{\gamma_o}{2}\right)(\tilde{\sigma}^2)^{\frac{n}{2}}}\left[1+\frac{n-1}{\gamma_o}\right]^{-\frac{1}{2}(n+\gamma_o)}.\end{aligned}$$

Thus, the LR is given by

$$\begin{aligned}\Lambda &= a\left\{\frac{\widetilde{\boldsymbol{E}}'\mathbf{V}_n^{-1}\widetilde{\boldsymbol{E}}/(n-1)}{\hat{\boldsymbol{E}}'\mathbf{V}_n^{-1}\hat{\boldsymbol{E}}/m}\right\}^{\frac{n}{2}}\\ &= a\left\{\frac{m\widetilde{\boldsymbol{E}}'\mathbf{V}_n^{-1}\widetilde{\boldsymbol{E}}}{(n-1)\left[\widetilde{\boldsymbol{E}}'\mathbf{V}_n^{-1}\widetilde{\boldsymbol{E}}+\tilde{\boldsymbol{\beta}}_n'\boldsymbol{H}'\boldsymbol{C}_{11}\boldsymbol{H}\tilde{\boldsymbol{\beta}}_n\right]}\right\}^{\frac{n}{2}}\\ &= a\left(\frac{1}{1+\frac{p-1}{m}\mathcal{L}_n}\right)^{\frac{n}{2}},\end{aligned}$$

where

$$a = \left(\frac{n - 1 + \gamma_o}{m + \gamma_o} \right)^{\frac{1}{2}(n+\gamma_o)}.$$

Hence, $\mathcal{L}_n$ is the LR test for testing the null hypothesis $H_0 : \boldsymbol{\beta} = \beta_0 \mathbf{1}_p$.

For the non-null distribution of $\mathcal{L}_n$, consider that the conditioned distribution of $\mathcal{L}_n$ (under normal theory) is the central F. Hence, using the result of Saleh (2006), the unconditioned distribution of $\mathcal{L}_n$ is

$$g^*_{p,m}(\mathcal{L}_n) = \int_0^\infty W_o(t) \mathcal{G}_{p,m}(\mathcal{L}_n, \Delta_t^2)\, dt, \tag{6.2.10}$$

where $\mathcal{G}_{p,m}(\mathcal{L}_n, \Delta_t^2)$ denotes the distribution of $\mathcal{L}_n$ under the (normality) assumption $\boldsymbol{E} \sim \mathcal{N}_n(\mathbf{0}, \sigma^2 t^{-1} \mathbf{V}_n)$. Thus, we have

$$g^*_{p,m}(\mathcal{L}_n) = \sum_{r \geq 0} \frac{\left(\frac{p-1}{m}\right)^{\frac{1}{2}(p+2r-1)} \mathcal{L}_n^{\frac{1}{2}(p+2r-3)} K_r^{(0)}(\Delta^2)}{r!\, B\left(\frac{p+2r-1}{2}, \frac{m}{2}\right) \left(1 + \frac{p-1}{m}\mathcal{L}_n\right)^{\frac{1}{2}(p+m+2r-1)}}.$$

Corollary 6.2.2.1. *Under H_0, the pdf of $\mathcal{L}_n$ is given by*

$$g^*_{p,m}(\mathcal{L}_n) = \frac{\left(\frac{p-1}{m}\right)^{\frac{p-1}{2}} \mathcal{L}_n^{\frac{p-1}{2}-1}}{B\left(\frac{p-1}{2}, \frac{m}{2}\right) \left(1 + \frac{p-1}{m}\mathcal{L}_n\right)^{\frac{1}{2}(p+m-1)}}, \tag{6.2.11}$$

which is the central F-distribution with $(p-1, m)$ degrees of freedom.

Corollary 6.2.2.2. *The power function at γ-level of significance of $\mathcal{L}_n$, say generalized noncentral F cumulative distribution function of the statistic $\mathcal{L}_n$ is given by $1 - \mathcal{G}_{p,m}(l_\gamma; \Delta^2)$, where*

$$\mathcal{G}_{p,m}(l_\gamma; \Delta^2) = \sum_{r \geq 0} K_r^{(0)}(\Delta^2) I_{l_\gamma}\left[\frac{1}{2}(p+2r), \frac{m}{2}\right], \tag{6.2.12}$$

and $l_\gamma = \frac{(p-1)F_{p-1,m}(\gamma)}{m+(p-1)F_{p-1,m}(\gamma)}$.

Then, we propose the following theorem.

Theorem 6.2.3. *Under the assumed regularity conditions, the following holds:*

$$\begin{pmatrix}\hat{\boldsymbol{\Theta}}_n \\ \tilde{\boldsymbol{\beta}}_n\end{pmatrix} \sim M_t^{(2p)}\left\{\begin{pmatrix}\boldsymbol{\Theta}+\boldsymbol{K}_1^{-1}\boldsymbol{K}_3\boldsymbol{H}\boldsymbol{\beta} \\ \boldsymbol{\beta}\end{pmatrix};\sigma^2\begin{pmatrix}\boldsymbol{D}_{11} & \boldsymbol{D}_{12} \\ \boldsymbol{D}_{21} & \boldsymbol{C}_{11}^{-1}\end{pmatrix},\gamma_o\right\},$$

$$\begin{pmatrix}\tilde{\boldsymbol{\beta}}_n \\ \tilde{\boldsymbol{\beta}}_n-\hat{\boldsymbol{\beta}}_n\end{pmatrix} \sim M_t^{(2p)}\left\{\begin{pmatrix}\boldsymbol{\beta} \\ \boldsymbol{H}\boldsymbol{\beta}\end{pmatrix};\sigma^2\begin{pmatrix}\boldsymbol{C}_{11}^{-1} & \boldsymbol{H}\boldsymbol{C}_{11}^{-1} \\ \boldsymbol{C}_{11}^{-1}\boldsymbol{H}' & \boldsymbol{H}\boldsymbol{C}_{11}^{-1}\end{pmatrix},\gamma_o\right\},$$

$$\begin{pmatrix}\hat{\boldsymbol{\beta}}_n-\beta_0\mathbf{1}_p \\ \tilde{\boldsymbol{\beta}}_n-\hat{\boldsymbol{\beta}}_n\end{pmatrix} \sim M_t^{(2p)}\left\{\begin{pmatrix}(\bar{\beta}-\beta_0)\mathbf{1}_p \\ \boldsymbol{H}\boldsymbol{\beta}\end{pmatrix};\sigma^2\begin{pmatrix}\frac{\mathbf{1}_p\mathbf{1}_p'}{C} & \mathbf{0} \\ \mathbf{0} & \boldsymbol{H}\boldsymbol{C}_{11}^{-1}\end{pmatrix},\gamma_o\right\},$$

where

$$\bar{\beta}\mathbf{1}_p=\frac{\mathbf{1}_p\mathbf{1}_p'\boldsymbol{C}_{11}^{-1}}{\mathbf{1}_p'\boldsymbol{C}_{11}^{-1}\mathbf{1}_p},\quad \boldsymbol{D}_{12}=\frac{-\mathbf{1}_p\mathbf{1}_p'\boldsymbol{K}_1^{-1}\boldsymbol{K}_3}{\mathbf{1}_p'\boldsymbol{C}_{11}^{-1}\mathbf{1}_p},\quad \boldsymbol{D}_{11}=\boldsymbol{K}_1^{-1}+\frac{\boldsymbol{K}_1^{-1}\boldsymbol{K}_3\mathbf{1}_p\mathbf{1}_p'\boldsymbol{K}_3\boldsymbol{K}_1^{-1}}{\mathbf{1}_p'\boldsymbol{C}_{11}^{-1}\mathbf{1}_p}.$$

Now, we propose three more estimators and consider the preliminary test estimators (PTEs), Stein-type shrinkage estimators (SEs), and the positive-rule estimators (PRSEs). Accordingly, the PT estimators of $\boldsymbol{\beta}$ and $\boldsymbol{\Theta}$, respectively, are given by

$$\hat{\boldsymbol{\beta}}_n^{PT}=\tilde{\boldsymbol{\beta}}_n-\boldsymbol{H}\tilde{\boldsymbol{\beta}}_n I(\mathcal{L}_n<F_{p-1,m}(\gamma)), \tag{6.2.13}$$

and

$$\hat{\boldsymbol{\Theta}}_n^{PT}=\tilde{\boldsymbol{\Theta}}_n+\boldsymbol{K}_1^{-1}\boldsymbol{K}_3\boldsymbol{H}\tilde{\boldsymbol{\beta}}_n I(\mathcal{L}_n<F_{p-1,m}(\gamma)), \tag{6.2.14}$$

where $F_{p-1,m}(\gamma)$ is the upper γ-quantile of the central F distribution with $p-1$ and m degrees of freedom.

The PTE has the disadvantage that it is a discrete process and depends on $\gamma(0<\gamma<1)$, the level of significance. To eliminate such dependency, we define the Stein-type estimators of $\boldsymbol{\beta}$ and $\boldsymbol{\Theta}$, which is continoues version of $\hat{\boldsymbol{\beta}}_n^{PT}$ and $\hat{\boldsymbol{\Theta}}_n^{PT}$ and free of level of significance. They are given by

$$\hat{\boldsymbol{\beta}}_n^{S}=\tilde{\boldsymbol{\beta}}_n-c\boldsymbol{H}\tilde{\boldsymbol{\beta}}_n\mathcal{L}_n^{-1} \tag{6.2.15}$$

and

$$\hat{\boldsymbol{\Theta}}_n^{S}=\tilde{\boldsymbol{\Theta}}_n+c\boldsymbol{K}_1^{-1}\boldsymbol{K}_3\boldsymbol{H}\tilde{\boldsymbol{\beta}}_n\mathcal{L}_n^{-1}, \tag{6.2.16}$$

where

$$c=\frac{(p-3)m}{(p-1)(m+2)}. \tag{6.2.17}$$

Note that the two foregoing estimators are not convex combinations, so they may exceed the values $\hat{\boldsymbol{\Theta}}_n$ and $\hat{\boldsymbol{\beta}}_n$. Thus, in order to shrink these estimators towards $\hat{\boldsymbol{\Theta}}_n$ and $\hat{\boldsymbol{\beta}}_n$, we consider some convex combinations of $\hat{\boldsymbol{\Theta}}_n$ and $\hat{\boldsymbol{\Theta}}_n^S$, and $\hat{\boldsymbol{\beta}}_n$ and $\hat{\boldsymbol{\beta}}_n^S$, via the preliminary test procedure to obtain positive-rule Stein-type (PRS) estimators of $\boldsymbol{\beta}$ and $\boldsymbol{\Theta}$ for $p \geq 4$, respectively, via preliminary test procedure. Thus, we obtain the following positive-rule estimators (for $p > 3$) as

$$\begin{aligned}\hat{\boldsymbol{\beta}}_n^{S+} &= \hat{\boldsymbol{\beta}}_n I(\mathcal{L}_n < c) + \hat{\boldsymbol{\beta}}_n^S I(\mathcal{L}_n > c) \\ &= \hat{\boldsymbol{\beta}}_n + (1 - c\mathcal{L}_n^{-1}) I(\mathcal{L}_n > c) \boldsymbol{H}\tilde{\boldsymbol{\beta}}_n \end{aligned} \quad (6.2.18)$$

and

$$\begin{aligned}\hat{\boldsymbol{\Theta}}_n^{S+} &= \hat{\boldsymbol{\Theta}}_n I(\mathcal{L}_n < c) + \hat{\boldsymbol{\Theta}}_n^S I(\mathcal{L}_n > c) \\ &= \hat{\boldsymbol{\Theta}}_n - (\hat{\boldsymbol{\Theta}}_n - \hat{\boldsymbol{\Theta}}_n^S) I(\mathcal{L}_n > c) \\ &= \hat{\boldsymbol{\Theta}}_n - (1 - c\mathcal{L}_n^{-1}) I(\mathcal{L}_n > c) \boldsymbol{K}_1^{-1}\boldsymbol{K}_3\boldsymbol{H}\tilde{\boldsymbol{\beta}}_n \\ &= \tilde{\boldsymbol{\Theta}}_n + \boldsymbol{K}_1^{-1}\boldsymbol{K}_3\boldsymbol{H}\tilde{\boldsymbol{\beta}}_n\{1 - (1 - c\mathcal{L}_n^{-1}) I(\mathcal{L}_n > c)\}. \end{aligned} \quad (6.2.19)$$

Thus, we have defined five estimators of $\boldsymbol{\Theta}$ and $\boldsymbol{\beta}$. In the next section we obtain the bias, MSE matrix, and risk expressions of these estimators.

6.3 Bias, MSE, and Risk Expressions

The bias, the weighted quadratic risk, and the MSE-matrix of the estimators are given in this section, based on the notations that are given in Chapter 4.

6.3.1 Expressions of Bias, MSE Matrix, and Risks of $\tilde{\boldsymbol{\beta}}_n$, $\tilde{\boldsymbol{\Theta}}_n$, $\hat{\boldsymbol{\beta}}_n$, and $\hat{\boldsymbol{\Theta}}_n$

Using Theorem 6.2.1, we have

$$\begin{aligned} \boldsymbol{b}(\tilde{\boldsymbol{\beta}}_n) &= \mathbf{0}, \\ \boldsymbol{M}(\tilde{\boldsymbol{\beta}}_n) &= \sigma_{\varepsilon}^2 \boldsymbol{C}_{11}^{-1}, \\ R(\tilde{\boldsymbol{\beta}}_n; \boldsymbol{W}) &= \sigma_{\varepsilon}^2 tr(\boldsymbol{W}\boldsymbol{C}_{11}^{-1}), \end{aligned} \quad (6.3.1)$$

and

$$\begin{aligned} \boldsymbol{b}(\tilde{\boldsymbol{\Theta}}_n) &= \mathbf{0}, \\ \boldsymbol{M}(\tilde{\boldsymbol{\Theta}}_n) &= \sigma_{\varepsilon}^2 \boldsymbol{C}_{22}^{-1}, \\ R(\tilde{\boldsymbol{\Theta}}_n, \boldsymbol{W}) &= \sigma_{\varepsilon}^2 tr(\boldsymbol{W}\boldsymbol{C}_{22}^{-1}). \end{aligned} \quad (6.3.2)$$

By Theorem 6.2.3,

$$\begin{aligned}
\boldsymbol{b}(\hat{\boldsymbol{\beta}}_n) &= -\boldsymbol{H}\boldsymbol{\beta}, \\
\boldsymbol{M}(\hat{\boldsymbol{\beta}}_n) &= \frac{\sigma_\varepsilon^2 \mathbf{1}_p \mathbf{1}_p'}{\mathbf{1}_p' \boldsymbol{C}_{11}^{-1} \mathbf{1}_p} + \boldsymbol{H}\boldsymbol{\beta}\boldsymbol{\beta}'\boldsymbol{H}', \\
R(\hat{\boldsymbol{\beta}}_n; \boldsymbol{W}) &= \frac{\sigma_\varepsilon^2 \mathbf{1}_p' \boldsymbol{W} \mathbf{1}_p}{\mathbf{1}_p' \boldsymbol{C}_{11}^{-1} \mathbf{1}_p} + \boldsymbol{\beta}'\boldsymbol{H}'\boldsymbol{W}\boldsymbol{H}\boldsymbol{\beta}, \qquad (6.3.3)
\end{aligned}$$

and

$$\begin{aligned}
\boldsymbol{b}(\hat{\boldsymbol{\Theta}}_n) &= \boldsymbol{K}_1^{-1}\boldsymbol{K}_3\boldsymbol{H}\boldsymbol{\beta}, \\
\boldsymbol{M}(\hat{\boldsymbol{\Theta}}_n) &= \sigma_\varepsilon^2 \boldsymbol{D}_{11} + \boldsymbol{K}_1^{-1}\boldsymbol{K}_3\boldsymbol{H}\boldsymbol{\beta}\boldsymbol{\beta}'\boldsymbol{H}'\boldsymbol{K}_3\boldsymbol{K}_1^{-1}, \\
R(\hat{\boldsymbol{\Theta}}_n, \boldsymbol{W}) &= \sigma_\varepsilon^2 tr(\boldsymbol{W}\boldsymbol{D}_{11}) + \boldsymbol{\beta}'\boldsymbol{H}'\boldsymbol{K}_1^{-1}\boldsymbol{K}_3\boldsymbol{W}\boldsymbol{K}_3\boldsymbol{K}_1^{-1}\boldsymbol{H}\boldsymbol{\beta}. \qquad (6.3.4)
\end{aligned}$$

6.3.2 Expressions of Bias, MSE Matrix, and Risks of the PTEs of $\boldsymbol{\beta}$ and $\boldsymbol{\Theta}$

$$\begin{aligned}
\boldsymbol{b}(\hat{\boldsymbol{\beta}}_n^{PT}) &= -\boldsymbol{H}\boldsymbol{\beta} G_{p+1,m}^{(2)}(l_\gamma; \Delta^2), \\
\boldsymbol{M}(\hat{\boldsymbol{\beta}}_n^{PT}) &= \sigma_\varepsilon^2 \boldsymbol{C}_{11}^{-1} - \sigma_\varepsilon^2 \boldsymbol{H}\boldsymbol{C}_{11}^{-1} G_{p+1,m}^{(1)}(l_\gamma; \Delta^2) + \boldsymbol{H}\boldsymbol{\beta}\boldsymbol{\beta}'\boldsymbol{H}' \\
&\quad \times \{2G_{p+1,m}^{(2)}(l_\gamma; \Delta^2) - G_{p+3,m}^{(2)}(l_\gamma; \Delta^2)\}, \\
R(\hat{\boldsymbol{\beta}}_n^{PT}; \boldsymbol{W}) &= \sigma_\varepsilon^2 tr(\boldsymbol{W}\boldsymbol{C}_{11}^{-1}) - \sigma_\varepsilon^2 tr(\boldsymbol{W}\boldsymbol{H}\boldsymbol{C}_{11}^{-1}) G_{p+1,m}^{(1)}(l_\gamma; \Delta^2) \\
&\quad + \boldsymbol{\beta}'\boldsymbol{H}'\boldsymbol{W}\boldsymbol{H}\boldsymbol{\beta}\{2G_{p+1,m}^{(2)}(l_\gamma; \Delta^2) - G_{p+3,m}^{(2)}(l_\gamma; \Delta^2)\}, \\
&\qquad (6.3.5)
\end{aligned}$$

and

$$\begin{aligned}
\boldsymbol{b}(\hat{\boldsymbol{\Theta}}_n^{PT}) &= \boldsymbol{K}_1^{-1}\boldsymbol{K}_3\boldsymbol{H}\boldsymbol{\beta} G_{p+1,m}^{(2)}(l_\gamma; \Delta^2), \\
\boldsymbol{M}(\hat{\boldsymbol{\Theta}}_n^{PT}) &= \sigma_\varepsilon^2 \boldsymbol{C}_{22}^{-1} - \sigma_\varepsilon^2 (\boldsymbol{C}_{22}^{-1} - \boldsymbol{D}_{11}) G_{p+1,m}^{(1)}(l_\gamma; \Delta^2) \\
&\quad + \boldsymbol{K}_1^{-1}\boldsymbol{K}_3\boldsymbol{H}\boldsymbol{\beta}\boldsymbol{\beta}'\boldsymbol{H}'\boldsymbol{K}_3\boldsymbol{K}_1^{-1} \\
&\quad \times \{2G_{p+1,m}^{(2)}(l_\gamma; \Delta^2) - G_{p+3,m}^{(2)}(l_\gamma; \Delta^2)\}, \\
R(\hat{\boldsymbol{\Theta}}_n^{PT}, \boldsymbol{W}) &= \sigma_\varepsilon^2 tr(\boldsymbol{W}\boldsymbol{C}_{22}^{-1}) - \sigma_\varepsilon^2 tr[\boldsymbol{W}(\boldsymbol{C}_{22}^{-1} - \boldsymbol{D}_{11})] G_{p+1,m}^{(1)}(l_\gamma; \Delta^2) \\
&\quad + \boldsymbol{\beta}'\boldsymbol{H}'\boldsymbol{K}_1^{-1}\boldsymbol{K}_3'\boldsymbol{W}\boldsymbol{K}_3\boldsymbol{K}_1^{-1}\boldsymbol{H}\boldsymbol{\beta} \\
&\quad \times \{2G_{p+1,m}^{(2)}(l_\gamma; \Delta^2) - G_{p+3,m}^{(2)}(l_\gamma; \Delta^2)\}, \qquad (6.3.6)
\end{aligned}$$

where $G_{p,m}^{(j)}(l_\gamma; \Delta^2)$ is given by (5.3.4).

Note that $G_{p,m}^{(2)}(l_\gamma; \Delta^2) = \mathcal{G}_{p,m}(l_\gamma; \Delta^2)$.

6.3.3 Expressions of Bias, MSE Matrix, and Risks of the SSEs of β and Θ

$$
\begin{aligned}
\boldsymbol{b}(\hat{\boldsymbol{\beta}}_n^S) &= -c(p-1)\boldsymbol{H\beta}E^{(2)}[\chi_{p+1}^{-2}(\Delta^2)], \\
\boldsymbol{M}(\hat{\boldsymbol{\beta}}_n^S) &= \sigma_{\varepsilon}^2\boldsymbol{C}_{11}^{-1} - c(p-1)\sigma_{\varepsilon}^2\boldsymbol{HC}_{11}^{-1}\{2E^{(1)}[\chi_{p+1}^{-2}(\Delta^2)] \\
&\quad -(p-3)E^{(1)}[\chi_{p+1}^{-4}(\Delta^2)]\} \\
&\quad +c(p^2-1)\boldsymbol{H\beta\beta}'\boldsymbol{H}'E^{(2)}[\chi_{p+3}^{-4}(\Delta^2)], \\
R(\hat{\boldsymbol{\beta}}_n^S;\boldsymbol{W}) &= \sigma_{\varepsilon}^2 tr(\boldsymbol{WC}_{11}^{-1}) - c(p-1)\sigma_{\varepsilon}^2 tr(\boldsymbol{WHC}_{11}^{-1})\{2E^{(1)}[\chi_{p+1}^{-2}(\Delta^2)] \\
&\quad -(p-3)E^{(1)}[\chi_{p+1}^{-4}(\Delta^2)]\} \\
&\quad +c(p^2-1)\boldsymbol{\beta}'\boldsymbol{H}'\boldsymbol{WH\beta}E^{(2)}[\chi_{p+3}^{-4}(\Delta^2)],
\end{aligned}
$$

and

$$
\begin{aligned}
\boldsymbol{b}(\hat{\boldsymbol{\Theta}}_n^S) &= c(p-1)\boldsymbol{K}_1^{-1}\boldsymbol{K}_3\boldsymbol{H\beta}E^{(2)}[\chi_{p+1}^{-2}(\Delta^2)], \\
\boldsymbol{M}(\hat{\boldsymbol{\Theta}}_n^S) &= \sigma_{\varepsilon}^2\boldsymbol{C}_{22}^{-1} - c(p-1)\sigma_{\varepsilon}^2\boldsymbol{K}_1^{-1}\boldsymbol{K}_3\boldsymbol{HC}_{11}^{-1}\boldsymbol{H}'\boldsymbol{K}_3\boldsymbol{K}_1^{-1} \\
&\quad \times\{(p-3)E^{(1)}[\chi_{p+1}^{-4}(\Delta^2)] + 2\Delta^2E^{(1)}[\chi_{p+3}^{-4}(\Delta^2)]\} \\
&\quad +c(p^2-1)\boldsymbol{K}_1^{-1}\boldsymbol{K}_3\boldsymbol{H\beta\beta}'\boldsymbol{H}'\boldsymbol{K}_3\boldsymbol{K}_1^{-1}E^{(2)}[\chi_{p+3}^{-4}(\Delta^2)], \\
R(\hat{\boldsymbol{\Theta}}_n^S;\boldsymbol{W}) &= \sigma_{\varepsilon}^2 tr(\boldsymbol{WC}_{22}^{-1}) \\
&\quad -c(p-1)\sigma_{\varepsilon}^2 tr(\boldsymbol{WK}_1^{-1}\boldsymbol{K}_3\boldsymbol{HC}_{11}^{-1}\boldsymbol{H}'\boldsymbol{K}_3\boldsymbol{K}_1^{-1}) \\
&\quad \times\{(p-3)E^{(1)}[\chi_{p+1}^{-4}(\Delta^2)] + 2\Delta^2E^{(1)}[\chi_{p+3}^{-4}(\Delta^2)]\} \\
&\quad +c(p^2-1)\boldsymbol{\beta}'\boldsymbol{H}'\boldsymbol{K}_1^{-1}\boldsymbol{K}_3\boldsymbol{WK}_3\boldsymbol{K}_1^{-1}\boldsymbol{H\beta}E^{(2)}[\chi_{p+3}^{-4}(\Delta^2)],
\end{aligned}
\tag{6.3.7}
$$

where $E^{(j)}[\chi_{p+s}^{-2}(\Delta^2)]$ and $E^{(j)}[\chi_{p+s}^{-4}(\Delta^2)]$ are given by (5.3.6) and (5.3.7), respectively.

6.3.4 Expressions of Bias, MSE Matrix, and Risks of the PRSEs of β and Θ

$$
\begin{aligned}
\boldsymbol{b}(\hat{\boldsymbol{\beta}}_n^{S+}) &= -\boldsymbol{H}\beta\{G_{p+1,m}^{(2)}(c_1;\Delta^2) \\
&\qquad +c_1E^{(2)}[F_{p+1,m}^{-1}(\Delta^2)I(F_{p+1,m}(\Delta^2) > c_1)]]\}, \\
\boldsymbol{M}(\hat{\boldsymbol{\beta}}_n^{S+}) &= \boldsymbol{M}(\hat{\boldsymbol{\beta}}_n^S)
\end{aligned}
$$

$$
\begin{aligned}
& -\sigma_{\varepsilon}^{2}\boldsymbol{H}\boldsymbol{C}_{11}^{-1}E^{(1)}[(1-c_1F_{p+1,m}^{-1}(\Delta^2))^2 I(F_{p+1,m}(\Delta^2)<c_1)] \\
& +2\boldsymbol{H}\boldsymbol{\beta}\boldsymbol{\beta}'\boldsymbol{H}'E^{(2)}[(1-c_1F_{p+1,m}^{-1}(\Delta^2))I(F_{p+1,m}(\Delta^2)<c_1)] \\
& -E^{(2)}[(1-c_2F_{p+3,m}^{-1}(\Delta^2))^2 I(F_{p+3,m}(\Delta^2)<c_2)],
\end{aligned}
$$

$$
\begin{aligned}
R(\hat{\boldsymbol{\beta}}_n^{S+};\boldsymbol{W}) &= \boldsymbol{R}(\hat{\boldsymbol{\beta}}_n^{S};\boldsymbol{W}) - \sigma_{\varepsilon}^{2}tr(\boldsymbol{W}\boldsymbol{H}\boldsymbol{C}_{11}^{-1}) \\
&\qquad \times E^{(1)}[(1-c_1F_{p+1,m}^{-1}(\Delta^2))^2 I(F_{p+1,m}(\Delta^2)<c_1)] \\
&\quad +2\boldsymbol{\beta}'\boldsymbol{H}'\boldsymbol{W}\boldsymbol{H}\boldsymbol{\beta}E^{(2)}[(1-c_1F_{p+1,m}^{-1}(\Delta^2))I(F_{p+1,m}(\Delta^2)<c_1)] \\
&\quad -E^{(2)}[(1-c_2F_{p+3,m}^{-1}(\Delta^2))^2 I(F_{p+3,m}(\Delta^2)<c_2)], \qquad (6.3.8)
\end{aligned}
$$

and

$$
\begin{aligned}
\boldsymbol{b}(\hat{\boldsymbol{\Theta}}_n^{S+}) &= c_1\boldsymbol{K}_1^{-1}\boldsymbol{K}_3\boldsymbol{H}\boldsymbol{\beta}\{E^{(2)}[F_{p+1,m}^{-1}(\Delta^2)] \\
&\quad -E^{(2)}[F_{p+1,m}^{-1}(\Delta^2)I(F_{p+1,m}^{-1}(\Delta^2)<c_1)]\} \\
&\quad +\boldsymbol{K}_1^{-1}\boldsymbol{K}_3\boldsymbol{H}\boldsymbol{\beta}G_{p+1,m}^{(2)}(c_1;\Delta^2), \\
\boldsymbol{M}(\hat{\boldsymbol{\Theta}}_n^{S+}) &= \boldsymbol{M}(\hat{\boldsymbol{\Theta}}_n^{S}) - \sigma_{\varepsilon}^{2}\boldsymbol{K}_1^{-1}\boldsymbol{K}_3\boldsymbol{H}\boldsymbol{C}_{11}^{-1}\boldsymbol{H}'\boldsymbol{K}_3\boldsymbol{K}_1^{-1} \\
&\quad \times E^{(1)}[(1-c_1F_{p+1,m}^{-1}(\Delta^2))^2 I(F_{p+1,m}(\Delta^2)<c_1)] \\
&\quad +\boldsymbol{K}_1^{-1}\boldsymbol{K}_3\boldsymbol{H}\boldsymbol{C}_{11}^{-1}\boldsymbol{H}'\boldsymbol{K}_3\boldsymbol{K}_1^{-1} \\
&\quad \times\{2E^{(2)}[(1-c_1F_{p+1,m}^{-1}(\Delta^2))I(F_{p+1,m}(\Delta^2)<c_1)] \\
&\quad -E^{(2)}[(1-c_2F_{p+3,m}^{-1}(\Delta^2))^2 I(F_{p+3,m}(\Delta^2)<c_2)]\}, \\
R(\hat{\boldsymbol{\Theta}}_n^{S+};\boldsymbol{W}) &= \boldsymbol{R}(\hat{\boldsymbol{\Theta}}_n^{S};\boldsymbol{W}) - \sigma_{\varepsilon}^{2}tr(\boldsymbol{W}\boldsymbol{K}_1^{-1}\boldsymbol{K}_3\boldsymbol{H}\boldsymbol{C}_{11}^{-1}\boldsymbol{H}'\boldsymbol{K}_3\boldsymbol{K}_1^{-1}) \\
&\quad \times E^{(1)}[(1-c_1F_{p+1,m}^{-1}(\Delta^2))^2 I(F_{p+1,m}(\Delta^2)<c_1)] \\
&\quad +\boldsymbol{\beta}'\boldsymbol{H}'\boldsymbol{K}_1^{-1}\boldsymbol{K}_3'\boldsymbol{W}\boldsymbol{K}_3\boldsymbol{K}_1^{-1}\boldsymbol{H}\boldsymbol{\beta} \\
&\quad \times\{2E^{(2)}[(1-c_1F_{p+1,m}^{-1}(\Delta^2))I(F_{p+1,m}(\Delta^2)<c_1)] \\
&\quad -E^{(2)}[(1-c_2F_{p+3,m}^{-1}(\Delta^2))^2 I(F_{p+3,m}(\Delta^2)<c_2)]\}, \\
& \qquad\qquad (6.3.9)
\end{aligned}
$$

where $E^{(j)}[F_{p+s,m}^{-1}(\Delta^2)I(F_{p+s,m}(\Delta^2)<c_i)]$ is given by equation (5.3.9) and, similar to Example 2.6.2, for $h=0,1$

$$
\begin{aligned}
& E^{(2-h)}[F_{p+s,m}^{-2}(\Delta^2)I(F_{p+s,m}(\Delta^2)<c)] = \\
& \sum_{r\geq 0} K_r^{(h)}(\Delta^2)m(p+s)^2[(p+s-2+2r)(p+s-4+2r)]^{-1} \\
& \quad \times I_y\left(\frac{1}{2}(p+s-4+2r);\frac{1}{2}(m+4)\right),
\end{aligned}
$$

under which for $s=1$, we use $c_1=\frac{c(p-1)}{p+1}$; for $s=3$, we use $c_2=\frac{c(p-1)}{p+3}$ and $y=\frac{c(p-1)}{m+c(p-1)}$.

6.4 Risk Analysis

In this section, for the sake of brevity, we only consider the analysis of the properties of the intercept estimators.

Taking $\ddot{\boldsymbol{B}} = \boldsymbol{\beta}'\boldsymbol{H}'\boldsymbol{K}_1^{-1}\boldsymbol{K}_3\boldsymbol{W}\boldsymbol{K}_3\boldsymbol{K}_1^{-1}\boldsymbol{H}\boldsymbol{\beta}$ from equations (6.3.2) and (6.3.4), we get

$$\boldsymbol{R}(\tilde{\boldsymbol{\Theta}}_n;\boldsymbol{W}) - \boldsymbol{R}(\hat{\boldsymbol{\Theta}}_n;\boldsymbol{W}) = \sigma_{\boldsymbol{\varepsilon}}^2 tr[\boldsymbol{W}(\boldsymbol{C}_{22}^{-1} - \boldsymbol{D}_{11})] - \ddot{\boldsymbol{B}}. \tag{6.4.1}$$

Consider that

$$\sigma_{\boldsymbol{\varepsilon}}^2 tr[\boldsymbol{W}(\boldsymbol{C}_{22}^{-1} - \boldsymbol{D}_{11})] \le \sigma_{\boldsymbol{\varepsilon}}^2 tr(\boldsymbol{C}_{22}^{-1}\boldsymbol{W}). \tag{6.4.2}$$

Also

$$\boldsymbol{K}_3\boldsymbol{K}_1^{-1}\boldsymbol{C}_{11}^{-1}\boldsymbol{K}_1^{-1}\boldsymbol{K}_3 = \boldsymbol{C}_{22}^{-1} - \boldsymbol{K}_1^{-1}. \tag{6.4.3}$$

Therefore, by (6.4.1)-(6.4.3), we obtain

$$\begin{aligned}
\frac{\ddot{\boldsymbol{B}}}{\xi} &\le \lambda_1(\boldsymbol{C}_{11}^{-1}\boldsymbol{K}_1^{-1}\boldsymbol{K}_3\boldsymbol{W}\boldsymbol{K}_3\boldsymbol{K}_1^{-1}), \ \ \lambda_1 = Ch_{\max}(\boldsymbol{A}), \text{ for the matrix } \boldsymbol{A} \\
&= \lambda_1(\boldsymbol{K}_3\boldsymbol{K}_1^{-1}\boldsymbol{C}_{11}^{-1}\boldsymbol{K}_1^{-1}\boldsymbol{K}_3\boldsymbol{W}) \\
&= \lambda_1[(\boldsymbol{C}_{22}^{-1} - \boldsymbol{K}_1^{-1})\boldsymbol{W}] \\
&< \lambda_1(\boldsymbol{C}_{22}^{-1}\boldsymbol{W}) \\
&< tr(\boldsymbol{C}_{22}^{-1}\boldsymbol{W}).
\end{aligned} \tag{6.4.4}$$

Thus, from (6.4.4) it can be seen that the risk difference (6.4.1) is non-negative, i.e., $\hat{\boldsymbol{\Theta}}_n$ dominates $\tilde{\boldsymbol{\Theta}}_n$ ($\hat{\boldsymbol{\Theta}}_n \succeq \tilde{\boldsymbol{\Theta}}_n$) provided

$$\Delta^2 \le \frac{tr(\boldsymbol{C}_{22}^{-1}\boldsymbol{W})}{\lambda_1(\boldsymbol{C}_{22}^{-1}\boldsymbol{W})}. \tag{6.4.5}$$

Also, using the inequality

$$\ddot{\boldsymbol{B}} \ge \xi\lambda_p[(\boldsymbol{C}_{22}^{-1} - \boldsymbol{K}_1^{-1})\boldsymbol{W}], \tag{6.4.6}$$

$\tilde{\boldsymbol{\Theta}}_n \succeq \hat{\boldsymbol{\Theta}}_n$ whenever

$$\Delta^2 \ge \frac{tr(\boldsymbol{C}_{22}^{-1}\boldsymbol{W})}{\lambda_p(\boldsymbol{C}_{22}^{-1}\boldsymbol{W})}. \tag{6.4.7}$$

Comparing $\tilde{\boldsymbol{\Theta}}_n$ and $\hat{\boldsymbol{\Theta}}_n$ with $\hat{\boldsymbol{\Theta}}_n^{PT}$, using equations (6.3.2), (6.3.4), and (6.3.6), we obtain

$$\hat{\boldsymbol{\Theta}}_n^{PT} \succeq \tilde{\boldsymbol{\Theta}}_n \quad \Leftrightarrow \quad \Delta^2 \le \frac{tr(\boldsymbol{C}_{22}^{-1}\boldsymbol{W})G_{p+1,m}^{(1)}(l_\gamma;\Delta^2)}{\lambda_1(\boldsymbol{C}_{22}^{-1}\boldsymbol{W})[2H_{p+1,m}^{(2)}(l_\gamma;\Delta^2) - H_{p+3,m}^{(2)}(l_\gamma;\Delta^2)]}.$$

$$\hat{\boldsymbol{\Theta}}_n^{PT} \succeq \hat{\boldsymbol{\Theta}}_n \quad \Leftrightarrow \quad \Delta^2 \ge \frac{tr(\boldsymbol{C}_{22}^{-1}\boldsymbol{W})[1 - G_{p+1,m}^{(1)}(l_\gamma;\Delta^2)]}{\lambda_1(\boldsymbol{C}_{22}^{-1}\boldsymbol{W})[1 - 2G_{p+1,m}^{(2)}(l_\gamma;\Delta^2) + G_{p+3,m}^{(2)}(l_\gamma;\Delta^2)]}.$$

Comparing $\hat{\boldsymbol{\Theta}}_n^S$ relative to $\tilde{\boldsymbol{\Theta}}_n$ using the risk difference for $p \geq 4$, we have

$$\begin{aligned}
R(\tilde{\boldsymbol{\Theta}}_n; \boldsymbol{W}) - R(\hat{\boldsymbol{\Theta}}_n^S; \boldsymbol{W}) &= c(p-1)\sigma_{\boldsymbol{\varepsilon}}^2 tr(\boldsymbol{W}\boldsymbol{K}_1^{-1}\boldsymbol{K}_3\boldsymbol{H}\boldsymbol{C}_{11}^{-1}\boldsymbol{H}'\boldsymbol{K}_3\boldsymbol{K}_1^{-1}) \\
&\quad \times \{(p-3)E^{(1)}[\chi_{p+1}^{-4}(\Delta^2)] + 2\Delta^2 E^{(1)}[\chi_{p+3}^{-4}(\Delta^2)]\} \\
&\quad - c(p^2-1)\ddot{\boldsymbol{B}}E^{(2)}[\chi_{p+3}^{-4}(\Delta^2)] \\
&\quad (p-1)\sigma_{\boldsymbol{\varepsilon}}^2 tr(\boldsymbol{W}\boldsymbol{K}_1^{-1}\boldsymbol{K}_3\boldsymbol{H}\boldsymbol{C}_{11}^{-1}\boldsymbol{H}'\boldsymbol{K}_3\boldsymbol{K}_1^{-1}) \\
&\quad \times \{(p-3)E^{(1)}[\chi_{p+1}^{-4}(\Delta^2)] + 2\Delta^2 E^{(1)}[\chi_{p+3}^{-4}(\Delta^2)]\} \\
&\quad - c(p^2-1)\Delta^2 tr(\boldsymbol{C}_{22}^{-1}\boldsymbol{W})E^{(2)}[\chi_{p+3}^{-4}(\Delta^2)].
\end{aligned} \tag{6.4.8}$$

The R.H.S. of equation (6.4.8) is a nonincreasing function w.r.t. Δ^2 and maximizes at $\Delta^2 = 0$. Thus, $\hat{\boldsymbol{\Theta}}_n^S \succeq \tilde{\boldsymbol{\Theta}}_n$.

In the comparison of $\hat{\boldsymbol{\Theta}}_n^{S+}$ with $\hat{\boldsymbol{\Theta}}_n^S$, using the risk difference, we may write

$$\begin{aligned}
R(\hat{\boldsymbol{\Theta}}_n^S; \boldsymbol{W}) - R(\hat{\boldsymbol{\Theta}}_n^{S+}) &= \sigma_{\boldsymbol{\varepsilon}}^2 tr(\boldsymbol{W}\boldsymbol{K}_1^{-1}\boldsymbol{K}_3\boldsymbol{H}\boldsymbol{C}_{11}^{-1}\boldsymbol{H}'\boldsymbol{K}_3\boldsymbol{K}_1^{-1}) \\
&\quad \times E^{(1)}\left[\left(1 - c_1 F_{p+1,m}^{-1}(\Delta^2)\right)^2 I\left(F_{p+1,m}(\Delta^2) < c_1\right)\right] \\
&\quad -2\ddot{\boldsymbol{B}}E^{(2)}\left[\left(1 - c_1 F_{p+1,m}^{-1}(\Delta^2)\right) I\left(F_{p+1,m}(\Delta^2) < c_1\right)\right] \\
&\quad +\ddot{\boldsymbol{B}}E^{(2)}\left[\left(1 - c_2 F_{p+3,m}^{-1}(\Delta^2)\right)^2 I\left(F_{p+3,m}(\Delta^2) < c_2\right)\right].
\end{aligned} \tag{6.4.9}$$

The R.H.S. of (6.4.9) is definitely positive because, for $F_{p+1,m}(\Delta^2) < c_1$, we have $\left(1 - c_1 F_{p+1,m}^{-1}(\Delta^2)\right) < 0$ and, therefore,

$$-2\ddot{\boldsymbol{B}}E^{(2)}\left[\left(1 - c_1 F_{p+1,m}^{-1}(\Delta^2)\right) I\left(F_{p+1,m}(\Delta^2) < c_1\right)\right] > 0.$$

Also, the expectation of a positive random variable is positive, so

$$E^{(1)}\left[\left(1 - c_1 F_{p+1,m}^{-1}(\Delta^2)\right)^2 I\left(F_{p+1,m}(\Delta^2) < c_1\right)\right] > 0.$$

Thus, we conclude $\hat{\boldsymbol{\Theta}}_n^{S+}$ strictly performs better than $\hat{\boldsymbol{\Theta}}_n^S$ and uniformly $\hat{\boldsymbol{\Theta}}_n^{S+} \succeq \hat{\boldsymbol{\Theta}}_n^S$.

This analysis shows that if we can demonstrate the superiority of $\hat{\boldsymbol{\Theta}}_n^S$ over each of the other estimators, then we will conclude the same result for $\hat{\boldsymbol{\Theta}}_n^{S+}$. Therefore, we continue with the comparison of $\hat{\boldsymbol{\Theta}}_n^S$ with the others.

In order to compare $\hat{\Theta}_n^S$ with $\hat{\Theta}_n$, first note that under H_0, $\Delta^2 = 0$ and, therefore,

$$\begin{aligned}
R(\hat{\Theta}_n; \boldsymbol{W}) - R(\hat{\Theta}_n^S; \boldsymbol{W}) &= \sigma_{\varepsilon}^2 tr(\boldsymbol{D}_{11}\boldsymbol{W}) - \sigma_{\varepsilon}^2 tr(\boldsymbol{C}_{22}^{-1}\boldsymbol{W}) + c(p-1)(p-3)\sigma_{\varepsilon}^2 \\
&\times tr(\boldsymbol{W}\boldsymbol{K}_1^{-1}\boldsymbol{K}_3\boldsymbol{H}\boldsymbol{C}_{11}^{-1}\boldsymbol{H}'\boldsymbol{K}_3\boldsymbol{K}_1^{-1})E^{(1)}[\chi_{p+1}^{*^{-4}}(0)] \\
&= \sigma_{\varepsilon}^2 tr(\boldsymbol{D}_{11}\boldsymbol{W}) - \sigma_{\varepsilon}^2 tr(\boldsymbol{C}_{22}^{-1}\boldsymbol{W}) + c(p-1)(p-3)\sigma_{\varepsilon}^2 \\
&\times tr(\boldsymbol{W}[\boldsymbol{C}_{22}^{-1} - \boldsymbol{D}_{11}])E^{(1)}[\chi_{p+1}^{*^{-4}}(0)] \\
&= \sigma_{\varepsilon}^2 \left\{ \frac{m(p-3)^2 E^{(1)}[\chi_{p+1}^{*^{-4}}(0)]}{m+2} - 1 \right\} tr(\boldsymbol{W}[\boldsymbol{C}_{22}^{-1} - \boldsymbol{D}_{11}]) \\
&\leq 0.
\end{aligned} \tag{6.4.10}$$

Hence, $\hat{\Theta}_n \succeq \hat{\Theta}_n^S$ under H_0 provided $E^{(1)}[\chi_{p+1}^{-4}(0)] \leq \frac{m+2}{m(p-3)^2}$. But as Δ^2 departs from zero, the risk of $\hat{\Theta}_n$ becomes unbounded while the risk of $\hat{\Theta}_n^S$ remains below $\tilde{\Theta}_n$. In this case, $\hat{\Theta}_n^S$ performs better than $\hat{\Theta}_n$.

Finally, comparing $\hat{\Theta}_n^S$ with $\hat{\Theta}_n^{PT}$, under H_0, using the risk difference we have

$$\begin{aligned}
R(\hat{\Theta}_n^{PT}; \boldsymbol{W}) - R(\hat{\Theta}_n^S; \boldsymbol{W}) &= c(p-1)(p-3)\sigma_{\varepsilon}^2 \\
&\quad \times tr(\boldsymbol{W}\boldsymbol{K}_1^{-1}\boldsymbol{K}_3\boldsymbol{H}\boldsymbol{C}_{11}^{-1}\boldsymbol{H}'\boldsymbol{K}_3\boldsymbol{K}_1^{-1})E^{(1)}[\chi_{p+1}^{*^{-4}}(0)] \\
&\quad - \sigma_{\varepsilon}^2 tr(\boldsymbol{W}[\boldsymbol{C}_{22}^{-1} - \boldsymbol{D}_{11}])G_{p+1,m}^{(1)}(l_\gamma; 0) \\
&= \sigma_{\varepsilon}^2 \left\{ \frac{(p-3)^2 m}{m+2} tr(\boldsymbol{W}[\boldsymbol{C}_{22}^{-1} - \boldsymbol{D}_{11}])E^{(1)}[\chi_{p+1}^{*^{-4}}(0)] \right. \\
&\quad \left. - tr(\boldsymbol{W}[\boldsymbol{C}_{22}^{-1} - \boldsymbol{D}_{11}])G_{p+1,m}^{(1)}(l_\gamma; 0) \right\}.
\end{aligned} \tag{6.4.11}$$

The RHS of (6.4.11) is non-positive, that is $\hat{\Theta}_n^{PT} \succeq \hat{\Theta}_n^S$, for all γ satisfying

$$m(p-3)^2 E^{(1)}[\chi_{p+1}^{*^{-4}}(0)] \leq (m+2)G_{p+1,m}^{(1)}(l_\gamma; 0). \tag{6.4.12}$$

However, as Δ^2 moves away from zero, the risk of $\hat{\Theta}_n^{PT}$ exceeds the risks of $\tilde{\Theta}_n$ while the risk of $\hat{\Theta}_n^S$ remains below the risk of $\tilde{\Theta}_n$ and both ultimately merge with the risk of $\tilde{\Theta}_n$ as $\Delta^2 \to \infty$. Thus, $\hat{\Theta}_n^S$ dominates $\hat{\Theta}_n^{PT}$ outside a small neighborhood of the origin.

Similar to Theorem 5.4.1, in the following theorem, we show that the shrinkage factor c of the Stein-type estimator is robust with respect to the slope parameters and the unknown mixing distributions.

Theorem 6.4.1. *Consider the model* (6.1.1) *where the error-vector belongs to the* $M_t^{(n)}(\mathbf{0}, \sigma^2\mathbf{V}_n, \gamma_o)$, $\gamma_o > 4$. *Then, the Stein-type estimator,* $\hat{\boldsymbol{\beta}}_n^S$ *of the slope given by*

$$\begin{aligned}\hat{\boldsymbol{\beta}}_n^S &= \tilde{\boldsymbol{\beta}}_n - c\boldsymbol{Q}^{-1}\boldsymbol{H}\tilde{\boldsymbol{\beta}}_n\mathcal{L}_n^{-1} \\ &= \tilde{\boldsymbol{\beta}}_n - c^*(ms_\varepsilon^2)\boldsymbol{Q}^{-1}\boldsymbol{H}\tilde{\boldsymbol{\beta}}_n(\tilde{\boldsymbol{\beta}}_n'\boldsymbol{H}'\boldsymbol{Q}^{-1}\boldsymbol{H}\tilde{\boldsymbol{\beta}}_n)^{-1} \\ &= \tilde{\boldsymbol{\beta}}_n - c^*(ms_\varepsilon^2)\boldsymbol{Q}^{-1}\mathbf{Z}(\mathbf{Z}'\boldsymbol{Q}^{-1}\mathbf{Z})^{-1},\end{aligned}$$

where $\mathbf{Z} = \boldsymbol{H}\tilde{\boldsymbol{\beta}}_n$, *uniformly dominates the unrestricted estimator* $\tilde{\boldsymbol{\beta}}_n$ *with respect to a quadratic loss-function with the positive-definite matrix* $\boldsymbol{Q}$ *iff* $0 < c^* \le \frac{2(p-3)}{m+2}$. *The largest reduction of the risk is attained when* $c^* = \frac{p-3}{m+2}$.

Proof: Consider the risk-difference of the SE and the UE, given by

$$\begin{aligned}& E(\hat{\boldsymbol{\beta}}_n^S - \boldsymbol{\beta})'\boldsymbol{Q}(\hat{\boldsymbol{\beta}}_n^S - \boldsymbol{\beta}) - E(\tilde{\boldsymbol{\beta}}_n - \boldsymbol{\beta})'\boldsymbol{Q}(\tilde{\boldsymbol{\beta}}_n - \boldsymbol{\beta}) \\ = \; & (c^*)^2E[(ms_\varepsilon^2)^2(\mathbf{Z}'\boldsymbol{Q}^{-1}\mathbf{Z})^{-1}] - 2c^*E[(ms_\varepsilon^2)(\tilde{\boldsymbol{\beta}}_n - \boldsymbol{\beta})'\boldsymbol{H}\tilde{\boldsymbol{\beta}}_n(\mathbf{Z}'\boldsymbol{Q}^{-1}\mathbf{Z})^{-1})] \\ = \; & (c^*)^2m(m+2)E_t[E_N\mathbf{Z}'\boldsymbol{Q}^{-1}\mathbf{Z}^{-1}] - 2c^*m(p-3)E_t[E_N(\mathbf{Z}'\boldsymbol{Q}^{-1}\mathbf{Z})^{-1}] \le 0,\end{aligned}$$

if and only if $0 < c* \le \frac{2(p-3)}{m+2}$.

Maximum reduction occurs when $c^* = \frac{p-3}{m+2}$. Thus, we choose $c = \frac{(p-3)m}{(p-1)(m+2)}$.

6.5 Problems

1. Verify $\hat{\boldsymbol{\Theta}}_n$ and $\hat{\boldsymbol{\beta}}_n$ in equation (6.2.6).
2. Prove Theorem 6.2.3.
3. Find the joint distribution of $\left((\hat{\boldsymbol{\Theta}}_n - \boldsymbol{\Theta})', (\tilde{\boldsymbol{\Theta}}_n, \boldsymbol{\Theta})'\right)'$ based on the Theorems 6.2.1 and 6.2.3.
4. Verify the bias, MSE matrices, and risk expressions of section 6.3.
5. Verify equation (6.4.3).
6. Refer to problem 8 in Chapter 5. Consider ridge-type and Liu-type estimators obtained by premultiplying each estimators of $\boldsymbol{\Theta}$ and $\boldsymbol{\beta}$ by $\left(\boldsymbol{C}_{11}^{-1} + k_o\boldsymbol{I}_p\right)^{-1}$, for $k_o \ge 0$ and $\left(\boldsymbol{C}_{11}^{-1} + d\boldsymbol{I}_p\right)\left(\boldsymbol{C}_{11}^{-1} + \boldsymbol{I}_p\right)^{-1}$, for $0 < d \le 1$, respectively. Define the estimators and then derive bias, MSE matrix, and risk expressions.

CHAPTER 7

MULTIPLE REGRESSION MODEL

Outline

In this chapter, we discuss the multiple regression model in detail, beginning with the estimation of the model parameters and test of hypothesis. In the continuation, we introduce improved estimators of the regression coefficients with their dominance properties. In this case, we derive the characteristic properties based on the so-called symmetric balanced loss function.

A. K. Md. Ehsanes Saleh, M. Arashi, S.M.M. Tabatabaey.

7.1 Model Specification

The multiple regression model is arguably the most widely used statistical tool applied in almost every discipline of the modern era. The estimation of parameters of the multiple regression model is of interest to many users. To deal with a common multiple regression equation, consider the linear model

$$\boldsymbol{y} = \boldsymbol{X\beta} + \boldsymbol{\varepsilon}, \tag{7.1.1}$$

where $\boldsymbol{y} = (y_1, \ldots, y_n)'$ is an $(n \times 1)$ vector of observations, $\boldsymbol{X} = (\boldsymbol{x}_1', \ldots, \boldsymbol{x}_n')'$ is a nonstochastic $(n \times p)$ matrix of full rank p, $\boldsymbol{\beta}$ is a $(p \times 1)$ vector of unknown regression coefficients, and $\boldsymbol{\varepsilon}$ is an $(n \times 1)$ random error-vector distributed as $M_t^{(n)}(\mathbf{0}, \sigma^2 \mathbf{V}_n, \gamma_o)$ (in this case).

7.2 Shrinkage Estimators and Testing

Using standard conditions, it is well known that the LSE of $\boldsymbol{\beta}$ is (see Ravishanker and Dey, 2001 for details)

$$\tilde{\boldsymbol{\beta}}_n = (\boldsymbol{X}'\mathbf{V}_n^{-1}\boldsymbol{X})^{-1}\boldsymbol{X}'\mathbf{V}_n^{-1}\boldsymbol{y} = \boldsymbol{C}^{-1}\boldsymbol{X}'\mathbf{V}_n^{-1}\boldsymbol{y}, \qquad \boldsymbol{C} = \boldsymbol{X}'\mathbf{V}_n^{-1}\boldsymbol{X}. \tag{7.2.1}$$

According to Theorem 2.6.3, its distribution is $M_t^{(p)}(\boldsymbol{\beta}, \sigma^2\boldsymbol{C}^{-1}, \gamma_o)$.

From Corollary 2.6.3.1, the 1st moment is the $\boldsymbol{\beta}$ vector and the 2nd central moment is

$$E(\tilde{\boldsymbol{\beta}}_n - \boldsymbol{\beta})'(\tilde{\boldsymbol{\beta}}_n - \boldsymbol{\beta}) = \sigma_{\boldsymbol{\varepsilon}}^2 tr(\boldsymbol{C}^{-1}), \quad \sigma_{\boldsymbol{\varepsilon}}^2 = \frac{\gamma_o \sigma^2}{\gamma_o - 2}.$$

Similarly, the estimate of the σ^2 is

$$\tilde{\sigma}^2 = \frac{1}{n}(\boldsymbol{y} - \boldsymbol{X}\tilde{\boldsymbol{\beta}}_n)'\mathbf{V}_n^{-1}(\boldsymbol{y} - \boldsymbol{X}\tilde{\boldsymbol{\beta}}_n). \tag{7.2.2}$$

Now, consider an estimator of σ^2 with the form

$$S^2 = \frac{(\boldsymbol{y} - \boldsymbol{X}\tilde{\boldsymbol{\beta}}_n)'\mathbf{V}_n^{-1}(\boldsymbol{y} - \boldsymbol{X}\tilde{\boldsymbol{\beta}}_n)}{n-p}. \tag{7.2.3}$$

Let

$$\begin{aligned} \boldsymbol{Z} &= \mathbf{V}_n^{-\frac{1}{2}}(\boldsymbol{y} - \boldsymbol{X}\tilde{\boldsymbol{\beta}}_n) \\ &= \mathbf{V}_n^{-\frac{1}{2}}\left(\boldsymbol{I} - \boldsymbol{X}\boldsymbol{C}^{-1}\boldsymbol{X}'\mathbf{V}_n^{-1}\right)\boldsymbol{y}. \end{aligned}$$

Then, under normality (mimicking (2.6.3)),

$$\boldsymbol{Z}|t \sim \mathcal{N}_n\left(\mathbf{0}, \sigma^2 t^{-1}(\boldsymbol{I} - \boldsymbol{A})\right), \quad \boldsymbol{A} = \mathbf{V}_n^{-\frac{1}{2}}\boldsymbol{X}\boldsymbol{C}^{-1}\boldsymbol{X}'\mathbf{V}_n^{-\frac{1}{2}}.$$

$\boldsymbol{A}$ is a symmetric idempotent matrix and so is $\boldsymbol{I}-\boldsymbol{A}$. Thus, $rank(\boldsymbol{I}-\boldsymbol{A}) = tr(\boldsymbol{I}-\boldsymbol{A}) = n-p$. Therefore,

$$\boldsymbol{Z}'\boldsymbol{Z}|t = \boldsymbol{Z}'(\boldsymbol{I}-\boldsymbol{A})^{-\frac{1}{2}}(\boldsymbol{I}-\boldsymbol{A})(\boldsymbol{I}-\boldsymbol{A})^{-\frac{1}{2}}\boldsymbol{Z}|t \sim \sigma^2 t^{-1}\chi^2_{n-p}.$$

In conclusion, along with using (2.6.4), we have

$$(n-p)S^2 \sim \sigma^2\kappa^{(1)}\chi^2_{n-p} \equiv \sigma^2_{\boldsymbol{\varepsilon}}\chi^2_{n-p},$$

becuase $\kappa^{(1)} = \frac{\gamma_o}{\gamma_o-2}$; this demonstrates S^2 is an unbiased estimator of $\sigma^2_{\boldsymbol{\varepsilon}}$.

Based on the unrestricted estimator $\tilde{\boldsymbol{\beta}}_n$, a plausible estimator of γ_o can also be given as (see Singh, 1988)

$$\tilde{\gamma}_o = \frac{2(2\tilde{\nu}-3)}{\tilde{\nu}-3},$$

where

$$\tilde{\nu} = n\frac{\sum_{i=1}^n (y_i - \boldsymbol{x}_i'\tilde{\boldsymbol{\beta}}_n)^4}{(\sum_{i=1}^n (y_i - \boldsymbol{x}_i'\tilde{\boldsymbol{\beta}}_n)^2)^2}.$$

For test of $\boldsymbol{H\beta} = \boldsymbol{h}$ (where $q < p$), we first consider the restricted estimator given by

$$\hat{\boldsymbol{\beta}}_n = \tilde{\boldsymbol{\beta}}_n - \boldsymbol{C}^{-1}\boldsymbol{H}'\mathbf{V}_1(\boldsymbol{H}\tilde{\boldsymbol{\beta}}_n - \boldsymbol{h}), \qquad \mathbf{V}_1 = [\boldsymbol{HC}^{-1}\boldsymbol{H}']^{-1}. \tag{7.2.4}$$

By making use of equation (7.2.1) one can easily see that $\hat{\boldsymbol{\beta}}_n \sim E_p\left(\boldsymbol{\beta}-\boldsymbol{\delta}, \sigma^2\mathbf{V}_2, f\right)$ for $\boldsymbol{\delta} = \boldsymbol{C}^{-1}\boldsymbol{H}'\mathbf{V}_1(\boldsymbol{H\beta}-\boldsymbol{h})$ and $\mathbf{V}_2 = \boldsymbol{C}^{-1}(\boldsymbol{I}_p - \boldsymbol{H}'\mathbf{V}_1\boldsymbol{HC}^{-1})$.

Now, we consider the linear hypothesis $\boldsymbol{H\beta} = \boldsymbol{h}$ and obtain the test statistic for the null hypothesis $H_0 : \boldsymbol{H\beta} = \boldsymbol{h}$.

Theorem 7.2.1. *Let*

$$\begin{aligned}
\Omega_1 &= \{\boldsymbol{\beta} : \boldsymbol{\beta} \in \mathbb{R}^p, \sigma \in \mathbb{R}^+, \mathbf{V}_n > 0\} \\
\Omega_0 &= \{\boldsymbol{\beta} : \boldsymbol{\beta} \in \mathbb{R}^p, \boldsymbol{H\beta} = \boldsymbol{h}, \sigma \in \mathbb{R}^+, \mathbf{V}_n > 0\}.
\end{aligned}$$

The LRC for testing the null hypothesis is given by

$$\mathcal{L}_n = \frac{(\boldsymbol{H}\tilde{\boldsymbol{\beta}}_n - \boldsymbol{h})'\mathbf{V}_1(\boldsymbol{H}\tilde{\boldsymbol{\beta}}_n - \boldsymbol{h})}{qS^2}, \tag{7.2.5}$$

having the following generalized noncentral F distribution:

$$g^*_{q,m}(\mathcal{L}_n) = \sum_{r\geq 0} \frac{\left(\frac{q}{m}\right)^{\frac{1}{2}(q+2r)} \mathcal{L}_n^{\frac{1}{2}(q+2r-2)} K_r^{(0)}(\Delta^2)}{r!\, B\left(\frac{q+2r}{2}, \frac{m}{2}\right)\left(1+\frac{q}{m}\mathcal{L}_n\right)^{\frac{1}{2}(q+m+2r)}}, \tag{7.2.6}$$

where $m = n - p$, $\Delta^2 = \theta/\sigma^2_{\varepsilon}$ *for* $\theta = (\boldsymbol{H\beta} - \boldsymbol{h})'\mathbf{V}_1(\boldsymbol{H\beta} - \boldsymbol{h})$.

Proof: The likelihood ratio is $\Lambda = L_0/L_1$, where L_i, $i = 0, 1$ is the largest value that the likelihood function takes in region Ω_i, $i = 0, 1$. From the fact that $\boldsymbol{y} \sim M_t^{(n)}(\boldsymbol{X\beta}, \sigma^2\boldsymbol{V}_n, \gamma_o)$, we obtain

$$\begin{aligned}
L_0 &= \frac{\Gamma\left(\frac{n+\gamma_o}{2}\right)}{(\pi\gamma_o)^{\frac{n}{2}}\Gamma\left(\frac{\gamma_o}{2}\right)(\hat{\sigma}^2_{\varepsilon})^{\frac{n}{2}}}\left[1 + \frac{(\boldsymbol{y} - \boldsymbol{X}\hat{\boldsymbol{\beta}})'\boldsymbol{V}_n^{-1}(\boldsymbol{y} - \boldsymbol{X}\hat{\boldsymbol{\beta}})}{\gamma_o\hat{\sigma}^2_{\varepsilon}}\right]^{-\frac{1}{2}(n+\gamma_o)} \\
&= \frac{\Gamma\left(\frac{n+\gamma_o}{2}\right)}{(\pi\gamma_o)^{\frac{n}{2}}\Gamma\left(\frac{\gamma_o}{2}\right)(\hat{\sigma}^2_{\varepsilon})^{\frac{n}{2}}}\left[1 + \frac{n-p+q}{\gamma_o}\right]^{-\frac{1}{2}(n+\gamma_o)}
\end{aligned}$$

and

$$\begin{aligned}
L_1 &= \frac{\Gamma\left(\frac{n+\gamma_o}{2}\right)}{(\pi\gamma_o)^{\frac{n}{2}}\Gamma\left(\frac{\gamma_o}{2}\right)(\tilde{\sigma}^2_{\varepsilon})^{\frac{n}{2}}}\left[1 + \frac{(\boldsymbol{y} - \boldsymbol{X}\tilde{\boldsymbol{\beta}})'\boldsymbol{V}_n^{-1}(\boldsymbol{y} - \boldsymbol{X}\tilde{\boldsymbol{\beta}})}{\gamma_o\tilde{\sigma}^2_{\varepsilon}}\right]^{-\frac{1}{2}(n+\gamma_o)} \\
&= \frac{\Gamma\left(\frac{n+\gamma_o}{2}\right)}{(\pi\gamma_o)^{\frac{n}{2}}\Gamma\left(\frac{\gamma_o}{2}\right)(\tilde{\sigma}^2_{\varepsilon})^{\frac{n}{2}}}\left[1 + \frac{n-p}{\gamma_o}\right]^{-\frac{1}{2}(n+\gamma_o)},
\end{aligned}$$

giving

$$\Lambda = \frac{L_0}{L_1} = a\left(\frac{1}{1 + \frac{(\boldsymbol{H}\tilde{\boldsymbol{\beta}}_n - \boldsymbol{h})'(\boldsymbol{HC}^{-1}\boldsymbol{H}')^{-1}(\boldsymbol{H}\tilde{\boldsymbol{\beta}}_n - \boldsymbol{h})}{(\boldsymbol{y} - \boldsymbol{X}\tilde{\boldsymbol{\beta}}_n)'\mathbf{V}_n^{-1}(\boldsymbol{y} - \boldsymbol{X}\tilde{\boldsymbol{\beta}}_n)}}\right)^{\frac{n}{2}},$$

where

$$a = \left(\frac{n-p+q}{n-p}\right)^{\frac{n}{2}}\left(\frac{n-p+\gamma_o}{n-p+q+\gamma_o}\right)^{\frac{1}{2}(n+\gamma_o)}.$$

Therefore,

$$\Lambda^{\frac{n}{2}} = a^{\frac{2}{n}}\left(\frac{1}{1 + \frac{q}{n-p}\mathcal{L}_n}\right).$$

Λ is a decreasing function of $\mathcal{L}_n$. Thus, $\Lambda < \Lambda_0$ is equivalent to $\mathcal{L}_n > F_\alpha$ for a certain constant F_α. That is, the likelihood ration test statistic is equivalent to the test statistic

$$\mathcal{L}_n = \frac{(\boldsymbol{H}\tilde{\boldsymbol{\beta}}_n - \boldsymbol{h})'\mathbf{V}_1(\boldsymbol{H}\tilde{\boldsymbol{\beta}}_n - \boldsymbol{h})}{qS^2}.$$

Now, consider that

$$\begin{aligned}
(n-p)S^2|t &= (\boldsymbol{y} - \boldsymbol{X}\tilde{\boldsymbol{\beta}}_n)'\mathbf{V}^{-1}(\boldsymbol{y} - \boldsymbol{X}\tilde{\boldsymbol{\beta}}_n)|t \\
&= \boldsymbol{y}'\left[\left(\mathbf{V}^{-1} - \mathbf{V}^{-1}\boldsymbol{X}(\boldsymbol{X}'\mathbf{V}^{-1}\boldsymbol{X})^{-1}\boldsymbol{X}'\mathbf{V}^{-1}\right)\right]\boldsymbol{y}|t \sim \sigma^2 t^{-1}\chi^2_{n-p}
\end{aligned}$$

also $(\boldsymbol{HC}^{-1}\boldsymbol{H}')^{-1/2}(\boldsymbol{H}\tilde{\boldsymbol{\beta}}_n)|t \sim \mathcal{N}_q((\boldsymbol{HC}^{-1}\boldsymbol{H}')^{-1/2}(\boldsymbol{H\beta}-\boldsymbol{h}), t^{-1}\sigma^2\boldsymbol{I}_q)$. Then

$$\tilde{\boldsymbol{\beta}}_n'\boldsymbol{H}'(\boldsymbol{HC}^{-1}\boldsymbol{H}')^{-1}\boldsymbol{H}\tilde{\boldsymbol{\beta}}_n|t \sim \chi^2_{q,\Delta_t^2},$$

where $\Delta_t^2 = \frac{t\theta}{\sigma^2}$.

Using the fact that $(\boldsymbol{y}-\boldsymbol{X}\tilde{\boldsymbol{\beta}}_n)'\mathbf{V}^{-1}(\boldsymbol{y}-\boldsymbol{X}\tilde{\boldsymbol{\beta}}_n)|t$ and $\tilde{\boldsymbol{\beta}}_n'\boldsymbol{H}'(\boldsymbol{HC}^{-1}\boldsymbol{H}')^{-1}\boldsymbol{H}\tilde{\boldsymbol{\beta}}_n|t$ are independent, we get

$$\mathcal{L}_n|t = \frac{\tilde{\boldsymbol{\beta}}_n'\boldsymbol{H}'(\boldsymbol{HC}^{-1}\boldsymbol{H}')^{-1}\boldsymbol{H}\tilde{\boldsymbol{\beta}}_n|t}{qS^2|t} \sim F_{q,n-p,\Delta_t^2}.$$

Hence,

$$\begin{aligned} g^*_{q,n-p}(\mathcal{L}_n) &= \int_0^\infty W_o(t)F_{q,n-p,\Delta_t^2}(\mathcal{L}_n|t)dt \\ &= \int_0^\infty W_o(t)\sum_{r=0}^\infty \frac{e^{\frac{-\Delta_t^2}{2}}(\frac{\Delta_t^2}{2})^r}{\Gamma(r+1)}\left(\frac{q}{n-p}\right)^{\frac{q}{2}+r} \\ &\quad \times \frac{\mathcal{L}_n^{\frac{q}{2}+r-1}}{B(\frac{q}{2}+r,\frac{n-p}{2})(1+\frac{q}{n-p}\mathcal{L}_n)^{\frac{q+n-p}{2}+r}}dt \\ &= \sum_{r=0}^\infty \left(\frac{q}{n-p}\right)^{\frac{q}{2}+r}\frac{\mathcal{L}_n^{\frac{q}{2}+r-1}K_r^{(0)}(\Delta^2)}{B(\frac{q}{2}+r,\frac{n-p}{2})(1+\frac{q}{n-p}\mathcal{L}_n)^{\frac{q+n-p}{2}+r}}. \end{aligned}$$

In the rest, we propose the shrinkage estimators that are discussed in previous chapters. First, we consider the PT estimator of $\boldsymbol{\beta}$, which is a convex combination of $\tilde{\boldsymbol{\beta}}_n$ and $\hat{\boldsymbol{\beta}}_n$:

$$\hat{\boldsymbol{\beta}}_n^{PT} = \tilde{\boldsymbol{\beta}}_n I(\mathcal{L}_n \geq F_\alpha) + \hat{\boldsymbol{\beta}}_n I(\mathcal{L}_n < F_\alpha). \tag{7.2.7}$$

The S estimator of $\boldsymbol{\beta}$, as

$$\hat{\boldsymbol{\beta}}_n^S = \hat{\boldsymbol{\beta}}_n + (1-d\mathcal{L}_n^{-1})(\tilde{\boldsymbol{\beta}}_n - \hat{\boldsymbol{\beta}}_n) = \tilde{\boldsymbol{\beta}}_n - d\mathcal{L}_n^{-1}(\tilde{\boldsymbol{\beta}}_n - \hat{\boldsymbol{\beta}}_n), \tag{7.2.8}$$

where

$$d = \frac{(q-2)m}{q(m+2)} \quad \text{and} \quad q \geq 3. \tag{7.2.9}$$

Up until this part, except in Theorem 4.5.1, no reason is given for why we select such d in (7.2.9). A full analysis of why we choose the shrinkage factor to be equal to (7.2.9) is given in Saleh (2006) under normal theory. In the following, we extend the stated result for ECDs under balanced loss function.

Finally, the PRS estimator of $\boldsymbol{\beta}$ is defined by

$$\hat{\boldsymbol{\beta}}_n^{S+} = \hat{\boldsymbol{\beta}}_n + (1-d\mathcal{L}_n^{-1})I[\mathcal{L}_n > d](\tilde{\boldsymbol{\beta}}_n - \hat{\boldsymbol{\beta}}_n). \tag{7.2.10}$$

In the forthcoming section, we derive the risk function of proposed estimators under balanced loss function. Arashi, Tabatabaey and Khan (2008) considered the proposed problem under LINEX loss for elliptical models. Also, Arashi and Tabatabaey

(2008) and Arashi, Tabatabaey and Iranmanesh (2010) formulated the use of stochastic constraints $\boldsymbol{h} = \boldsymbol{H}\boldsymbol{\beta} + \boldsymbol{v}$, where $\boldsymbol{v}$ is a random error term independent of ε, in regression with elliptical errors. More recently, Arashi (2012) extended the materials of this chapter for a seemingly unrelated regression model under the assumption of elliptical symmetry.

7.3 Bias and Risk Expressions

In this section, we first discuss balanced loss function, and then the underlying characteristic will be derived.

7.3.1 Balanced Loss Function

Let $\boldsymbol{\beta}^*$ denote any estimator of $\boldsymbol{\beta}$; then the quadratic loss function that reflects the goodness of fit of the model is

$$(\boldsymbol{X}\boldsymbol{\beta}^* - \boldsymbol{y})'(\boldsymbol{X}\boldsymbol{\beta}^* - \boldsymbol{y}).$$

Similarly, the precision of estimation of $\boldsymbol{\beta}^*$ is measured by the weighted loss function

$$(\boldsymbol{\beta}^* - \boldsymbol{\beta})'\boldsymbol{X}'\boldsymbol{X}(\boldsymbol{\beta}^* - \boldsymbol{\beta}).$$

Generally, both of the previous criteria are used to judge the performance of any estimator. Throughout this section, we shall consider the estimation problem through the following loss function:

$$\begin{aligned} L^{\boldsymbol{W}}_{\omega,\boldsymbol{\delta}_0}(\boldsymbol{\delta};\boldsymbol{\beta}) &= \omega r\left(\|\boldsymbol{\beta}\|^2\right)(\boldsymbol{\delta} - \boldsymbol{\delta}_0)'\boldsymbol{W}(\boldsymbol{\delta} - \boldsymbol{\delta}_0) \\ &\quad +(1-\omega) r\left(\|\boldsymbol{\beta}\|^2\right)(\boldsymbol{\delta} - \boldsymbol{\beta})'\boldsymbol{W}(\boldsymbol{\delta} - \boldsymbol{\beta}), \end{aligned} \tag{7.3.1}$$

where $\omega \in [0, 1]$, $r(.)$ is a positive weight function, $\boldsymbol{W}$ is a weight matrix, and $\boldsymbol{\delta}_0$ is a target estimator.

This loss is proposed by Jozani et al. (2006), inspired by Zellner's (1994) balanced loss function. This loss function takes both goodness of fit and error of estimation into account. The

$$\omega r\left(\|\boldsymbol{\beta}\|^2\right)(\boldsymbol{\delta} - \boldsymbol{\delta}_0)'(\boldsymbol{\delta} - \boldsymbol{\delta}_0)$$

part of the loss is analogous to a penalty term for lack of smoothness in nonparametric regression. The weight ω in (6.3.1) calibrates the relative importance of these two criteria. Dey et al. (1999) also considered issues of admissibility and dominance, under the loss (7.3.1) ignoring the term $r(.)$ when $\boldsymbol{W} = \boldsymbol{I}_p$. For the case $\omega = 0$, we will simply write $L^{\boldsymbol{W}}_0(\boldsymbol{\delta};\boldsymbol{\beta})$ as the quadratic loss function. Of course, duty of the weight function $r(.)$ is clearly apparent in the Bayesian viewpoint. In this chapter, we take it into consideration for the sake of generality. As will be seen later, the structure of $r(.)$ does not alter the whole superiority conclusions. The following result plays a deterministic role in deriving comparative studies. The following result can be found in Jozani et al. (2006) and Arashi (2012).

Assume $\boldsymbol{h}_i : \mathbb{R}^p \to \mathbb{R}^p$, $i = 1, 2$ are measurable functions.

Lemma 7.3.1.

(i) The estimator $\boldsymbol{\delta}_0(\boldsymbol{X}) + (1-\omega)\boldsymbol{h}_1(\boldsymbol{X})$ *dominates* $\boldsymbol{\delta}_0(\boldsymbol{X}) + (1-\omega)\boldsymbol{h}_2(\boldsymbol{X})$ *under the balanced loss function* $L^{\boldsymbol{W}}_{\omega,\boldsymbol{\delta}_0}(\boldsymbol{\delta};\boldsymbol{\beta})$ *if and only if* $\boldsymbol{\delta}_0(\boldsymbol{X}) + \boldsymbol{h}_1(\boldsymbol{X})$ *dominates* $\boldsymbol{\delta}_0(\boldsymbol{X}) + \boldsymbol{h}_2(\boldsymbol{X})$ *under the quadratic loss function* $L^{\boldsymbol{W}}_0(\boldsymbol{\delta};\boldsymbol{\beta})$.

(ii) Suppose the estimator $\boldsymbol{\delta}_0(\boldsymbol{X})$ *has constant risk* γ *under the quadratic loss function* $L^{\boldsymbol{W}}_0(\boldsymbol{\delta};\boldsymbol{\beta})$. *Then* $\boldsymbol{\delta}_0(\boldsymbol{X})$ *is minimax under the balanced loss function* $L^{\boldsymbol{W}}_{\omega,\boldsymbol{\delta}_0}(\boldsymbol{\delta};\boldsymbol{\beta})$ *with constant (and minimax) risk* $(1-\omega)\gamma$ *if and only if* $\boldsymbol{\delta}_0(\boldsymbol{X})$ *is minimax under the quadratic loss function* $L^{\boldsymbol{W}}_0(\boldsymbol{\delta};\boldsymbol{\beta})$ *with constant (and minimax) risk* γ.

The risk function for any estimator $\boldsymbol{\beta}^*$ of $\boldsymbol{\beta}$ associated with (7.3.1) is defined as

$$\boldsymbol{R}^{\boldsymbol{W}}_{\omega,\boldsymbol{\delta}_0}(\boldsymbol{\beta}^*;\boldsymbol{\beta}) = E[L^{\boldsymbol{W}}_{\omega,\boldsymbol{\delta}_0}(\boldsymbol{\beta}^*;\boldsymbol{\beta})]. \tag{7.3.2}$$

In the next section, we determine the bias and, using the risk function (7.3.2) with $\boldsymbol{\delta}_0 = \tilde{\boldsymbol{\beta}}_n$ as the target estimator and $\boldsymbol{W} = \boldsymbol{C}$, evaluate the risks of the five different estimators under study. For the case $\omega = 0$, we will simply write $\boldsymbol{R}^{\boldsymbol{W}}_0(\boldsymbol{\beta}^*;\boldsymbol{\beta})$. First, we consider bias expressions of the estimators.

7.3.2 Properties

Directly

$$\boldsymbol{b}_1 = E[\tilde{\boldsymbol{\beta}}_n - \boldsymbol{\beta}] = \boldsymbol{0}, \qquad \boldsymbol{b}_2 = E[\hat{\boldsymbol{\beta}}_n - \boldsymbol{\beta}] = -\boldsymbol{\delta}. \tag{7.3.3}$$

$$\begin{aligned} \boldsymbol{b}_3 &= E(\hat{\boldsymbol{\beta}}_n^{PT} - \boldsymbol{\beta}) = E[\tilde{\boldsymbol{\beta}}_n - I(\mathcal{L}_n \le F_\alpha)(\tilde{\boldsymbol{\beta}}_n - \hat{\boldsymbol{\beta}}_n) - \boldsymbol{\beta}] \\ &= -\boldsymbol{C}\boldsymbol{H}'\mathbf{V}_1^{1/2}\, E[I(\mathcal{L}_n \le F_\alpha)\mathbf{V}_1^{1/2}(\boldsymbol{H}\tilde{\boldsymbol{\beta}}_n - \boldsymbol{h})] \\ &= -\boldsymbol{\delta} G^{(2)}_{q+2,m}\left(F_\alpha;\Delta^2\right). \end{aligned} \tag{7.3.4}$$

Also,

$$\begin{aligned} \boldsymbol{b}_4 &= E(\hat{\boldsymbol{\beta}}_n^{S} - \boldsymbol{\beta}) = E[\tilde{\boldsymbol{\beta}}_n - d\mathcal{L}_n^{-1}(\tilde{\boldsymbol{\beta}}_n - \hat{\boldsymbol{\beta}}_n) - \boldsymbol{\beta}] \\ &= -d\boldsymbol{C}^{-1}\boldsymbol{H}'\mathbf{V}_1^{1/2}E[\mathcal{L}_n^{-1}\mathbf{V}_1^{1/2}(\boldsymbol{H}\tilde{\boldsymbol{\beta}}_n - \boldsymbol{h})] = -dq\boldsymbol{\delta}E^{(2)}_N[\chi^{*\,-2}_{q+2}(\Delta^2)], \end{aligned}$$

(7.3.5)

For the final bias expression we have

$$\begin{aligned}
\boldsymbol{b}_5 &= E(\hat{\boldsymbol{\beta}}_n^S - \boldsymbol{\beta}) - E[I(\mathcal{L}_n \leq d)(\tilde{\boldsymbol{\beta}}_n - \hat{\boldsymbol{\beta}}_n)] + dE[\mathcal{L}_n^{-1} I(\mathcal{L}_n \leq d)(\tilde{\boldsymbol{\beta}}_n - \hat{\boldsymbol{\beta}}_n)] \\
&= -dq\boldsymbol{\delta} E_N^{(2)}[\chi_{q+2}^{*^{-4}}(\Delta^2)] + \boldsymbol{\delta} G_{q+2,m}^{(2)}\left(d;\Delta^2\right) \\
&\quad + \frac{qd}{q+2}\boldsymbol{\delta} E_N^{(2)}\left[F_{q+2,m}^{-1}(\Delta^2) I\left(F_{q+2,m}(\Delta^2) \leq \frac{qd}{q+2}\right)\right],
\end{aligned} \tag{7.3.6}$$

where $x' = \frac{dq}{m+dq}$.

Note that as the noncentrality parameter $\Delta^2 \to \infty$, $\boldsymbol{b}_1 = \boldsymbol{b}_3 = \boldsymbol{b}_4 = \boldsymbol{b}_5 = \boldsymbol{0}$, while $\boldsymbol{b}_2$ becomes unbounded. However, under $H_0 : \boldsymbol{H\beta} = \boldsymbol{h}$, because $\boldsymbol{\delta} = \boldsymbol{0}$, $\boldsymbol{b}_1 = \boldsymbol{b}_2 = \boldsymbol{b}_3 = \boldsymbol{b}_4 = \boldsymbol{b}_5 = \boldsymbol{0}$.

Arashi, Saleh, and Tabatabaey (2013) derived the risk of the estimators using squared error loss function. Here we derive the risk expressions using the balanced loss function. For the risks of the estimators, taking $\boldsymbol{R}^{\boldsymbol{C}}_{\omega,\tilde{\boldsymbol{\beta}}_n}(.;\boldsymbol{\beta})$ given by (7.3.2), we have

$$\begin{aligned}
\boldsymbol{R}^{\boldsymbol{C}}_{\omega,\tilde{\boldsymbol{\beta}}_n}(\tilde{\boldsymbol{\beta}}_n;\boldsymbol{\beta}) &= (1-\omega)r\left(\|\boldsymbol{\beta}\|^2\right) E[(\tilde{\boldsymbol{\beta}}_n - \boldsymbol{\beta})'\boldsymbol{C}(\tilde{\boldsymbol{\beta}}_n - \boldsymbol{\beta})] \\
&= p\,\sigma_{\boldsymbol{\beta}}^2(1-\omega)r\left(\|\boldsymbol{\beta}\|^2\right).
\end{aligned} \tag{7.3.7}$$

Afterward, using the fact that $\mathbf{V}_1^{\frac{1}{2}}(\boldsymbol{H}\tilde{\boldsymbol{\beta}}_n - \boldsymbol{h}) \sim M_t^{(q)}(\mathbf{V}_1^{\frac{1}{2}}(\boldsymbol{H\beta} - \boldsymbol{h}), \sigma^2\boldsymbol{I}_q, f)$, it can be concluded that

$$\begin{aligned}
\boldsymbol{R}^{\boldsymbol{C}}_{\omega,\tilde{\boldsymbol{\beta}}_n}(\hat{\boldsymbol{\beta}}_n;\boldsymbol{\beta}) &= \omega r\left(\|\boldsymbol{\beta}\|^2\right) E[(\boldsymbol{H}\tilde{\boldsymbol{\beta}}_n - \boldsymbol{h})'\mathbf{V}_1(\boldsymbol{H}\tilde{\boldsymbol{\beta}}_n - \boldsymbol{h})] \\
&\quad + (1-\omega)r\left(\|\boldsymbol{\beta}\|^2\right) E[(\hat{\boldsymbol{\beta}}_n - \boldsymbol{\beta})'\boldsymbol{C}(\hat{\boldsymbol{\beta}}_n - \boldsymbol{\beta})] \\
&= -q\omega\sigma_{\boldsymbol{\varepsilon}}^2 r\left(\|\boldsymbol{\beta}\|^2\right) + (1-\omega)r\left(\|\boldsymbol{\beta}\|^2\right)[\sigma_{\boldsymbol{\varepsilon}}^2 \boldsymbol{tr}(\mathbf{V}_2\boldsymbol{C}) + \boldsymbol{\delta}'\boldsymbol{C\delta}] \\
&= -q\omega\sigma_{\boldsymbol{\varepsilon}}^2 r\left(\|\boldsymbol{\beta}\|^2\right) + \boldsymbol{R}^{\boldsymbol{C}}_{\omega,\tilde{\boldsymbol{\beta}}_n}(\tilde{\boldsymbol{\beta}}_n;\boldsymbol{\beta}) \\
&\quad + (1-\omega)r\left(\|\boldsymbol{\beta}\|^2\right)\left(-\sigma_{\boldsymbol{\varepsilon}}^2 \boldsymbol{tr}[H'\mathbf{V}_1 HC] + \delta'\boldsymbol{W}\delta\right) \\
&= -q\sigma_{\boldsymbol{\varepsilon}}^2 r\left(\|\boldsymbol{\beta}\|^2\right) + \boldsymbol{R}^{\boldsymbol{C}}_{\omega,\tilde{\boldsymbol{\beta}}_n}(\tilde{\boldsymbol{\beta}}_n;\boldsymbol{\beta}) + (1-\omega)r\left(\|\boldsymbol{\beta}\|^2\right)\theta,
\end{aligned} \tag{7.3.8}$$

where $\theta = \boldsymbol{\delta}'\boldsymbol{C\delta} = (\boldsymbol{H\beta}-\boldsymbol{h})'\mathbf{V}_1(\boldsymbol{H\beta}-\boldsymbol{h})$. Note that $\boldsymbol{R} = \boldsymbol{C}_1^{-1/2}\boldsymbol{H}'\mathbf{V}_1\boldsymbol{H}\boldsymbol{C}_1^{-1/2}$ is a symmetric idempotent matrix of rank $q \leq p$. Thus, there exists an orthonormal matrix $\boldsymbol{Q}$ ($\boldsymbol{Q}'\boldsymbol{Q} = \boldsymbol{I}_p$) such that $\boldsymbol{QRQ}' = \begin{bmatrix} \boldsymbol{I}_q & \boldsymbol{0} \\ \boldsymbol{0} & \boldsymbol{0} \end{bmatrix}$. Now we define random variable $\boldsymbol{w} = \boldsymbol{QC}_1^{1/2}\tilde{\boldsymbol{\beta}}_n - \boldsymbol{QC}_1^{-1/2}\boldsymbol{H}'\mathbf{V}_1\boldsymbol{h}$, then $\boldsymbol{w} \sim M_t^{(p)}(\boldsymbol{\eta}, \sigma^2\boldsymbol{I}_p, f)$, where $\boldsymbol{\eta} = \boldsymbol{QC}_1^{1/2}\boldsymbol{\beta} - \boldsymbol{QC}_1^{-1/2}\boldsymbol{H}'\mathbf{V}_1\boldsymbol{h}$. Partitioning the vectors $\boldsymbol{w} = (\boldsymbol{w}_1', \boldsymbol{w}_2')'$ and $\boldsymbol{\eta} = (\boldsymbol{\eta}_1', \boldsymbol{\eta}_2')'$, where $\boldsymbol{w}_1$ and $\boldsymbol{w}_2$ are subvectors of order q and $p-q$, respectively, we can represent the test statistic $\mathcal{L}_n$ as

$$\mathcal{L}_n = \frac{\boldsymbol{w}_1'\boldsymbol{w}_1}{qS^2}, \qquad \theta = \boldsymbol{\eta}_1'\boldsymbol{\eta}_1. \tag{7.3.9}$$

Consequently, for the risk of PTE, noting that $\hat{\boldsymbol{\beta}}_n - \tilde{\boldsymbol{\beta}}_n = \boldsymbol{C}^{-1}\boldsymbol{H}'\mathrm{V}_1\boldsymbol{H}\boldsymbol{C}^{-\frac{1}{2}}\boldsymbol{w}$, simplifying the equations we can write

$$\begin{aligned}
\boldsymbol{R}^{\boldsymbol{C}}_{\omega,\tilde{\boldsymbol{\beta}}_n}(\hat{\boldsymbol{\beta}}^{PT}_n;\boldsymbol{\beta}) &= \omega r\left(\|\boldsymbol{\beta}\|^2\right) E[I(\mathcal{L}_n < F_\alpha)(\hat{\boldsymbol{\beta}}_n - \tilde{\boldsymbol{\beta}}_n)'\boldsymbol{C}(\hat{\boldsymbol{\beta}}_n - \tilde{\boldsymbol{\beta}}_n)] \\
&\quad +(1-\omega)r\left(\|\boldsymbol{\beta}\|^2\right) E[(\hat{\boldsymbol{\beta}}^{PT}_n - \boldsymbol{\beta})'\boldsymbol{C}(\hat{\boldsymbol{\beta}}^{PT}_n - \boldsymbol{\beta})] \\
&= \omega r\left(\|\boldsymbol{\beta}\|^2\right) E[\boldsymbol{w}_1'\boldsymbol{w}_1 I(\mathcal{L}_n < F_\alpha)] \\
&\quad +\boldsymbol{R}^{\boldsymbol{C}}_{\omega,\tilde{\boldsymbol{\beta}}_n}(\tilde{\boldsymbol{\beta}}_n;\boldsymbol{\beta}) - (1-\omega)r\left(\|\boldsymbol{\beta}\|^2\right) \\
&\quad \times\left(E[\boldsymbol{w}_1'\boldsymbol{w}_1 I(\mathcal{L}_n \le F_\alpha)] - 2\boldsymbol{\eta}_1' E[\boldsymbol{w}_1 I(\mathcal{L}_n \le F_\alpha)]\right) \\
&= \boldsymbol{R}^{\boldsymbol{C}}_{\omega,\tilde{\boldsymbol{\beta}}_n}(\tilde{\boldsymbol{\beta}}_n;\boldsymbol{\beta}) - (1-2\omega)r\left(\|\boldsymbol{\beta}\|^2\right) E[\boldsymbol{w}_1'\boldsymbol{w}_1 I(\mathcal{L}_n \le F_\alpha)] \\
&\quad +2(1-\omega)r\left(\|\boldsymbol{\beta}\|^2\right) \boldsymbol{\eta}_1' E[\boldsymbol{w}_1 I(\mathcal{L}_n \le F_\alpha)] \\
&= \boldsymbol{R}^{\boldsymbol{C}}_{\omega,\tilde{\boldsymbol{\beta}}_n}(\tilde{\boldsymbol{\beta}}_n;\boldsymbol{\beta}) - (1-2\omega)q\sigma^2_{\boldsymbol{\varepsilon}} r\left(\|\boldsymbol{\beta}\|^2\right) G^{(1)}_{q+2,m}\left(F_\alpha;\Delta^2\right) \\
&\quad +2\theta(1-\omega)r\left(\|\boldsymbol{\beta}\|^2\right)\Big[2G^{(2)}_{q+2,m}\left(F_\alpha;\Delta^2\right) \\
&\qquad\qquad -G^{(2)}_{q+4,m}\left(F_\alpha;\Delta^2\right)\Big].
\end{aligned}$$

Similarly, for the risk of SE after simplifying the equations, we have

$$\begin{aligned}
\boldsymbol{R}^{\boldsymbol{C}}_{\omega,\tilde{\boldsymbol{\beta}}_n}(\hat{\boldsymbol{\beta}}^{S}_n;\boldsymbol{\beta}) &= \omega r\left(\|\boldsymbol{\beta}\|^2\right) d^2 E[\mathcal{L}_n^{-2}(\hat{\boldsymbol{\beta}}_n - \tilde{\boldsymbol{\beta}}_n)'\boldsymbol{C}(\hat{\boldsymbol{\beta}}_n - \tilde{\boldsymbol{\beta}}_n)] \\
&\quad +(1-\omega)r\left(\|\boldsymbol{\beta}\|^2\right) E[(\hat{\boldsymbol{\beta}}^{S}_n - \boldsymbol{\beta})'\boldsymbol{C}(\hat{\boldsymbol{\beta}}^{S}_n - \boldsymbol{\beta})] \\
&= \omega r\left(\|\boldsymbol{\beta}\|^2\right) d^2 E[\mathcal{L}_n^{-1}\boldsymbol{w}_1'\boldsymbol{w}_1] + \boldsymbol{R}^{\boldsymbol{C}}_{\omega,\tilde{\boldsymbol{\beta}}_n}(\tilde{\boldsymbol{\beta}}_n;\boldsymbol{\beta}) \\
&\quad -2d(1-\omega)r\left(\|\boldsymbol{\beta}\|^2\right) E[\mathcal{L}_n^{-1}(\boldsymbol{w}_1'\boldsymbol{w}_1 - \boldsymbol{\eta}_1'\boldsymbol{w}_1)] \\
&\quad +d^2(1-\omega)r\left(\|\boldsymbol{\beta}\|^2\right) E[\mathcal{L}_n^{-2}\boldsymbol{w}_1'\boldsymbol{w}_1] \\
&= \boldsymbol{R}^{\boldsymbol{C}}_{\omega,\tilde{\boldsymbol{\beta}}_n}(\tilde{\boldsymbol{\beta}}_n;\boldsymbol{\beta}) + qr\left(\|\boldsymbol{\beta}\|^2\right)\Big\{\left[d^2\omega - 2d(1-\omega)\right] \\
&\quad \times E^{(1)}[\chi^{*^{-2}}_{q+2}(\Delta^2)] + d^2(1-\omega)E^{(1)}[\chi^{*^{-4}}_{q+2}(\Delta^2)]\Big\} \\
&\quad +\theta r\left(\|\boldsymbol{\beta}\|^2\right)\Big\{\left[d^2\omega - 2d(1-\omega)\right] E^{(2)}[\chi^{*^{-2}}_{q+4}(\Delta^2)] \\
&\quad -2d(1-\omega)E^{(2)}[\chi^{*^{-2}}_{q+2}(\Delta^2)] \\
&\quad +d^2(1-\omega)E^{(2)}[\chi^{*^{-4}}_{q+4}(\Delta^2)]\Big\}. \qquad (7.3.10)
\end{aligned}$$

Finally, for the risk of PRSE we can obtain

$$\begin{aligned}
\boldsymbol{R}^{\boldsymbol{C}}_{\omega,\tilde{\boldsymbol{\beta}}_n}(\hat{\boldsymbol{\beta}}^{S+}_n;\boldsymbol{\beta}) &= \omega r\left(\|\boldsymbol{\beta}\|^2\right)\Big\{E[(\hat{\boldsymbol{\beta}}^{S}_n - \tilde{\boldsymbol{\beta}}_n)'\boldsymbol{C}(\hat{\boldsymbol{\beta}}^{S}_n - \tilde{\boldsymbol{\beta}}_n)] - E[(1-d\mathcal{L}_n^{-1})^2 \\
&\quad \times I(\mathcal{L}_n \le d)(\tilde{\boldsymbol{\beta}}_n - \hat{\boldsymbol{\beta}}_n)'\boldsymbol{C}(\tilde{\boldsymbol{\beta}}_n - \hat{\boldsymbol{\beta}}_n)] \\
&\quad -2E[(1-d\mathcal{L}_n^{-1})I(\mathcal{L}_n \le d)(\hat{\boldsymbol{\beta}}_n - \boldsymbol{\beta})'\boldsymbol{C}(\tilde{\boldsymbol{\beta}}_n - \hat{\boldsymbol{\beta}}_n)]\Big\} \\
&\quad +(1-\omega)r\left(\|\boldsymbol{\beta}\|^2\right) E[(\hat{\boldsymbol{\beta}}^{S+}_n - \boldsymbol{\beta})'\boldsymbol{C}(\hat{\boldsymbol{\beta}}^{S+}_n - \boldsymbol{\beta})]
\end{aligned}$$

$$
\begin{aligned}
&= \boldsymbol{R}^{\boldsymbol{C}}_{\omega,\tilde{\boldsymbol{\beta}}_n}(\hat{\boldsymbol{\beta}}^S_n;\boldsymbol{\beta}) - r\left(\|\boldsymbol{\beta}\|^2\right) E[(1-d\mathcal{L}_n^{-1})^2 I(\mathcal{L}_n \le d) \\
&\quad \times(\tilde{\boldsymbol{\beta}}_n - \hat{\boldsymbol{\beta}}_n)'\boldsymbol{C}(\tilde{\boldsymbol{\beta}}_n - \hat{\boldsymbol{\beta}}_n)] - 2r\left(\|\boldsymbol{\beta}\|^2\right) \\
&\quad \times E\{[(1-d\mathcal{L}_n^{-1})I(\mathcal{L}_n \le d)(\hat{\boldsymbol{\beta}}_n - \boldsymbol{\beta})'\boldsymbol{C}(\tilde{\boldsymbol{\beta}}_n - \hat{\boldsymbol{\beta}}_n)]\} \\
&= \boldsymbol{R}^{\boldsymbol{C}}_{\omega,\tilde{\boldsymbol{\beta}}_n}(\hat{\boldsymbol{\beta}}^S_n;\boldsymbol{\beta}) - r\left(\|\boldsymbol{\beta}\|^2\right) \\
&\quad \times E[(1-d\mathcal{L}_n^{-1})^2 I(\mathcal{L}_n \le d)\boldsymbol{w}_1'\boldsymbol{w}_1] \\
&\quad -2r\left(\|\boldsymbol{\beta}\|^2\right) E[(1-d\mathcal{L}_n^{-1})I(\mathcal{L}_n \le d)(\boldsymbol{w}_1'\boldsymbol{w}_1 - \boldsymbol{\eta}_1'\boldsymbol{w}_1)] \\
&= \boldsymbol{R}^{\boldsymbol{C}}_{\omega,\tilde{\boldsymbol{\beta}}_n}(\hat{\boldsymbol{\beta}}^S_n;\boldsymbol{\beta}) - \sigma^2_{\boldsymbol{\varepsilon}}\Bigg\{qE^{(1)}\left[\left(1 - \frac{qd}{q+2}F^{-1}_{q+2,m}(\Delta^2)\right)^2\right. \\
&\quad \left.\times I\left(F_{q+2,m}(\Delta^2) \le \frac{qd}{q+2}\right)\right] + \frac{\theta}{\sigma^2_{\boldsymbol{\varepsilon}}} \\
&\quad \times E^{(2)}\left[\left(1 - \frac{qd}{q+2}F^{-1}_{q+2,m}(\Delta^2)\right)^2\right. \\
&\qquad \left.\times I\left(F_{q+2,m}(\Delta^2) \le \frac{qd}{q+2}\right)\right]\Bigg\} \\
&\quad -2\theta E^{(2)}\left[\left(1 - \frac{qd}{q+2}F^{-1}_{q+2,m}(\Delta^2)\right)\right. \\
&\quad \times \qquad \left. I\left(F_{q+2,m}(\Delta^2) \le \frac{qd}{q+2}\right)\right].
\end{aligned} \tag{7.3.11}
$$

7.4 Comparison

Providing risk analysis of the underlying estimators with the weight matrix $\boldsymbol{C}$ by making use of equations (7.3.7) and (7.3.8), the risk difference is given by

$$
\boldsymbol{D}_1 = \boldsymbol{R}^{\boldsymbol{C}}_{\omega,\tilde{\boldsymbol{\beta}}_n}(\hat{\boldsymbol{\beta}}_n;\boldsymbol{\beta}) - \boldsymbol{R}^{\boldsymbol{C}}_{\omega,\tilde{\boldsymbol{\beta}}_n}(\tilde{\boldsymbol{\beta}}_n;\boldsymbol{\beta}) = r\left(\|\boldsymbol{\beta}\|^2\right)\left[(1-\omega)\theta - q\sigma^2_{\boldsymbol{\varepsilon}}\right]. \tag{7.4.1}
$$

Then it can be directly considered that $\hat{\boldsymbol{\beta}}_n$ performs better than $\tilde{\boldsymbol{\beta}}_n$, say $\hat{\boldsymbol{\beta}}_n$ dominates $\tilde{\boldsymbol{\beta}}_n$ ($\hat{\boldsymbol{\beta}}_n \succeq \tilde{\boldsymbol{\beta}}_n$) provided that $0 \le \theta \le \frac{q\sigma^2_{\boldsymbol{\varepsilon}}}{1-\omega}$, for $\omega \neq 1$. Taking $r\left(\|\boldsymbol{\beta}\|^2\right) = (\boldsymbol{H\beta} - \boldsymbol{h})'\mathbf{V}_1(\boldsymbol{H\beta} - \boldsymbol{h}) = \theta$ into account gives the same result.

Comparing $\hat{\boldsymbol{\beta}}^{PT}_n$ versus $\tilde{\boldsymbol{\beta}}_n$, using risk difference,

$$
\begin{aligned}
\boldsymbol{D}_2 &= \boldsymbol{R}^{\boldsymbol{C}}_{\omega,\tilde{\boldsymbol{\beta}}_n}(\tilde{\boldsymbol{\beta}}_n;\boldsymbol{\beta}) - \boldsymbol{R}^{\boldsymbol{C}}_{\omega,\tilde{\boldsymbol{\beta}}_n}(\hat{\boldsymbol{\beta}}^{PT}_n;\boldsymbol{\beta}) \\
&= (1-2\omega)q\sigma^2_{\boldsymbol{\varepsilon}} r\left(\|\boldsymbol{\beta}\|^2\right) G^{(1)}_{q+2,m}\left(F_\alpha;\Delta^2\right) \\
&\quad -2\theta(1-\omega)[2G^{(2)}_{q+2,m}\left(F_\alpha;\Delta^2\right) - G^{(2)}_{q+4,m}\left(F_\alpha;\Delta^2\right)].
\end{aligned} \tag{7.4.2}
$$

It can be followed that the R.H.S. of (7.4.2) is non-negative, i.e., $\hat{\boldsymbol{\beta}}^{PT}_n \succeq \tilde{\boldsymbol{\beta}}_n$ for $\omega \neq 1$ whenever

$$
\theta \le \frac{(1-2\omega)q\sigma^2_{\boldsymbol{\varepsilon}} G^{(1)}_{q+2,m}\left(F_\alpha;\Delta^2\right)}{2(1-\omega)\left[2G^{(2)}_{q+2,m}\left(F_\alpha;\Delta^2\right) - G^{(2)}_{q+4,m}\left(F_\alpha;\Delta^2\right)\right]}. \tag{7.4.3}
$$

Moreover, under $H_0 : \boldsymbol{H}\boldsymbol{\beta} = \boldsymbol{h}$, because of $\theta = 0$, $\hat{\boldsymbol{\beta}}_n^{PT} \succeq \tilde{\boldsymbol{\beta}}_n$ for values ω such that $\omega \leq \frac{1}{2}$. Now we compare $\hat{\boldsymbol{\beta}}_n$ and $\hat{\boldsymbol{\beta}}_n^{PT}$ by the risk difference as follows:

$$\begin{aligned}
\boldsymbol{D}_3 &= \boldsymbol{R}^{\boldsymbol{C}}_{\omega,\tilde{\boldsymbol{\beta}}_n}(\hat{\boldsymbol{\beta}}_n;\boldsymbol{\beta}) - \boldsymbol{R}^{\boldsymbol{C}}_{\omega,\tilde{\boldsymbol{\beta}}_n}(\hat{\boldsymbol{\beta}}_n^{PT};\boldsymbol{\beta}) \\
&= -q\sigma_{\boldsymbol{\varepsilon}}^2 r\left(\|\boldsymbol{\beta}\|^2\right)\left[1-(1-2\omega)G^{(1)}_{q+2,m}\left(F_\alpha;\Delta^2\right)\right] \\
&\quad +\theta(1-\omega)r\left(\|\boldsymbol{\beta}\|^2\right)\left[1-2G^{(2)}_{q+2,m}\left(F_\alpha;\Delta^2\right)+G^{(2)}_{q+4,m}\left(F_\alpha;\Delta^2\right)\right].
\end{aligned} \tag{7.4.4}$$

Thus, $\hat{\boldsymbol{\beta}}_n^{PT} \succeq \hat{\boldsymbol{\beta}}_n$ whenever

$$\theta \geq \frac{q\sigma_{\boldsymbol{\varepsilon}}^2\left[1-(1-2\omega)G^{(1)}_{q+2,m}\left(F_\alpha;\Delta^2\right)\right]}{(1-\omega)\left[1-2G^{(2)}_{q+2,m}\left(F_\alpha;\Delta^2\right)+G^{(2)}_{q+4,m}\left(F_\alpha;\Delta^2\right)\right]}, \tag{7.4.5}$$

and vice versa. However, under H_0, the dominance order of $\tilde{\boldsymbol{\beta}}_n$, $\hat{\boldsymbol{\beta}}_n$, and $\hat{\boldsymbol{\beta}}_n^{PT}$ is as follows:

$$\hat{\boldsymbol{\beta}}_n \succeq \hat{\boldsymbol{\beta}}_n^{PT} \succeq \tilde{\boldsymbol{\beta}}_n. \tag{7.4.6}$$

In order to determine the superiority of $\hat{\boldsymbol{\beta}}_n^{S}$ to $\tilde{\boldsymbol{\beta}}_n$, we give the following results.

Theorem 7.4.1. *Consider the model* (7.1.1) *where the error-vector is,* $M_t^{(n)}(\mathbf{0},\sigma^2\mathbf{V}_n,\gamma_o)$, $\gamma_o > 4$. *Then the Stein-type shrinkage estimator,* $\hat{\boldsymbol{\beta}}_n^{S}$ *of* $\boldsymbol{\beta}$ *given by*

$$\hat{\boldsymbol{\beta}}_n^{S} = \tilde{\boldsymbol{\beta}}_n - d^*\mathcal{L}_n^{-1}(\tilde{\boldsymbol{\beta}}_n - \hat{\boldsymbol{\beta}}_n)$$

uniformly dominates the unrestricted estimator $\tilde{\boldsymbol{\beta}}_n$ *with respect to the quadratic loss function* $L_0^{\boldsymbol{C}}(\boldsymbol{\delta};\boldsymbol{\beta})$ *and is minimax if and only if* $0 < d^* \leq \frac{2m}{m+2}$. *The largest reduction of the risk is attained when* $d^* = \frac{m}{m+2}$.

Proof: By making use of $\dot{\mathbf{Z}} = \boldsymbol{H}'\mathbf{V}_1(\boldsymbol{H}\tilde{\boldsymbol{\beta}}_n - \boldsymbol{h})$, the SE can be rewritten as

$$\begin{aligned}
\hat{\boldsymbol{\beta}}_n^{S} &= \tilde{\boldsymbol{\beta}}_n - qd^*S^2\left[(\boldsymbol{H}\tilde{\boldsymbol{\beta}}_n-\boldsymbol{h})'\mathbf{V}_1(\boldsymbol{H}\tilde{\boldsymbol{\beta}}_n-\boldsymbol{h})\right]^{-1}\boldsymbol{C}^{-1}\boldsymbol{H}'\mathbf{V}_1(\boldsymbol{H}\tilde{\boldsymbol{\beta}}_n-\boldsymbol{h}) \\
&= \tilde{\boldsymbol{\beta}}_n - qd^*S^2\left(\dot{\mathbf{Z}}'\boldsymbol{C}^{-1}\dot{\mathbf{Z}}\right)^{-1}\boldsymbol{C}^{-1}\dot{\mathbf{Z}}.
\end{aligned}$$

Then, the risk difference of the SE and the UE under quadratic loss function is given by

$$\begin{aligned} \boldsymbol{D}_4 &= E(\hat{\boldsymbol{\beta}}_n^S - \boldsymbol{\beta})'\boldsymbol{C}(\hat{\boldsymbol{\beta}}_n^S - \boldsymbol{\beta}) - E(\tilde{\boldsymbol{\beta}}_n - \boldsymbol{\beta})'\boldsymbol{C}(\tilde{\boldsymbol{\beta}}_n - \boldsymbol{\beta}) \\ &= (d^*)^2 E\left[q^2 S^4 \left(\dot{\mathbf{Z}}'\boldsymbol{C}^{-1}\dot{\mathbf{Z}}\right)^{-1}\right] - 2d^* E\left[qS^2 \left(\dot{\mathbf{Z}}'\boldsymbol{C}^{-1}\dot{\mathbf{Z}}\right)^{-1} (\tilde{\boldsymbol{\beta}}_n - \boldsymbol{\beta})'\dot{\mathbf{Z}}\right] \\ &= (d^*)^2 E_t\left\{E_N\left[q^2 S^4 \left(\dot{\mathbf{Z}}'\boldsymbol{C}^{-1}\dot{\mathbf{Z}}\right)^{-1}\Big|t\right]\right\} \\ &\quad -2d^* E_t\left\{E_N\left[qS^2\left(\dot{\mathbf{Z}}'\boldsymbol{C}^{-1}\dot{\mathbf{Z}}\right)^{-1}(\tilde{\boldsymbol{\beta}}_n - \boldsymbol{\beta})'\boldsymbol{H}'\mathbf{V}_1(\boldsymbol{H}\tilde{\boldsymbol{\beta}}_n - \boldsymbol{h})\Big|t\right]\right\} \\ &= \frac{q^2(m+2)}{m}(d^*)^2 E_t\left(\frac{t^{-2}}{\dot{\mathbf{Z}}'\boldsymbol{C}^{-1}\dot{\mathbf{Z}}}\right) - 2q^2 d^* E_t\left(\frac{t^{-2}}{\dot{\mathbf{Z}}'\boldsymbol{C}^{-1}\dot{\mathbf{Z}}}\right), \end{aligned}$$

since $\left(\frac{mS^2}{\sigma^2}\right)\Big|t \sim t^{-1}\chi^2_m$ and $\tilde{\boldsymbol{\beta}}_n'\boldsymbol{H}'\mathbf{V}_1\boldsymbol{H}\tilde{\boldsymbol{\beta}}_n \mid t \sim t^{-2}\sigma^4\chi^2_q(\dot{\boldsymbol{\delta}})$, where $\dot{\boldsymbol{\delta}} = \boldsymbol{\beta}'\boldsymbol{H}'\mathbf{V}_1\boldsymbol{H}\boldsymbol{\beta}$.

Therefore, $\boldsymbol{D}_4 \leq 0$ if and only if $0 < d^* \leq \frac{2m}{m+2}$, where we used (5.4.10).

Remark 7.4.1. *Consider the coefficient d given by* (7.2.10). *From* $q \geq 3$, *we get* $0 < d = \frac{(q-2)m}{q(m+2)} < \frac{2m}{m+2}$ *and thus using Theorem 7.4.1,* $\hat{\boldsymbol{\beta}}_n^S$ *in equation* (7.2.9) *uniformly dominates* $\tilde{\boldsymbol{\beta}}_n$ *on the whole parameter space under quadratic loss function.*

Theorem 7.4.2. *Suppose in the model* (7.1.1), $\boldsymbol{\varepsilon} \sim M_t^{(n)}(\mathbf{0}, \sigma^2\mathbf{V}, \gamma_o)$. *Then the Stein-type shrinkage estimator*

$$\hat{\boldsymbol{\beta}}_n^S = \tilde{\boldsymbol{\beta}}_n - d(1-\omega)\mathcal{L}_n^{-1}(\tilde{\boldsymbol{\beta}}_n - \hat{\boldsymbol{\beta}}_n) \tag{7.4.7}$$

uniformly dominates $\tilde{\boldsymbol{\beta}}_n$ *under the balanced loss function* $L^{\boldsymbol{C}}_{\omega,\tilde{\boldsymbol{\beta}}_n}(.;\boldsymbol{\beta})$.

The proof directly follows using Lemma 7.3.1 (i) and Theorem 7.4.1.

Corollary 7.4.2.1. *Suppose in the model* (6.1.1), $\boldsymbol{\varepsilon} \sim M_t^{(n)}(\mathbf{0}, \sigma^2\mathbf{V}, \gamma_o)$. *Then* $\hat{\boldsymbol{\beta}}_n^S \succeq \tilde{\boldsymbol{\beta}}_n$ *under the balanced loss function* $L^{\boldsymbol{C}}_{\omega,\tilde{\boldsymbol{\beta}}_n}(.;\boldsymbol{\beta})$.

The proof directly follows from Theorem 7.4.2 for the special case $\omega = 0$.

Lemma 7.4.1. *Suppose in the model* (7.1.1), $\boldsymbol{\varepsilon} \sim M_t^{(n)}(\mathbf{0}, \sigma^2\mathbf{V}_n, \gamma_o)$. *Then the estimator* $\tilde{\boldsymbol{\beta}}_n$ *of* $\boldsymbol{\beta}$ *is minimax under the balanced loss function* $L^{\boldsymbol{C}}_{\omega,\tilde{\boldsymbol{\beta}}_n}(.;\boldsymbol{\beta})$.

The proof directly follows by knowing that $\tilde{\beta}_n$ is minimax under quadratic loss function and applying Lemma 7.3.1 (ii).

Remark 7.4.2. *Using Corollary 7.4.1 and Lemma 7.4.1, the Stein-type shrinkage estimator* $\hat{\beta}_n^S$ *of* β *is minimax.*

To compare $\hat{\beta}_n$ and $\hat{\beta}_n^S$, it is easy to show that

$$\begin{aligned} \boldsymbol{R}_0^{\boldsymbol{C}}(\hat{\boldsymbol{\beta}}_n^S;\boldsymbol{\beta}) &= \boldsymbol{R}_0^{\boldsymbol{C}}(\hat{\boldsymbol{\beta}}_n;\boldsymbol{\beta}) + q\sigma_\varepsilon^2 - \theta - dq^2\sigma_\varepsilon^2\Big\{(q-2)E[\chi_{q+2}^{*\,-4}(\Delta^2)] \\ &+ \left[1 - \frac{(q+2)\theta}{2q\sigma_\varepsilon^2\Delta^2}\right](2\Delta^2)E[\chi_{q+4}^{*\,-4}(\Delta^2)]\Big\}. \end{aligned} \tag{7.4.8}$$

Under H_0, this becomes

$$\boldsymbol{R}_0^{\boldsymbol{C}}(\hat{\boldsymbol{\beta}}_n^S;\boldsymbol{\beta}) = \boldsymbol{R}_0^{\boldsymbol{C}}(\hat{\boldsymbol{\beta}}_n;\boldsymbol{\beta}) + q\sigma_\varepsilon^2(1-d) \geq \boldsymbol{R}_0^{\boldsymbol{C}}(\hat{\boldsymbol{\beta}}_n;\boldsymbol{\beta}), \tag{7.4.9}$$

while

$$\boldsymbol{R}_0^{\boldsymbol{C}}(\hat{\boldsymbol{\beta}}_n;\boldsymbol{\beta}) = \boldsymbol{R}_0^{\boldsymbol{C}}(\tilde{\boldsymbol{\beta}}_n;\boldsymbol{\beta}) - q\sigma_\varepsilon^2 \leq \boldsymbol{R}_0^{\boldsymbol{C}}(\tilde{\boldsymbol{\beta}}_n;\boldsymbol{\beta}). \tag{7.4.10}$$

Therefore, $\hat{\boldsymbol{\beta}}_n \succeq \hat{\boldsymbol{\beta}}_n^S$ under H_0 with the quadratic loss $L_0^{\boldsymbol{C}}(.,\boldsymbol{\beta})$. Therefore using Lemma 7.3.1 (i), under H_0, $\hat{\boldsymbol{\beta}}_n \succeq \hat{\boldsymbol{\beta}}_n^S$ with the balanced loss $L_{\omega,\tilde{\boldsymbol{\beta}}_n}^{\boldsymbol{C}}(.;\boldsymbol{\beta})$. However, as $\boldsymbol{\eta}_1$ moves away from 0, θ increases and the risk of $\hat{\boldsymbol{\beta}}_n$ becomes unbounded while the risk of $\hat{\boldsymbol{\beta}}_n^S$ remains below the risk of $\tilde{\boldsymbol{\beta}}_n$; thus, for similar reasons, $\hat{\boldsymbol{\beta}}_n^S$ dominates $\hat{\boldsymbol{\beta}}_n$ outside an interval around the origin under the balanced loss $L_{\omega,\tilde{\boldsymbol{\beta}}_n}^{\boldsymbol{C}}(.;\boldsymbol{\beta})$. This scenario repeats when we compare $\hat{\boldsymbol{\beta}}_n^S$ and $\hat{\boldsymbol{\beta}}_n^{PT}$. Consider under H_0

$$\boldsymbol{R}_0^{\boldsymbol{C}}(\hat{\boldsymbol{\beta}}_n^S;\boldsymbol{\beta}) = \boldsymbol{R}_0^{\boldsymbol{C}}(\hat{\boldsymbol{\beta}}_n^{PT};\boldsymbol{\beta}) + q\sigma_\varepsilon^2[1-\alpha-d] \geq \boldsymbol{R}_0^{\boldsymbol{C}}(\hat{\boldsymbol{\beta}}_n^{PT};\boldsymbol{\beta})$$

for all α such that $F_{q+2,m}^{-1}(d,0) \leq \frac{qF_\alpha}{q+2}$. This means the estimator $\hat{\boldsymbol{\beta}}_n^S$ does not always dominate $\hat{\boldsymbol{\beta}}_n^{PT}$ under H_0. Thus, under H_0 with α satisfying $F_{q+2,m}^{-1}(d,0) \leq \frac{qF_\alpha}{q+2}$ taking the balanced loss function, we have $\hat{\boldsymbol{\beta}}_n \succeq \hat{\boldsymbol{\beta}}_n^{PT} \succeq \hat{\boldsymbol{\beta}}_n^S \succeq \tilde{\boldsymbol{\beta}}_n$. Then we compare the risks of $\hat{\boldsymbol{\beta}}_n^{S+}$ and $\hat{\boldsymbol{\beta}}_n^S$. Subsequently, the risk difference is given by

$$\begin{aligned} \boldsymbol{D}_5 &= \boldsymbol{R}_{\omega,\tilde{\boldsymbol{\beta}}_n}^{\boldsymbol{C}}(\hat{\boldsymbol{\beta}}_n^{S+};\boldsymbol{\beta}) - \boldsymbol{R}_{\omega,\tilde{\boldsymbol{\beta}}_n}^{\boldsymbol{C}}(\hat{\boldsymbol{\beta}}_n^S;\boldsymbol{\beta}) \\ &= -\sigma_\varepsilon^2\Big\{qE^{(1)}\left[\left(1 - \frac{qd}{q+2}F_{q+2,m}^{-1}(\Delta^2)\right)^2 I\left(F_{q+2,m}(\Delta^2) \leq \frac{qd}{q+2}\right)\right] \\ &+ \frac{\theta}{\sigma_\varepsilon^2}E^{(2)}\left[\left(1 - \frac{qd}{q+2}F_{q+2,m}^{-1}(\Delta^2)\right)^2 I\left(F_{q+2,m}(\Delta^2) \leq \frac{qd}{q+2}\right)\right]\Big\} \\ &- 2\theta E^{(2)}\left[\left(1 - \frac{qd}{q+2}F_{q+2,m}^{-1}(\Delta^2)\right) I\left(F_{q+2,m}(\Delta^2) \leq \frac{qd}{q+2}\right)\right]. \end{aligned}$$

The R.H.S. of the above equality is negative since for $F_{q+2,m}(\Delta^2) \leq \frac{qd}{q+2}$, $(\frac{qd}{q+2}F_{q+2,m}(\Delta^2) - 1) \geq 0$ and also the expectation of a positive random variable is positive. That for all $\boldsymbol{\beta}$, $\hat{\boldsymbol{\beta}}_n^{S+} \succeq \hat{\boldsymbol{\beta}}_n^{S}$.

Remark 7.4.3. *The positive-rule shrinkage estimator $\hat{\boldsymbol{\beta}}_n^{S+}$ of $\boldsymbol{\beta}$ is minimax.*

In the rest, we continue the comparisons under $L_0^{\boldsymbol{C}}(.;\boldsymbol{\beta})$. The results are the same for the balanced loss $L^{\boldsymbol{C}}_{\omega,\tilde{\boldsymbol{\beta}}_n}(.;\boldsymbol{\beta})$. To compare $\hat{\boldsymbol{\beta}}_n$ and $\hat{\boldsymbol{\beta}}_n^{S+}$, first consider the case under H_0, i.e., $\boldsymbol{\eta}_1 = 0$. In this case,

$$\begin{aligned} \boldsymbol{R}_0^{\boldsymbol{C}}(\hat{\boldsymbol{\beta}}_n^{S+};\boldsymbol{\beta}) &= \boldsymbol{R}_0^{\boldsymbol{C}}(\hat{\boldsymbol{\beta}}_n;\boldsymbol{\beta}) + q\sigma_{\boldsymbol{\varepsilon}}^2\Big\{(1-d) - E\Big[(1 - \frac{qd}{q+2}F_{q+2,m}^{-1}(0))^2 \\ &\quad \times I(F_{q+2,m}(0) \leq \frac{qd}{q+2})\Big]\Big\} \geq \boldsymbol{R}_0^{\boldsymbol{C}}(\hat{\boldsymbol{\beta}}_n;\boldsymbol{\beta}), \end{aligned}$$

since $E\left[(1-\frac{qd}{q+2}F_{q+2,m}^{-1}(0))^2 I(F_{q+2,m}(0) \leq \frac{qd}{q+2})\right] \leq E\left[(1-\frac{qd}{q+2}F_{q+2,m}^{-1}(0))^2\right]$ $= 1-d$. Thus, under H_0, $\hat{\boldsymbol{\beta}}_n \succeq \hat{\boldsymbol{\beta}}_n^{S+}$. However, as $\boldsymbol{\eta}_1$ moves away from 0, θ increases and the risk of $\hat{\boldsymbol{\beta}}_n$ becomes unbounded while the risk of $\hat{\boldsymbol{\beta}}_n^{S+}$ remains below the risk of $\tilde{\boldsymbol{\beta}}_n$; thus, $\hat{\boldsymbol{\beta}}_n^{S+}$ dominates $\hat{\boldsymbol{\beta}}_n$ outside an interval around the origin.

Now we compare $\hat{\boldsymbol{\beta}}_n^{S+}$ and $\hat{\boldsymbol{\beta}}_n^{PT}$. When H_0 holds, because $G^*_{q+2,m}(F_\alpha, 0) = 1-\alpha$

$$\begin{aligned} \boldsymbol{R}_0^{\boldsymbol{C}}(\hat{\boldsymbol{\beta}}_n^{S+};\boldsymbol{\beta}) &= \boldsymbol{R}_0^{\boldsymbol{C}}(\hat{\boldsymbol{\beta}}_n^{PT};\boldsymbol{\beta}) + q\sigma_{\boldsymbol{\varepsilon}}^2\Big\{1-\alpha-d - E\Big[(1 - \frac{qd}{q+2}F_{q+2,m}^{-1}(0))^2 \\ &\quad \times I(F_{q+2,m}(0) \leq \frac{qd}{q+2})\Big]\Big\} \geq \boldsymbol{R}_0^{\boldsymbol{C}}(\hat{\boldsymbol{\beta}}_n^{PT};\boldsymbol{\beta}) \end{aligned}$$

for all α satisfying $E\left[(1-\frac{qd}{q+2}F_{q+2,m}^{-1}(0))^2 I(F_{q+2,m}(0) \leq \frac{qd}{q+2})\right] \leq 1-\alpha-d.$

Thus, $\hat{\boldsymbol{\beta}}_n^{S+}$ does not always dominate $\hat{\boldsymbol{\beta}}_n^{PT}$ when the null-hypothesis H_0 holds.

Therefore, the dominance order of five estimators under the balanced loss function $L^{\boldsymbol{C}}_{\omega,\tilde{\boldsymbol{\beta}}_n}(.;\boldsymbol{\beta})$ can be determined under the following two categories

1. $\hat{\boldsymbol{\beta}}_n \succeq \hat{\boldsymbol{\beta}}_n^{PT} \succeq \hat{\boldsymbol{\beta}}_n^{S+} \succeq \hat{\boldsymbol{\beta}}_n^{S} \succeq \tilde{\boldsymbol{\beta}}_n$, 2. $\hat{\boldsymbol{\beta}}_n \succeq \hat{\boldsymbol{\beta}}_n^{S+} \succeq \hat{\boldsymbol{\beta}}_n^{S} \succeq \hat{\boldsymbol{\beta}}_n^{PT} \succeq \tilde{\boldsymbol{\beta}}_n$.

To end this chapter, we display some graphical results for the risk of the proposed estimators. In this regard, we suppose that $n = 30$, $p = 5$, $q = 3$ and degree of freedom $\gamma_o = 5$ and $\omega \in \{0, 0.5, 0.9\}$ to cover all possible situations.

Figure 7.1 shows the risk behavior of the proposed estimators for a number of degrees of freedom and compares the risks of the PTE, SE, and PRSE for a selected number of degree of freedom ($\gamma_o = 5$) and varying values ω. The graphs in Figure 7.1 reaffirm the analytical comparison covered in this section. More important, as ω increases, the risk values decrease. In other words, based on the structure of BLF, it confirms that if the model fit is good, then the risk values are decreased as a natural consequence.

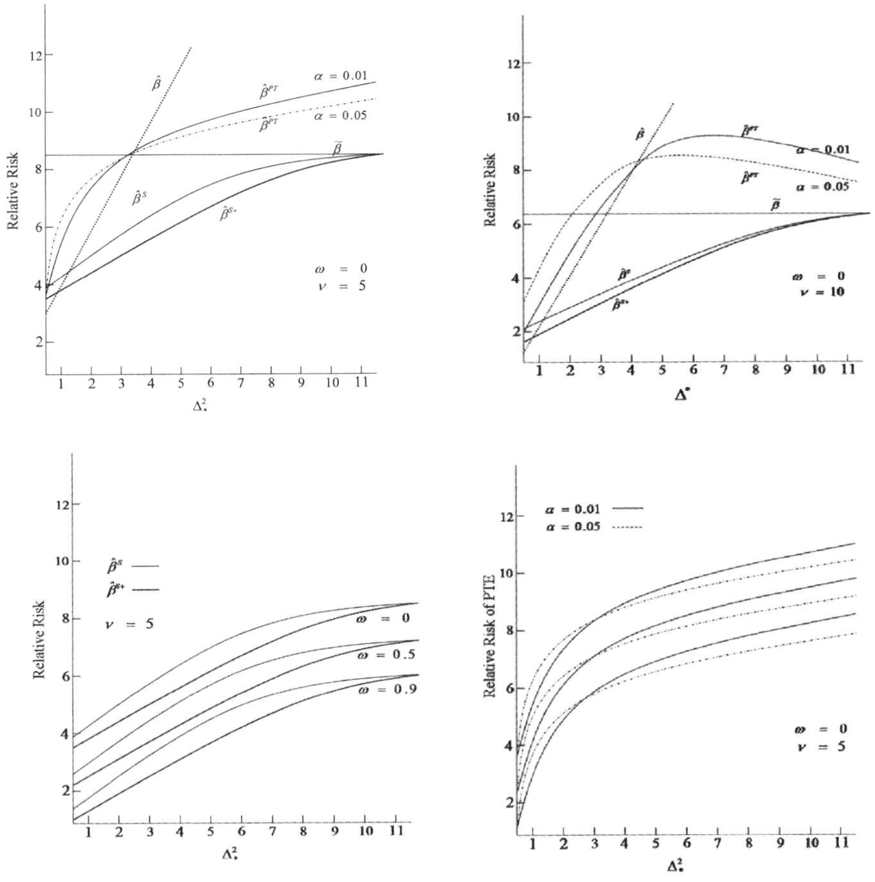

Figure 7.1 Risk performance

7.5 Problems

1. Define a shrinkage-type estimator of $\boldsymbol{\beta}$ as $\hat{\boldsymbol{\beta}}_n(c) = c\tilde{\boldsymbol{\beta}}_n$, $c > 0$, similar to (3.3.4). Derive the risk function based on the balanced loss function (7.3.1) and compare it with the other proposed estimators in this chapter.

2. Prove Lemma 7.3.1.

3. Verify equation (7.3.9).

4. Prove Theorem 7.4.2.

5. Prove Remark 7.4.3.

CHAPTER 8

RIDGE REGRESSION

Outline

The area of shrinkage estimators came to the forefront of statistical literature soon after Stein (1956) discovered that the sample mean in a multivariate model is not admissible, under a quadratic loss function, for dimension more than two. The idea took a decade to settle in to statistical methodology. Another class of shrinkage estimators appeared in the statistical literature in the 1970's due to Hoerl and Kennard (1970). The methodology is an advancement for linear models. The idea is simple

Statistical Inference for Models with Multivariate t-Distributed Errors, First Edition.

A. K. Md. Ehsanes Saleh, M. Arashi, S.M.M. Tabatabaey.

but its impact is great: that the ordinary least squares estimates (OLSEs) are unbiased and the covariance matrix is dependent on the design matrix $\boldsymbol{X}$, which may be ill-conditioned, that is, some of the eigenvalues may be zero or near zero, which impacts on the variance of the OLSEs very large to make the OLSEs useless. To overcome this problem, Hoerl and Kennard (1970) put forward the idea of using the $\left(\boldsymbol{X}'\boldsymbol{X} + k\boldsymbol{I}_p\right)^{-1}\boldsymbol{X}\boldsymbol{y}$ instead of $\left(\boldsymbol{X}'\boldsymbol{X}\right)^{-1}\boldsymbol{X}'\boldsymbol{y}$ for the estimation of the coefficients of a regression model, as in Chapter 7. These estimators are called "ridge regression estimators" (RREs), where k is called the tuning/biasing/ridge parameter, which is traditionally known as the "ridge parameter." In this chapter, we consider the regression model and apply the ridge regression methodology when the error distribution belongs to the class of multivariate t-distribution.

8.1 Model Specification

Consider the multiple regression model as in Chapter 7, given by

$$\boldsymbol{y} = \boldsymbol{X}\boldsymbol{\beta} + \boldsymbol{\varepsilon} \tag{8.1.1}$$

where $\boldsymbol{y}$ is an $(n \times 1)$ vector of observations, $\boldsymbol{X}$ is a nonstochastic $(n \times p)$ matrix of full rank p, $\boldsymbol{\beta}$ is a $(p \times 1)$ vector of unknown regression coefficients, and $\boldsymbol{\varepsilon}$ is an $(n \times 1)$ random error-vector distributed as $M_t^{(n)}(\boldsymbol{0}, \sigma^2\mathbf{V}_n, \gamma_o)$.

It is well-known that the OLS estimator of $\boldsymbol{\beta}$ is given by $\tilde{\boldsymbol{\beta}}_n = (\boldsymbol{X}'\boldsymbol{X})^{-1}\boldsymbol{X}'\boldsymbol{y}$. It is observed that the properties of the usual OLS estimator of $\boldsymbol{\beta}$ depend heavily on the characteristics of the design matrix, $\boldsymbol{X}$. If the matrix $\boldsymbol{X}'\boldsymbol{X}$ moves from a well-conditioned to an ill-conditioned one, then the OLS estimator is sensitive to a number of "errors." That is to say, if there is an "explosion" of any eigenvalues, there is an "explosion" of $(\boldsymbol{X}'\boldsymbol{X})^{-1}$, the corresponding estimator would loose its validity in the sense that some of the regression coefficients may be statistically insignificant with wrong signs and meaningful statistical inference becomes impossible for practitioners.

An adequate remedy for the effect of collinearity is to abandon the use of OLS and use a biased estimation method known as ridge regression. To overcome collinearity under ridge regression, Hoerl and Kennard (1970) suggested the use of $\boldsymbol{C}_{(k)} = \boldsymbol{X}'\boldsymbol{X} + k\boldsymbol{I}_p$, $(k \geq 0)$ rather than $(X'X)$ in the estimation of $\boldsymbol{\beta}$, thereby developing the idea of ridge estimation of β given by $\tilde{\boldsymbol{\beta}}_n(k) = \boldsymbol{C}_{(k)}^{(-1)}\boldsymbol{X}'\boldsymbol{y}$. The mathematical validity of RRE is obtained if we minimize $(\boldsymbol{y} - \boldsymbol{X}\boldsymbol{\beta})'(\boldsymbol{y} - \boldsymbol{X}\boldsymbol{\beta}) + k\boldsymbol{\beta}'\boldsymbol{\beta}$ w.r.t. k to obtain $-\boldsymbol{X}'(\boldsymbol{y} - \boldsymbol{X}\boldsymbol{\beta}) + k\boldsymbol{\beta} = \boldsymbol{0}$, which leads to the solution $\tilde{\boldsymbol{\beta}}_n(k) = \left(\boldsymbol{I}_p + k(X'X)^{-1}\right)^{-1}\tilde{\boldsymbol{\beta}}_n$.

The ridge regression approach has been studied by Hoerl and Kennard (1970), McDonald and Galarneau (1975), Lawless (1978), Gibbons (1981), Sarkar (1992), Saleh and Kibria (1993), Tabatabaey (1995), Kibria and Saleh (2004), Zhong and Yang (2007), Hassanzadeh Bashtian, Arashi, and Tabatabaey (2011), Saleh and Kibria (2011), Kibria, Mansson, and Shukur (2012), Kibria and Saleh (2012), Kibria, Kristofer, and Shukur (2013), Najarian, Arashi, and Kibria (2013), and Arashi, Tabatabaey and Hassanzadeh Bashtian (2014) to mention a few.

We are primarily interested in the estimation of $\boldsymbol{\beta}$ when the error distribution is multivariate t. It is a priori suspected that $\boldsymbol{\beta}$ may be restricted to the subspace $H_0 : \boldsymbol{H\beta} = \boldsymbol{h}$, where $\boldsymbol{H}$ is a $(q \times p)$ known matrix of rank $q(< p)$ and $\boldsymbol{h}$ is a $(q \times 1)$ vector of known constants.

The primary object of this study is to estimate $\boldsymbol{\beta}$, when a priori $H_0 : \boldsymbol{H\beta} = \boldsymbol{h}$ is suspected. We combine the idea of the preliminary test estimator and Stein-rule estimator with ridge regression approach estimation to obtain better estimators of $\boldsymbol{\beta}$. Accordingly, we consider five estimators, namely, the unrestricted ridge regression estimator (URRE), the restricted ridge regression estimator (RRRE), the preliminary test ridge regression estimator (PTRRE), the Stein-type ridge regression estimator (SRRE), and the positive-rule Stein ridge regression estimator (PRSRRE), an approach discussed in Saleh (2006).

8.2 Proposed Estimators

For the unrestricted estimator, we note that the OLSE of $\boldsymbol{\beta}$ under $M_t^{(n)}(\boldsymbol{0}, \sigma^2\mathbf{V}_n, \gamma_o)$ is given by $\tilde{\boldsymbol{\beta}}_n = (\boldsymbol{X}'\mathbf{V}_n^{-1}\boldsymbol{X})^{-1}\boldsymbol{X}'\mathbf{V}_n^{-1}\boldsymbol{y}$ with covariance $\frac{\gamma_o\sigma^2}{\gamma_o-2}(\boldsymbol{X}'\mathbf{V}_n^{-1}\boldsymbol{X})^{-1}$. Hence, we define the URRE of $\boldsymbol{\beta}$ under $M_t^{(n)}(\boldsymbol{0}, \sigma^2\mathbf{V}_n, \gamma_o)$ as

$$\begin{aligned}
\tilde{\boldsymbol{\beta}}_n(k) &= \boldsymbol{W}\tilde{\boldsymbol{\beta}}_n, \quad \boldsymbol{W} = (\boldsymbol{I}_p + k\boldsymbol{C}^{-1})^{-1}, \quad \boldsymbol{C} = (\boldsymbol{X}'\mathbf{V}_n^{-1}\boldsymbol{X}),\ k \geq 0, \\
&= \boldsymbol{W}\boldsymbol{C}^{-1}\boldsymbol{X}'\mathbf{V}_n^{-1}\boldsymbol{y} \\
&= (\boldsymbol{C} + k\boldsymbol{I}_p)^{-1}\boldsymbol{X}'\mathbf{V}_n^{-1}\boldsymbol{y} \\
&= (\boldsymbol{I}_p + k\boldsymbol{C}^{-1})^{-1}\tilde{\boldsymbol{\beta}}_n. \qquad (8.2.1)
\end{aligned}$$

Its distribution is $M_t^{(p)}(\boldsymbol{W\beta}, \sigma_\varepsilon^2\boldsymbol{W}\boldsymbol{C}^{-1}\boldsymbol{W}, \gamma_o)$. Thus, the 1st moment is $\boldsymbol{W\beta}$ and the 2nd central moment is

$$E(\tilde{\boldsymbol{\beta}}_n(k) - \boldsymbol{\beta})'(\tilde{\boldsymbol{\beta}}_n(k) - \boldsymbol{\beta}) = \sigma_\varepsilon^2 tr(\boldsymbol{W}\boldsymbol{C}^{-1}\boldsymbol{W}) + k^2\boldsymbol{\beta}'\boldsymbol{C}_{(k)}^{-2}\boldsymbol{\beta},$$

where $\boldsymbol{C}_{(k)} = \boldsymbol{X}'\boldsymbol{X} + k\boldsymbol{I}_p$.

The restricted ridge regression estimator (RRRE) is defined by

$$\hat{\boldsymbol{\beta}}_n(k) = \boldsymbol{W}\hat{\boldsymbol{\beta}}_n = \boldsymbol{W}\tilde{\boldsymbol{\beta}}_n - \boldsymbol{W}\boldsymbol{C}^{-1}\boldsymbol{H}'\mathbf{V}_1(\boldsymbol{H}\tilde{\boldsymbol{\beta}}_n - \boldsymbol{h}), \quad \mathbf{V}_1 = (\boldsymbol{H}\boldsymbol{C}^{-1}\boldsymbol{H}')^{-1}. \qquad (8.2.2)$$

This estimator has been considered by Sarkar (1992).

The test statistic for testing the null hypothesis $H_0 : \boldsymbol{H\beta} = \boldsymbol{h}$, is $\mathcal{L}_n$ given by (7.2.5).

In many practical situations, along with the model one may suspect that $\boldsymbol{\beta}$ belongs to the subspace defined by $\boldsymbol{H\beta} = \boldsymbol{h}$. In such a situation one combines the estimate of $\boldsymbol{\beta}$ and the test-statistic to obtain 3 or more estimators, in addition to the URR and RRR estimators of $\boldsymbol{\beta}$. First we consider the preliminary test ridge regression estimators (PTRRE) defined by

$$\hat{\boldsymbol{\beta}}_n^{PT}(k) = \boldsymbol{W}\hat{\boldsymbol{\beta}}_n^{PT} = \boldsymbol{W}\hat{\boldsymbol{\beta}}_n + [1 - I(\mathcal{L}_n \leq F_\alpha)]\boldsymbol{W}(\tilde{\boldsymbol{\beta}}_n - \hat{\boldsymbol{\beta}}_n). \tag{8.2.3}$$

This estimator has been considered by Saleh and Kibria (1993). The PTRRE has the disadvantage that it depends on α $(0 < \alpha < 1)$, the level of significance, and also it yields extreme results, namely $\hat{\boldsymbol{\beta}}_n(k)$ and $\tilde{\boldsymbol{\beta}}_n(k)$, depending on the outcome of the test. Afterward, we consider the shrinkage ridge regression estimator (SRRE) defined by

$$\hat{\boldsymbol{\beta}}_n^{S}(k) = \boldsymbol{W}\hat{\boldsymbol{\beta}}_n^{S} = \boldsymbol{W}\hat{\boldsymbol{\beta}}_n + (1 - d\mathcal{L}_n^{-1})\boldsymbol{W}(\tilde{\boldsymbol{\beta}}_n - \hat{\boldsymbol{\beta}}_n), \tag{8.2.4}$$

where d is given by (7.2.9). The SRRE has the disadvantage that it has strange behavior for small values of $\mathcal{L}_n$. Also, the shrinkage factor $(1 - d\mathcal{L}_n^{-1})$ becomes negative for $\mathcal{L}_n < d$. Hence, we consider the positive-rule shrinkage ridge regression estimator (PRSRRE) defined by

$$\hat{\boldsymbol{\beta}}_n^{S+}(k) = \boldsymbol{W}\hat{\boldsymbol{\beta}}_n^{S+} = \boldsymbol{W}\hat{\boldsymbol{\beta}}_n^{S} - (1 - d\mathcal{L}_n^{-1})I(\mathcal{L}_n \leq d)\boldsymbol{W}(\tilde{\boldsymbol{\beta}}_n - \hat{\boldsymbol{\beta}}_n). \tag{8.2.5}$$

Note that $\hat{\boldsymbol{\beta}}_n^{PT}(k)$ is a compromised estimator of $\boldsymbol{\beta}$ when we suspect $\boldsymbol{H}\boldsymbol{\beta} = \boldsymbol{h}$ and $\boldsymbol{C}$ is ill-conditioned, giving rise to multicollinearity.

8.3 Bias, MSE, and Risk Expressions

In this section, we provide the expressions for the bias, quadratic risk, and mean squared error matrix of the estimators $\tilde{\boldsymbol{\beta}}_n(k)$, $\hat{\boldsymbol{\beta}}_n(k)$, $\hat{\boldsymbol{\beta}}_n^{PT}(k)$, $\hat{\boldsymbol{\beta}}_n^{S}(k)$, and $\hat{\boldsymbol{\beta}}_n^{S+}(k)$.

8.3.1 Biases of the Estimators

In this subsection, we present expressions for the biases of all five estimators.

Theorem 8.3.1. *Biases of the URRE, RRRE, PTRRE, SRRE, and PRSRRE are given by (8.3.1)–(8.3.5), respectively:*

(i)

$$\boldsymbol{b}_1(k) = -k\boldsymbol{C}_{(k)}^{-1}\boldsymbol{\beta} \tag{8.3.1}$$

(ii)

$$\boldsymbol{b}_2(k) = -\boldsymbol{W}\boldsymbol{\delta} - k\boldsymbol{C}_{(k)}^{-1}\boldsymbol{\beta} \tag{8.3.2}$$

(iii)

$$\boldsymbol{b}_3(k) = -\boldsymbol{W}\boldsymbol{\delta}G^{(2)}_{q+2,m}(l_\gamma,\Delta^2) - k\boldsymbol{C}^{-1}_{(k)}\boldsymbol{\beta} \tag{8.3.3}$$

(iv)

$$\boldsymbol{b}_4(k) = -qd\boldsymbol{W}\boldsymbol{\delta}E^{(2)}[\chi^{-2}_{q+2}(\Delta^2)] - k\boldsymbol{C}^{-1}_{(k)}\boldsymbol{\beta} \tag{8.3.4}$$

(v)

$$\begin{aligned}\boldsymbol{b}_5(k) &= \boldsymbol{W}\boldsymbol{\delta}\Big\{\frac{qd}{q+2}E^{(2)}\left[F^{-1}_{q+2,n-p}(\Delta^2)I\left(F_{q+2,n-p}(\Delta^2)\le\frac{qd}{q+2}\right)\right]\\ &\quad -\frac{qd}{q+2}E^{(2)}[F^{-1}_{q+2,n-p}(\Delta^2)] - G^{(2)}_{q+2,n-p}(l_\gamma,\Delta^2)\Big\} - k\boldsymbol{C}^{-1}_{(k)}\boldsymbol{\beta},\end{aligned} \tag{8.3.5}$$

where $\boldsymbol{\delta} = \boldsymbol{C}^{-1}\boldsymbol{H}'\mathrm{V}_1(\boldsymbol{H}\boldsymbol{\beta}-\boldsymbol{h})$.

Proof. (i) By definition,

$$\boldsymbol{b}_1(k) = E[\tilde{\boldsymbol{\beta}}_n(k)-\boldsymbol{\beta}] = E(\boldsymbol{W}\tilde{\boldsymbol{\beta}}_n-\boldsymbol{\beta}) = (\boldsymbol{W}-\boldsymbol{I}_p)\boldsymbol{\beta}.$$

Using $\boldsymbol{W} = (\boldsymbol{I}_p + k\boldsymbol{C}^{-1})^{-1}, k\ge 0$, we get

$$\begin{aligned} & \boldsymbol{W}^{-1} = I_p + k\boldsymbol{C}^{-1}\\ \Leftrightarrow\quad & k\boldsymbol{W}\boldsymbol{C}^{-1} = \boldsymbol{I}_p - \boldsymbol{W}\\ \Leftrightarrow\quad & k(\boldsymbol{I}_p + k\boldsymbol{C}^{-1})^{-1}\boldsymbol{C}^{-1} = \boldsymbol{I}_p - \boldsymbol{W}\\ \Leftrightarrow\quad & k(\boldsymbol{C} + k\boldsymbol{I}_p)^{-1} = \boldsymbol{I}_p - \boldsymbol{W}\\ \Leftrightarrow\quad & k\boldsymbol{C}^{-1}_{(k)} = \boldsymbol{I}_p - \boldsymbol{W}.\end{aligned}$$

Thus, $\boldsymbol{b}_1(k) = -k\boldsymbol{C}^{-1}_{(k)}\boldsymbol{\beta}$.

(ii) Using (8.2.2),

$$\begin{aligned}\boldsymbol{b}_2(k) &= E(\hat{\boldsymbol{\beta}}_n(k)-\boldsymbol{\beta})\\ &= E(\tilde{\boldsymbol{\beta}}_n(k)-\boldsymbol{\beta}) - \boldsymbol{W}\boldsymbol{C}^{-1}\boldsymbol{H}'(\boldsymbol{H}\boldsymbol{C}^{-1}\boldsymbol{H}')^{-1}(\boldsymbol{H}\boldsymbol{\beta}-\boldsymbol{h})\\ &= -\boldsymbol{W}\boldsymbol{\delta} - k\boldsymbol{C}^{-1}_{(k)}\boldsymbol{\beta}.\end{aligned}$$

(iii) By making use of equation (3.3.3), we obtain

$$\begin{aligned}\boldsymbol{b}_3(k) &= E(\hat{\boldsymbol{\beta}}^{PT}_n(k)-\boldsymbol{\beta})\\ &= E(\tilde{\boldsymbol{\beta}}_n(k)-\boldsymbol{\beta}) - \boldsymbol{W}\boldsymbol{\delta}G^{(2)}_{q+2,m}(l_\gamma,\Delta^2)\\ &= -\boldsymbol{W}\boldsymbol{\delta}G^{(2)}_{q+2,m}(l_\gamma,\Delta^2) - k\boldsymbol{C}^{-1}_{(k)}\boldsymbol{\beta}.\end{aligned}$$

(iv) Also,

$$\begin{aligned}
\boldsymbol{b}_4(k) &= E(\hat{\boldsymbol{\beta}}_n^S(k) - \boldsymbol{\beta}) \\
&= E(\tilde{\boldsymbol{\beta}}_n(k) - \boldsymbol{\beta}) - qd\boldsymbol{W}\boldsymbol{\delta}E^{(2)}[\chi_{q+2}^{-2}(\Delta^2)] \\
&= -qd\boldsymbol{W}\boldsymbol{\delta}E^{(2)}[\chi_{q+2}^{-2}(\Delta^2)] - k\boldsymbol{C}_{(k)}^{-1}\boldsymbol{\beta}.
\end{aligned}$$

(v) Finally, we have that

$$\begin{aligned}
\boldsymbol{b}_5(k) &= E(\hat{\boldsymbol{\beta}}_n^{S+}(k) - \boldsymbol{\beta}) \\
&= E(\hat{\boldsymbol{\beta}}_n^S(k) - \boldsymbol{\beta}) - \boldsymbol{W}\boldsymbol{\delta}G_{q+2,m}^{(2)}(l_\gamma, \Delta^2) \\
&+ \frac{qd}{q+2}\boldsymbol{W}\boldsymbol{\delta}E^{(2)}\left[F_{q+2,n-p}^{-1}(\Delta^2)I\left(F_{q+2,n-p}(\Delta^2) \le \frac{qd}{q+2}\right)\right] \\
&= \boldsymbol{W}\boldsymbol{\delta}\left\{\frac{qd}{q+2}E^{(2)}\left[F_{q+2,n-p}^{-1}(\Delta^2)I\left(F_{q+2,n-p}(\Delta^2) \le \frac{qd}{q+2}\right)\right]\right. \\
&\left. - \frac{qd}{q+2}E^{(2)}[F_{q+2,n-p}^{-1}(\Delta^2)] - G_{q+2,n-p}^{(2)}(l_\gamma, \Delta^2)\right\} - k\boldsymbol{C}_{(k)}^{-1}\boldsymbol{\beta}.
\end{aligned}$$

8.3.2 MSE Matrices and Risks of the Estimators

For the risk of the estimators, we consider

$$\begin{aligned}
R(\boldsymbol{\beta}_{(k)}^*; \boldsymbol{\beta}) &= E[(\boldsymbol{\beta}_{(k)}^* - \boldsymbol{\beta})'(\boldsymbol{\beta}_{(k)}^* - \boldsymbol{\beta})] \\
&= tr[\boldsymbol{M}(\boldsymbol{\beta}_{(k)}^*)],
\end{aligned} \tag{8.3.6}$$

where

$$\boldsymbol{M}(\boldsymbol{\beta}_{(k)}^*) = E[(\boldsymbol{\beta}_{(k)}^* - \boldsymbol{\beta})(\boldsymbol{\beta}_{(k)}^* - \boldsymbol{\beta})'] \tag{8.3.7}$$

is the mean-squared error matrix of $\boldsymbol{\beta}_{(k)}^*$ and $\boldsymbol{\beta}_{(k)}^*$ can be any one of the five estimators. Similar computations as in Chapter 7 lead to the following theorem.

Theorem 8.3.2. *MSE matrices and risks of URRE, RRRE, PTRRE, SRRE, and PRSRRE are given by (8.3.8)–(8.3.17), respectively:*

$$\boldsymbol{M}_1(\tilde{\boldsymbol{\beta}}_n(k)) = \sigma_{\boldsymbol{\varepsilon}}^2\boldsymbol{W}\boldsymbol{C}^{-1}\boldsymbol{W} + k^2\boldsymbol{C}_{(k)}^{-1}\boldsymbol{\beta}\boldsymbol{\beta}'\boldsymbol{C}_{(k)}^{-1} \tag{8.3.8}$$

$$R_1(\tilde{\boldsymbol{\beta}}_n(k);\boldsymbol{\beta}) = \sigma_{\boldsymbol{\varepsilon}}^2 tr(\boldsymbol{W}\boldsymbol{C}^{-1}\boldsymbol{W}) + k^2\boldsymbol{\beta}'\boldsymbol{C}_{(k)}^{-2}\boldsymbol{\beta} \tag{8.3.9}$$

$$\begin{aligned}\boldsymbol{M}_2(\hat{\boldsymbol{\beta}}_n(k)) &= \sigma_{\boldsymbol{\varepsilon}}^2\boldsymbol{W}\boldsymbol{C}^{-1}\boldsymbol{W} - \sigma_{\boldsymbol{\varepsilon}}^2[\boldsymbol{W}\boldsymbol{C}^{-1}\boldsymbol{H}'(\boldsymbol{H}\boldsymbol{C}^{-1}\boldsymbol{H}')^{-1}\boldsymbol{H}\boldsymbol{C}^{-1}\boldsymbol{W}]\\ &\quad + [\boldsymbol{W}\boldsymbol{\delta} + k\boldsymbol{C}_{(k)}^{-1}\boldsymbol{\beta}][\boldsymbol{W}\boldsymbol{\delta} + k\boldsymbol{C}^{-1}\boldsymbol{\beta}]'\end{aligned} \tag{8.3.10}$$

$$\begin{aligned}R_2(\hat{\boldsymbol{\beta}}_n(k);\boldsymbol{\beta}) &= \sigma_{\boldsymbol{\varepsilon}}^2 tr(\boldsymbol{W}\boldsymbol{C}^{-1}\boldsymbol{W}) - \sigma_{\boldsymbol{\varepsilon}}^2 tr[\boldsymbol{W}\boldsymbol{C}^{-1}\boldsymbol{H}'(\boldsymbol{H}\boldsymbol{C}^{-1}\boldsymbol{H}')^{-1}\boldsymbol{H}\boldsymbol{C}^{-1}\boldsymbol{W}]\\ &\quad + \boldsymbol{\delta}'\boldsymbol{W}^2\boldsymbol{\delta} + k^2\boldsymbol{\beta}'\boldsymbol{C}_{(k)}^{2}\boldsymbol{\beta} + 2k\boldsymbol{\delta}'\boldsymbol{W}\boldsymbol{C}_{(k)}^{-1}\boldsymbol{\beta}\end{aligned} \tag{8.3.11}$$

$$\begin{aligned}\boldsymbol{M}_3(\hat{\boldsymbol{\beta}}_n^{PT}(k)) &= \sigma_e^2\boldsymbol{W}\boldsymbol{C}^{-1}\boldsymbol{W} + k^2\boldsymbol{C}_{(k)}^{-1}\boldsymbol{\beta}\boldsymbol{\beta}'\boldsymbol{C}_{(k)}^{-1}\\ &\quad - \sigma_{\boldsymbol{\varepsilon}}^2[\boldsymbol{W}\boldsymbol{C}^{-1}\boldsymbol{H}'(\boldsymbol{H}\boldsymbol{C}^{-1}\boldsymbol{H}')^{-1}\boldsymbol{H}\boldsymbol{C}^{-1}\boldsymbol{W}]G_{q+2,m}^{(1)}(l_\gamma,\Delta^2)\\ &\quad + \boldsymbol{W}\boldsymbol{\delta}\boldsymbol{\delta}'\boldsymbol{W}\left[2G_{q+2,m}^{(2)}(l_\gamma,\Delta^2) - G_{q+4,m}^{(2)}(l_\gamma,\Delta^2)\right]\\ &\quad + k\left[\boldsymbol{W}\boldsymbol{\delta}\boldsymbol{\beta}'\boldsymbol{C}_{(k)}^{-1} + \boldsymbol{C}_{(k)}^{-1}\boldsymbol{\beta}\boldsymbol{\delta}'\boldsymbol{W}\right]G_{q+2,m}^{(2)}(l_\gamma,\Delta^2)\end{aligned} \tag{8.3.12}$$

$$\begin{aligned}R_3(\hat{\boldsymbol{\beta}}_n^{PT}(k);\boldsymbol{\beta}) &= \sigma_{\boldsymbol{\varepsilon}}^2 tr(\boldsymbol{W}\boldsymbol{C}^{-1}\boldsymbol{W})\\ &\quad - \sigma_{\boldsymbol{\varepsilon}}^2 tr[\boldsymbol{W}\boldsymbol{C}^{-1}\boldsymbol{H}'(\boldsymbol{H}\boldsymbol{C}^{-1}\boldsymbol{H}')^{-1}\boldsymbol{H}\boldsymbol{C}^{-1}\boldsymbol{W}]G_{q+2,m}^{(1)}(l_\gamma,\Delta^2)\\ &\quad + \boldsymbol{\delta}'\boldsymbol{W}^2\boldsymbol{\delta}\left[2G_{q+2,m}^{(2)}(l_\gamma,\Delta^2) - G_{q+4,m}^{(2)}(l_\gamma,\Delta^2)\right]\\ &\quad + 2k\boldsymbol{\delta}'\boldsymbol{W}\boldsymbol{C}_{(k)}^{-1}\boldsymbol{\beta}G_{q+2,m}^{(2)}(l_\gamma,\Delta^2) + k^2\boldsymbol{\beta}'\boldsymbol{C}_{(k)}^{-2}\boldsymbol{\beta}\end{aligned} \tag{8.3.13}$$

$$\begin{aligned}\boldsymbol{M}_4(\hat{\boldsymbol{\beta}}_n^{S}(k)) &= \sigma_{\boldsymbol{\varepsilon}}^2\boldsymbol{W}\boldsymbol{C}^{-1}\boldsymbol{W} - dq\sigma_e^2\boldsymbol{W}\boldsymbol{C}^{-1}\boldsymbol{H}'(\boldsymbol{H}\boldsymbol{C}^{-1}\boldsymbol{H}')^{-1}\boldsymbol{H}\boldsymbol{C}^{-1}\boldsymbol{W}\\ &\quad \times\{(q-2)E^{(1)}[\chi_{q+2}^{-4}(\Delta^2)] + 2\Delta^2E^{(2)}[\chi_{q+4}^{-4}(\Delta^2)]\}\\ &\quad + dq(q+2)\boldsymbol{W}\boldsymbol{\delta}\boldsymbol{\delta}'\boldsymbol{W}E^{(2)}[\chi_{q+2}^{-4}(\Delta^2)]\\ &\quad + qd[\boldsymbol{W}\boldsymbol{\delta}\boldsymbol{\beta}'\boldsymbol{C}_{(k)}^{-1} + \boldsymbol{C}_{(k)}^{-1}\boldsymbol{\beta}\boldsymbol{\delta}'\boldsymbol{W}]E^{(2)}[\chi_{q+2}^{-2}(\Delta^2)]\\ &\quad + k^2\boldsymbol{C}_{(k)}^{-1}\boldsymbol{\beta}\boldsymbol{\beta}'\boldsymbol{C}_{(k)}^{-1}\end{aligned} \tag{8.3.14}$$

$$
\begin{aligned}
R_4(\hat{\boldsymbol{\beta}}_n^S(k);\boldsymbol{\beta}) &= \sigma_{\varepsilon}^2 tr(\boldsymbol{W}\boldsymbol{C}^{-1}\boldsymbol{W}) \\
&\quad - dq\sigma_{\varepsilon}^2 tr\left[\boldsymbol{W}\boldsymbol{C}^{-1}\boldsymbol{H}'(\boldsymbol{H}\boldsymbol{C}^{-1}\boldsymbol{H}')^{-1}\boldsymbol{H}\boldsymbol{C}^{-1}\boldsymbol{W}\right] \\
&\quad \times \left\{(q-2)E^{(1)}[\chi_{q+2}^{-4}(\Delta^2)]\right. \\
&\quad + \left[1 - \frac{(q+2)\boldsymbol{\delta}'\boldsymbol{W}^2\boldsymbol{\delta}}{2\sigma_{\varepsilon}^2\Delta^2 tr[\boldsymbol{W}\boldsymbol{C}^{-1}\boldsymbol{H}'(\boldsymbol{H}\boldsymbol{C}^{-1}\boldsymbol{H}')^{-1}\boldsymbol{H}\boldsymbol{C}^{-1}\boldsymbol{W}]}\right] \\
&\quad \left. \times (2\Delta^2)E^{(2)}[\chi_{q+4}^{-4}(\Delta^2)]\right\} + 2qdk\boldsymbol{\delta}'\boldsymbol{W}\boldsymbol{C}_{(k)}^{-1}\boldsymbol{\beta}E^{(2)}[\chi_{q+2}^{-2}(\Delta^2)] \\
&\quad + k^2\boldsymbol{\beta}'\boldsymbol{C}_{(k)}^{-2}\boldsymbol{\beta} \qquad (8.3.15)
\end{aligned}
$$

$$
\begin{aligned}
\boldsymbol{M}_5(\hat{\boldsymbol{\beta}}_n^{S+}(k)) &= \boldsymbol{M}_4(\hat{\boldsymbol{\beta}}_n^S(k)) - \sigma_e^2\left\{\boldsymbol{W}\boldsymbol{C}^{-1}\boldsymbol{H}'(\boldsymbol{H}\boldsymbol{C}^{-1}\boldsymbol{H}')^{-1}\boldsymbol{H}\boldsymbol{C}^{-1}\boldsymbol{W}\right. \\
&\quad \times E^{(1)}\left[\left(1 - \frac{qd}{q+2}F_{q+2,n-p}^{-1}(\Delta^2)\right)^2\right. \\
&\quad \left. \times I\left(F_{q+2,n-p}(\Delta^2) \le \frac{qd}{q+2}\right)\right] + \frac{\boldsymbol{W}\boldsymbol{\delta}\boldsymbol{\delta}'\boldsymbol{W}}{\sigma_{\varepsilon}^2} \\
&\quad \times E^{(2)}\left[\left(1 - \frac{qd}{q+4}F_{q+4,n-p}^{-1}(\Delta^2)\right)^2\right. \\
&\quad \left.\left. \times I\left(F_{q+4,n-p}(\Delta^2) \le \frac{qd}{q+4}\right)\right]\right\} - 2\boldsymbol{W}\boldsymbol{\delta}\boldsymbol{\delta}'\boldsymbol{W} \\
&\quad \times E^{(2)}\left[\left(\frac{qd}{q+2}F_{q+2,n-p}^{-1}(\Delta^2) - 1\right)\right. \\
&\quad \left. \times I\left(F_{q+2,n-p}(\Delta^2) \le \frac{qd}{q+2}\right)\right] \\
&\quad - k\left[\boldsymbol{W}\boldsymbol{\delta}\boldsymbol{\beta}'\boldsymbol{C}_{(k)}^{-1} + \boldsymbol{C}_{(k)}^{-1}\boldsymbol{\beta}\boldsymbol{\delta}'\boldsymbol{W}\right] \\
&\quad \times E^{(2)}\left[\left(\frac{qd}{q+2}F_{q+2,n-p}^{-1}(\Delta^2) - 1\right)\right. \\
&\quad \left. \times I\left(F_{q+2,n-p}(\Delta^2) \le \frac{qd}{q+2}\right)\right]
\end{aligned}
$$

(8.3.16)

$$
\begin{aligned}
R_5(\hat{\boldsymbol{\beta}}_n^{S+}(k);\boldsymbol{\beta}) &= R_4(\hat{\boldsymbol{\beta}}_n^S(k);\boldsymbol{\beta}) - \sigma_{\varepsilon}^2\left\{tr[\boldsymbol{W}\boldsymbol{C}^{-1}\boldsymbol{H}'(\boldsymbol{H}\boldsymbol{C}^{-1}\boldsymbol{H}')^{-1}\boldsymbol{H}\boldsymbol{C}^{-1}\boldsymbol{W}]\right. \\
&\quad \times E^{(1)}\left[\left(1 - \frac{qd}{q+2}F_{q+2,n-p}^{-1}(\Delta^2)\right)^2\right. \\
&\quad \left. \times I\left(F_{q+2,n-p}(\Delta^2) \le \frac{qd}{q+2}\right)\right] + \frac{\boldsymbol{\delta}'\boldsymbol{W}^2\boldsymbol{\delta}}{\sigma_{\varepsilon}^2}
\end{aligned}
$$

$$
\begin{aligned}
&\times E^{(2)}\left[\left(1-\frac{qd}{q+4}F^{-1}_{q+4,n-p}(\Delta^2)\right)^2\right.\\
&\qquad\left.\times I\left(F_{q+4,n-p}(\Delta^2)\le\frac{qd}{q+4}\right)\right]\Bigg\}-2\boldsymbol{\delta}'\boldsymbol{W}^2\boldsymbol{\delta}\\
&\times E^{(2)}\left[\left(\frac{qd}{q+2}F^{-1}_{q+2,n-p}(\Delta^2)-1\right)\right.\\
&\qquad\left.\times I\left(F_{q+2,n-p}(\Delta^2)\le\frac{qd}{q+2}\right)\right]-2k\boldsymbol{\delta}'\boldsymbol{W}\boldsymbol{C}^{-1}_{(k)}\boldsymbol{\beta}\\
&\times E^{(2)}\left[\left(\frac{qd}{q+2}F^{-1}_{q+2,n-p}(\Delta^2)-1\right)\right.\\
&\qquad\left.\times I\left(F_{q+2,n-p}(\Delta^2)\le\frac{qd}{q+2}\right)\right] \qquad (8.3.17)
\end{aligned}
$$

8.4 Performance of the Estimators

In this section, we compare the five estimators with the usual ones. In section 8.4.1, we show that URRE, SRRE, and PRSRRE have similar dominance properties as the URE, SRE, and PRSRE. Similarly, in section 8.4.2, we show that URRE, PTRRE, and RRRE have similar dominance properties as URE, PTRE, and RRE.

In the rest of the section, we find the dominance properties in pairs, namely, $(\tilde{\boldsymbol{\beta}}_n, \tilde{\boldsymbol{\beta}}_n(k))$, $(\hat{\boldsymbol{\beta}}_n, \hat{\boldsymbol{\beta}}_n(k))$, $(\hat{\boldsymbol{\beta}}_n^{PT}, \hat{\boldsymbol{\beta}}_n^{PT}(k))$, $(\hat{\boldsymbol{\beta}}_n^{S}, \hat{\boldsymbol{\beta}}_n^{S}(k))$, and $(\hat{\boldsymbol{\beta}}_n^{S+}, \hat{\boldsymbol{\beta}}_n^{S+}(k))$, respectively. For the analysis of the properties of the estimators, we need the following fact.

It is clear that $\boldsymbol{C}$ is a positive-definite matrix. So there exists an orthogonal matrix $\boldsymbol{\Gamma}$ such that $\boldsymbol{C}=\boldsymbol{\Gamma\Lambda\Gamma}'$ and $\boldsymbol{\Lambda}=\boldsymbol{\Gamma}'\boldsymbol{C\Gamma}=\text{Diag}\,(\lambda_1,\ldots,\lambda_p)$ where

$$\lambda_1\ge\lambda_2\ge\ldots\ge\lambda_p>0 \qquad (8.4.1)$$

are the eigenvalues of $\boldsymbol{C}$ (see Anderson, 1984). it is easy to see that the eigenvalues of $\boldsymbol{W}=(\boldsymbol{I}_p+k\boldsymbol{C}^{-1})^{-1}$ and $\boldsymbol{C}_{(k)}=\boldsymbol{C}+k\boldsymbol{I}_p$ are $\frac{\lambda_i}{\lambda_i+k}$ and $\lambda_i+k, i=1,\ldots,p$, respectively. We note that the eigenvectors of $\boldsymbol{C},\boldsymbol{W}$, and $\boldsymbol{C}_{(k)}$ are all the same. With this background, we get the following identities:

$$
\begin{aligned}
\boldsymbol{\beta}'\boldsymbol{C}^{-2}_{(k)}\boldsymbol{\beta} &= \boldsymbol{\beta}'\boldsymbol{\Gamma}(\boldsymbol{\Lambda}+kI_p)^{-2}\boldsymbol{\Gamma}'\boldsymbol{\beta}=\boldsymbol{\alpha}'(\boldsymbol{\Lambda}+k\boldsymbol{I}_p)^{-2}\boldsymbol{\alpha}\\
&= \sum_{i=1}^{p}\frac{\boldsymbol{\alpha}_i^2}{(\lambda_i+k)^2}, \qquad \boldsymbol{\alpha}=\boldsymbol{\Gamma}'\boldsymbol{\beta} \qquad (8.4.2)
\end{aligned}
$$

$$tr(\boldsymbol{W}\boldsymbol{C}^{-1}\boldsymbol{W})=tr(\boldsymbol{C}^{-1}\boldsymbol{W}^2)=\sum_{i=1}^{p}\frac{\lambda_i}{(\lambda_i+k)^2} \qquad (8.4.3)$$

$$tr[\boldsymbol{W}\boldsymbol{C}^{-1}\boldsymbol{H}'(\boldsymbol{H}\boldsymbol{C}^{-1}\boldsymbol{H}')^{-1}\boldsymbol{H}\boldsymbol{C}^{-1}\boldsymbol{W}]$$

$$
\begin{aligned}
&= tr[\boldsymbol{\Gamma}(\boldsymbol{\Lambda}+kI_p)^{-1}\boldsymbol{\Gamma}'\boldsymbol{H}'(\boldsymbol{H}\boldsymbol{C}^{-1}\boldsymbol{H}')^{-1}\boldsymbol{H}\boldsymbol{\Gamma}(\boldsymbol{\Lambda}+k\boldsymbol{I}_p)^{-1}\boldsymbol{\Gamma}'] \\
&= tr\{[\boldsymbol{\Gamma}'\boldsymbol{H}'(\boldsymbol{H}\boldsymbol{C}^{-1}\boldsymbol{H}')^{-1}\boldsymbol{\Gamma}](\boldsymbol{\Lambda}+kI_p)^{-2}\} \\
&= \sum_{i=1}^{p}\frac{h_{ii}^*}{(\lambda_i+k)^2},
\end{aligned} \tag{8.4.4}
$$

where $h_{ii}^* \geq 0$ is the ith diagonal element of $\boldsymbol{\Gamma}'\boldsymbol{H}'(\boldsymbol{H}\boldsymbol{C}^{-1}\boldsymbol{H}')^{-1}\boldsymbol{H}\boldsymbol{\Gamma} = \boldsymbol{H}^*$. Let us now compare the performance of the ridge estimator with each of the others and the usual ones.

8.4.1 Comparison between $\tilde{\boldsymbol{\beta}}_n(k)$, $\hat{\boldsymbol{\beta}}_n^{S}(k)$, and $\hat{\boldsymbol{\beta}}_n^{S+}(k)$

For the comparison of $\tilde{\boldsymbol{\beta}}_n(k)$, $\hat{\boldsymbol{\beta}}_n^{S}(k)$, and $\hat{\boldsymbol{\beta}}_n^{S+}(k)$, we consider the risk difference given by

$$
\begin{aligned}
R_1(\tilde{\boldsymbol{\beta}}_n(k);\boldsymbol{\beta}) - R_4(\hat{\boldsymbol{\beta}}_n^{S}(k);\boldsymbol{\beta}) &= dq\sigma_\varepsilon^2 tr\left[\boldsymbol{W}\boldsymbol{C}^{-1}\boldsymbol{H}'(\boldsymbol{H}\boldsymbol{C}^{-1}\boldsymbol{H}')^{-1}\boldsymbol{H}\boldsymbol{C}^{-1}\boldsymbol{W}\right] \\
&\quad \times\Bigg\{(q-2)E^{(1)}[\chi_{q+2}^{-4}(\Delta^2)] + \Bigg[1 \\
&\quad - \frac{(q+2)\boldsymbol{\delta}'\boldsymbol{W}^2\boldsymbol{\delta}}{2\sigma_\varepsilon^2\Delta^2 tr[\boldsymbol{W}\boldsymbol{C}^{-1}\boldsymbol{H}'(\boldsymbol{H}\boldsymbol{C}^{-1}\boldsymbol{H}')^{-1}\boldsymbol{H}\boldsymbol{C}^{-1}\boldsymbol{W}]}\Bigg] \\
&\quad \times(2\Delta^2)E^{(2)}[\chi_{q+4}^{-4}(\Delta^2)]\Bigg\} \\
&\quad -2qdk\boldsymbol{\delta}'\boldsymbol{W}\boldsymbol{C}_{(k)}^{-1}\boldsymbol{\beta}E^{(2)}[\chi_{q+2}^{-2}(\Delta^2)].
\end{aligned} \tag{8.4.5}
$$

For nonpositive bilinear form $\boldsymbol{\delta}'WC_{(k)}^{-1}\boldsymbol{\beta}$, the R.H.S. of (8.4.5) is non-negative if and only if (iff)

$$
\frac{tr\left(\boldsymbol{W}^2\boldsymbol{C}^{-1}\boldsymbol{H}'(\boldsymbol{H}\boldsymbol{C}^{-1}\boldsymbol{H}')^{-1}\boldsymbol{H}\boldsymbol{C}^{-1}\right)}{Ch_{\max}\left(\boldsymbol{W}^2\boldsymbol{C}^{-1}\boldsymbol{H}'(\boldsymbol{H}\boldsymbol{C}^{-1}\boldsymbol{H}')^{-1}\boldsymbol{H}\boldsymbol{C}^{-1}\right)} \geq \frac{q+2}{2}. \tag{8.4.6}
$$

Thus, if the bilinear form is nonpositive, then $\hat{\boldsymbol{\beta}}_n^{S}(k)$ dominates uniformly.

Next, consider the risk difference given by

$$
\begin{aligned}
&R_4(\hat{\boldsymbol{\beta}}_n^{S}(k);\boldsymbol{\beta}) - R_5(\hat{\boldsymbol{\beta}}_n^{S+}(k);\boldsymbol{\beta}) \\
= \;&\sigma_e^2\{tr[\boldsymbol{W}\boldsymbol{C}^{-1}\boldsymbol{H}'(\boldsymbol{H}\boldsymbol{C}^{-1}\boldsymbol{H}')^{-1}\boldsymbol{H}\boldsymbol{C}^{-1}\boldsymbol{W}] \\
&\times E^{(1)}\left[\left(1-\frac{qd}{q+2}F_{q+2,n-p}^{-1}(\Delta^2)\right)^2 I\left(F_{q+2,n-p}(\Delta^2) \leq \frac{qd}{q+2}\right)\right] \\
&+\frac{\boldsymbol{\delta}'\boldsymbol{W}^2\boldsymbol{\delta}}{\sigma_\varepsilon^2}
\end{aligned}
$$

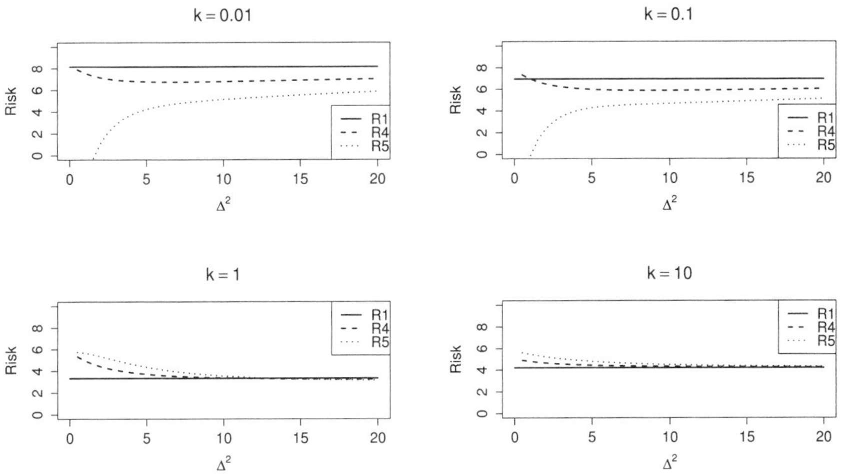

Figure 8.1 Risk of PRSRRE, SRRE and URRE

$$
\begin{aligned}
&\times E^{(2)}\left[\left(1-\frac{qd}{q+4}F_{q+4,n-p}^{-1}(\Delta^2)\right)^2 I\left(F_{q+4,n-p}(\Delta^2)\le\frac{qd}{q+4}\right)\right]\Bigg\}\\
&+2\boldsymbol{\delta}'\boldsymbol{W}^2\boldsymbol{\delta}\\
&\times\quad E^{(2)}\left[\left(\frac{qd}{q+2}F_{q+2,n-p}^{-1}(\Delta^2)-1\right)I\left(F_{q+2,n-p}(\Delta^2)\le\frac{qd}{q+2}\right)\right],\\
&+2k\boldsymbol{\delta}'\boldsymbol{W}\boldsymbol{C}_{(k)}^{-1}\boldsymbol{\beta}\\
&\times\quad E^{(2)}\left[\left(\frac{qd}{q+2}F_{q+2,n-p}^{-1}(\Delta^2)-1\right)I\left(F_{q+2,n-p}(\Delta^2)\le\frac{qd}{q+2}\right)\right].
\end{aligned}
\tag{8.4.7}
$$

If the bilinear form is non-negative, then R.H.S is non-negative. Hence the R.H.S. of (8.4.7) is non-negative and $\hat{\boldsymbol{\beta}}_n^{S+}(k)$ dominates uniformly over $\hat{\boldsymbol{\beta}}_n^{S}(k)$ for all $(\Delta^2, k) \in (0,\infty)\times(0,\infty)$.
Combining both results, we obtain

$$
R_5(\hat{\boldsymbol{\beta}}_n^{S+}(k);\boldsymbol{\beta}) \le R_4(\hat{\boldsymbol{\beta}}_n^{S}(k);\boldsymbol{\beta}) \le R_1(\tilde{\boldsymbol{\beta}}_n^{(k)};\boldsymbol{\beta}).
$$

For the nonpositive bilinear form, the uniform dominance results hold.

8.4.2 Comparison between $\tilde{\boldsymbol{\beta}}_n(\boldsymbol{k})$ and $\hat{\boldsymbol{\beta}}_n^{PT}(\boldsymbol{k})$

In this case, the risk difference is given by

$$\begin{aligned}R_1(\tilde{\boldsymbol{\beta}}_n(k);\boldsymbol{\beta}) - R_3(\hat{\boldsymbol{\beta}}_n^{PT}(k);\boldsymbol{\beta}) = \\ &\sigma_{\varepsilon}^2 tr[\boldsymbol{W}\boldsymbol{C}^{-1}\boldsymbol{H}'(\boldsymbol{H}\boldsymbol{C}^{-1}\boldsymbol{H}')^{-1}\boldsymbol{H}\boldsymbol{C}^{-1}\boldsymbol{W}]G_{q+2,m}^{(1)}(l_\gamma,\Delta^2) \\ &-\boldsymbol{\delta}'\boldsymbol{W}^2\boldsymbol{\delta}\left[2G_{q+2,m}^{(2)}(l_\gamma,\Delta^2) - G_{q+4,m}^{(2)}(l_\gamma,\Delta^2)\right] \\ &-2k\boldsymbol{\delta}'\boldsymbol{W}\boldsymbol{C}_{(k)}^{-1}\boldsymbol{\beta}G_{q+2,m}^{(2)}(l_\gamma,\Delta^2).\end{aligned} \quad (8.4.8)$$

For the nonpositive bilinear form $\boldsymbol{\delta}'WC_{(k)}^{-1}\boldsymbol{\beta}$, the R.H.S. of (8.4.8) is non-negative iff

$$\boldsymbol{\delta}'\boldsymbol{W}^2\boldsymbol{\delta} \le \frac{G_{q+2,m}^{(1)}(l_\gamma,\Delta^2)}{\left[2G_{q+2,m}^{(2)}(l_\gamma,\Delta^2) - G_{q+4,m}^{(2)}(l_\gamma,\Delta^2)\right]} \times \sigma_{\varepsilon}^2 tr[\boldsymbol{W}\boldsymbol{C}^{-1}\boldsymbol{H}'(\boldsymbol{H}\boldsymbol{C}^{-1}\boldsymbol{H}')^{-1}\boldsymbol{H}\boldsymbol{C}^{-1}\boldsymbol{W}]. \quad (8.4.9)$$

To check the accuracy of the proposed result of risk comparisons in the above, we display the graphs of the risk functions based on Δ^2 and different parameter values $n = 30$, $q = 3$, $p = 5$, and $k = 0.1, 0.5, 1, 10$. In this respect, for ease of computation, we assume $\boldsymbol{C} = \boldsymbol{X}'\boldsymbol{X} = I_p$,

$$\boldsymbol{H} = \begin{bmatrix} 1 & 0 & 0 & 0 & 0 \\ 0 & 1 & 0 & 0 & 0 \\ 0 & 0 & 1 & 0 & 0 \end{bmatrix}, \quad \boldsymbol{h} = [0,0,0]'.$$

The graphs of risk functions in Figure 8.3 clearly confirm the proposed results in sections 8.4.1 and 8.4.2.

8.4.3 Comparison between $\tilde{\boldsymbol{\beta}}_n(\boldsymbol{k})$ and $\tilde{\boldsymbol{\beta}}_n$

Theorem 8.4.1. *There always exists a $k > 0$ in range $0 < k < k^* = \frac{\sigma_{\varepsilon}^2}{\alpha_{\max}^2}$ such that the URE has a smaller risk value than the UE.*

Proof. Using Theorem 8.3.2 and equations (8.4.2) and (8.4.3), we obtain

$$\begin{aligned}R_1(\tilde{\boldsymbol{\beta}}_n(k);\boldsymbol{\beta}) &= \sigma_{\varepsilon}^2\sum_{i=1}^{p}\frac{\lambda_i}{(\lambda_i+k)^2} + k^2\sum_{i=1}^{p}\frac{\alpha_i^2}{(\lambda_i+k)^2} \\ &= \sigma_{\varepsilon}^2\sum_{i=1}^{p}\frac{\lambda_i}{(\lambda_i+k)^2} + \sum_{i=1}^{p}\frac{\alpha_i^2}{(1+\frac{\lambda_i}{k})^2}.\end{aligned} \quad (8.4.10)$$

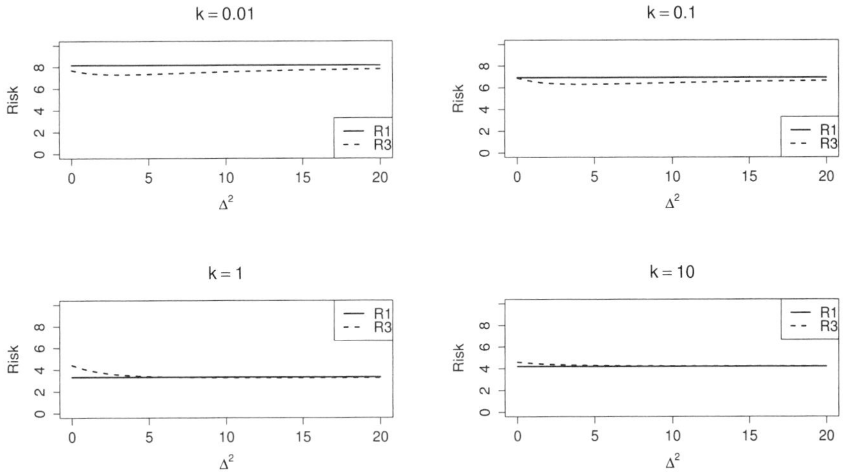

Figure 8.2 Risk of PTRRE and URRE

It is obvious that for $k = 0$, the first and second terms are equal to $\sigma_\varepsilon^2 \sum_{i=1}^{p} \frac{1}{\lambda_i}$ and zero, respectively. The first term is a continuous, monotonically decreasing function of k and its derivation w.r.t. k approaches $-\infty$ as $k \longrightarrow 0^+$ and $\lambda_p \longrightarrow 0$. The second term is also a continuous, monotonically increasing function of k and its derivative tends to zero as $k \longrightarrow 0^+$. We note that the second term approaches $\boldsymbol{\beta}'\boldsymbol{\beta}$ as $k \longrightarrow \infty$. Differentiating w.r.t. k, we get

$$\frac{\partial R_1(\tilde{\boldsymbol{\beta}}_n(k); \boldsymbol{\beta})}{\partial k} = 2\sum_{i=1}^{p} \frac{\lambda_i}{(\lambda_i + k)^3}(k\alpha_i^2 - \sigma_\varepsilon^2). \tag{8.4.11}$$

Thus, a sufficient condition for (8.4.11) to be negative is that $0 < k < k^*$ where

$$k^* = \frac{\sigma_\varepsilon^2}{\max_{1 \le i \le p} \alpha_i^2} = \frac{\sigma_\varepsilon^2}{\alpha_{\max}^2}, \tag{8.4.12}$$

where $\alpha_{\max}$ is the largest element of $\boldsymbol{\alpha}$. Hence, the theorem is proved.

8.4.4 Comparison between $\hat{\beta}_n(k)$ and $\hat{\beta}_n$

We consider two cases, namely, under H_0 and under H_A.

Theorem 8.4.2. *There always exists a $k > 0$ in range $0 < k < k_1^*$ such that RRRE has smaller risk value than RRE under $H_0 : \boldsymbol{H}\boldsymbol{\beta} = \boldsymbol{h}$.*

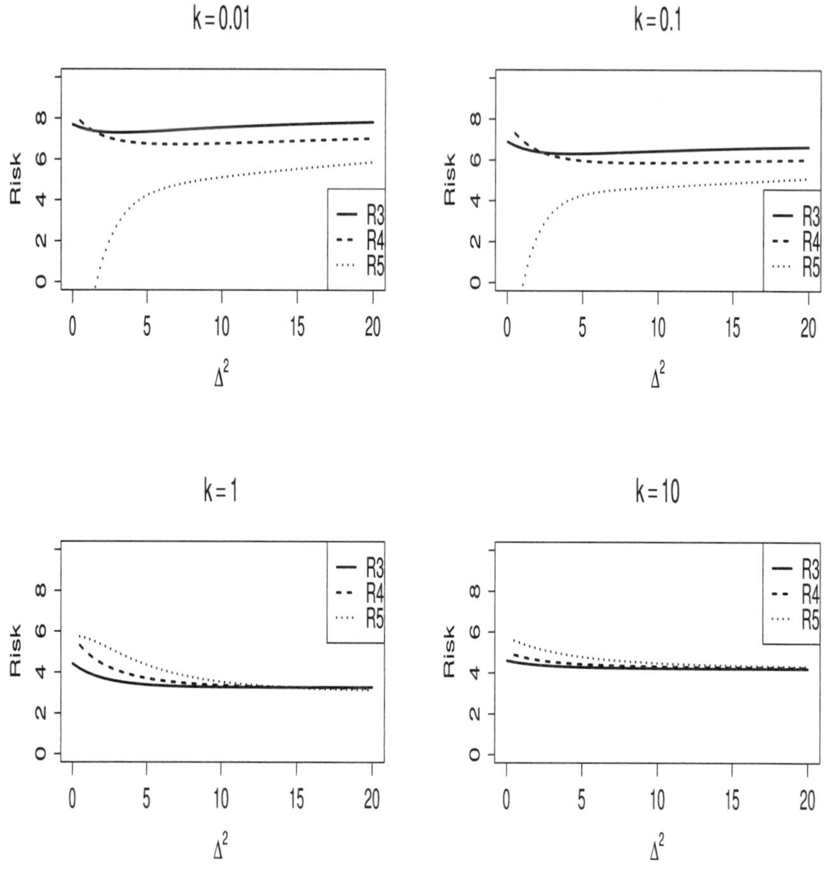

Figure 8.3 Risk of estimators

Proof. Using (8.3.11) and (8.4.2)-(8.4.4), we get

$$\begin{aligned} R_2(\hat{\boldsymbol{\beta}}_n(k);\boldsymbol{\beta}) &= \sigma_\varepsilon^2\sum_{i=1}^p\frac{\lambda_i}{(\lambda_i+k)^2}-\sigma_\varepsilon^2\sum_{i=1}^p\frac{h_{ii}^*}{(\lambda_i+k)^2}+k^2\sum_{i=1}^p\frac{\alpha_i^2}{(\lambda_i+k)^2} \\ &= \sigma_\varepsilon^2\sum_{i=1}^p\frac{\lambda_i-h_{ii}^*}{(\lambda_i+k)^2}+\sum_{i=1}^p\frac{\alpha_i^2}{(1+\frac{\lambda_i}{k})^2} \end{aligned} \quad (8.4.13)$$

following Theorem 8.4.1 and differentiating w.r.t. k, we find

$$\frac{\partial R_2(\hat{\boldsymbol{\beta}}_n(k);\boldsymbol{\beta})}{\partial k}=2\sum_{i=1}^p\frac{[k\alpha_i^2\lambda_i-\sigma_\varepsilon^2(\lambda_i-h_{ii}^*)]}{(\lambda_i+k)^2}. \quad (8.4.14)$$

It is clear from (8.4.14) that a sufficient condition for the first-order derivative of $R_2(\hat{\boldsymbol{\beta}}_n(k);\boldsymbol{\beta})$ to be negative is that $0 < k < k_1^*$ where

$$k_1^* = \frac{\sigma_{\boldsymbol{\varepsilon}}^2 \min_{1\le i\le p}(\lambda_i - h_{ii}^*)}{\max_{1\le i\le p}\alpha_i^2\lambda_i}. \tag{8.4.15}$$

It is obvious that $k_1^* \le k^*$ since $\frac{\lambda_i - h_{ii}^*}{\lambda_i} \le 1$ for all h_{ii}^* and $\lambda_i > 0, i = 1, ..., p$. Hence, we conclude that the range of values of k for the dominance of $\hat{\boldsymbol{\beta}}_n(k)$ over $\hat{\boldsymbol{\beta}}_n$ is smaller than the dominance of $\tilde{\boldsymbol{\beta}}_n(k)$ over $\tilde{\boldsymbol{\beta}}_n$.

Corollary 8.4.2.1. *The RRRE has smaller risk value than URRE under $H_0 : \boldsymbol{H\beta} = \boldsymbol{h}$.*

Proof. From (8.4.10) and (8.4.13), the result follows.

Theorem 8.4.3. *The RRRE is superior to the RRE for $0 < k < k_2^*$ under $H_A : \boldsymbol{H\beta} \neq \boldsymbol{h}$.*

Proof. Using (8.3.11) and (8.4.2)-(8.4.4), we get

$$R_2(\hat{\boldsymbol{\beta}}_n(k);\boldsymbol{\beta}) = \sum_{i=1}^{p} \frac{1}{(\lambda_i + k)^2}\left[\sigma_{\boldsymbol{\varepsilon}}^2(\lambda_i - h_{ii}^*) + k^2\alpha_i^2 + \lambda_i^2\Delta_i^2 + 2k\alpha_i\lambda_i\Delta_i^2\right], \tag{8.4.16}$$

where $\boldsymbol{\delta}^* = \boldsymbol{\Gamma}'\boldsymbol{\delta}$. Now, differentiating the above expression in (8.4.16) with respect to k, we find

$$\frac{\partial R_2(\hat{\boldsymbol{\beta}}_n(k);\boldsymbol{\beta})}{\partial k} = 2\sum_{i=1}^{p}\frac{1}{(\lambda_i + k)^2}\{k\lambda_i\alpha_i(\alpha_i - \Delta_i^2) - [\sigma_{\boldsymbol{\varepsilon}}^2(\lambda_i - h_{ii}^*) - \lambda_i^2\Delta_i^2(\alpha_i - \Delta_i^2)]\}. \tag{8.4.17}$$

Hence, a sufficient condition for (8.4.17) to be negative is that $0 < k < k_2^*$ where

$$k_2^* = \frac{\min_{1\le i\le p}[\sigma_{\boldsymbol{\varepsilon}}^2(\lambda_i - h_{ii}^*) - \lambda_i^2\Delta_i^2(\alpha_i - \Delta_i^2)]}{\max_{1\le i\le p}\lambda_i\alpha_i(\alpha_i - \Delta_i^2)}. \tag{8.4.18}$$

Hence, the proof is complete.

Corollary 8.4.3.1. *A sufficient condition for the RRRE to have smaller risk value than the URE is that there exists a value k such that $0 < k < k_1$ where k_1 is given by*

$$k_1 = \frac{\min_{1\le i\le p}(\sigma_{\boldsymbol{\varepsilon}}^2 h_{ii}^* - \lambda_i^2\delta_i^{*2})}{\max_{1\le i\le p}(2\alpha_i\lambda_i\Delta_i^2)}. \tag{8.4.19}$$

Proof. Using (8.4.10) and (8.4.16), we find

$$R_1(\tilde{\boldsymbol{\beta}}_n(k);\boldsymbol{\beta}) - R_2(\hat{\boldsymbol{\beta}}_n(k);\boldsymbol{\beta}) = \sum_{i=1}^{p} \frac{(\sigma_{\varepsilon}^2 h_{ii}^{*} - \lambda_i^2\delta_i^{*2} - 2k\alpha_i\lambda_i\Delta_i^2)}{(\lambda_i + k)^2}. \quad (8.4.20)$$

This difference is positive for all $k < k_1$, which is given in (8.4.19).

8.4.5 Comparison between $\hat{\boldsymbol{\beta}}_n^{PT}$ and $\hat{\boldsymbol{\beta}}_n^{PT}(k)$

We consider two cases, namely, under H_0 and under H_A.
Case 1. Under $H_0 : \boldsymbol{H\beta} = \boldsymbol{h}$ (i.e. $\Delta^2 = 0$).

Theorem 8.4.4. *There always exists a $k > 0$ in the region $0 < k < k_3^*$ such that PTRRE has smaller risk value than PTRE under $H_0 : \boldsymbol{H\beta} = \boldsymbol{h}$.*

Proof. Using (8.3.13) and (8.4.2)-(8.4.4), we find

$$R_3(\hat{\boldsymbol{\beta}}_n^{PT}(k);\boldsymbol{\beta}) = \sigma_{\varepsilon}^2 \sum_{i=1}^{p} \frac{\lambda_i - h_{ii}^{*} G_{q+2,n-p}^{(1)}(x',0)}{(\lambda_i + k)^2} + k^2 \sum_{i=1}^{p} \frac{\alpha_i^2}{(\lambda_i + k)^2}. \quad (8.4.21)$$

Differentiating w.r.t. k, we get

$$\frac{\partial R_3(\hat{\boldsymbol{\beta}}_n^{PT}(k);\boldsymbol{\beta})}{\partial k} = 2\sum_{i=1}^{p} \frac{1}{(\lambda_i + k)^3}\{k\alpha_i^2\lambda_i - \sigma_{\varepsilon}^2[\lambda_i - h_{ii}^{*} G_{q+2,n-p}^{(1)}(x',0)]\}. \quad (8.4.22)$$

Thus, a sufficient condition for (8.4.22) to be negative is that $0 < k < k_3^*$, where

$$k_3^* = \frac{\sigma_{\varepsilon}^2 \min_{1\le i\le p}[\lambda_i - h_{ii}^{*} G_{q+2,n-p}^{(1)}(x',0)]}{\max_{1\le i\le p} \alpha_i^2\lambda_i}. \quad (8.4.23)$$

Hence, the proof is complete.
We note that for $\alpha = 0, 1$, we obtain Theorems 8.4.1 and 8.4.2, respectively.

Corollary 8.4.4.1. *The PTRRE has smaller risk value than URRE under $H_0 : \boldsymbol{H\beta} = \boldsymbol{h}$.*

Proof Using (8.4.10) and (8.4.21), the result follows.
Case 2. Under $H_A : \boldsymbol{H\beta} \neq \boldsymbol{h}$ (i.e $\Delta^2 \neq 0$).

Theorem 8.4.5. *There always exists a $k > 0$ in the region $0 < k < k_4^*$ such that PTRRE has smaller risk value than PTRE under $H_A : \boldsymbol{H\beta} \neq \boldsymbol{h}$.*

Proof. Using (8.3.13) and (8.4.2)-(8.4.4), the risk expression is given by

$$R_3(\hat{\boldsymbol{\beta}}_n^{PT}(k);\boldsymbol{\beta}) = \sum_{i=1}^{p} \frac{1}{(\lambda_i+k)^2}\Big\{\sigma_{\boldsymbol{\varepsilon}}^2\left[\lambda_i - h_{ii}^* G_{q+2,n-p}^{(1)}(x',\Delta_*^2)\right]$$
$$+\lambda_i^2\delta_i^{*2}\left[2G_{q+2,n-p}^{(2)}(x',\Delta_*^2) - G_{q+4,n-p}^{(2)}(x',\Delta_*^2)\right]$$
$$+k^2\alpha_i^2 + 2k\alpha_i\lambda_i\Delta_i^2 G_{q+2,n-p}^{(2)}(x',\Delta_*^2)\Big\}. \quad (8.4.24)$$

Differentiating w.r.t. k, we obtain

$$\frac{\partial R_3(\hat{\boldsymbol{\beta}}_n^{PT}(k);\boldsymbol{\beta})}{\partial k} = 2\sum_{i=1}^{p} \frac{1}{(\lambda_i+k)^3}\Big\{k\alpha_i\lambda_i\left[\alpha_i - \Delta_i^2 G_{q+2,n-p}^{(1)}(x',\Delta_*^2)\right]$$
$$-\Big[\sigma_{\boldsymbol{\varepsilon}}^2\left(\lambda_i - h_{ii}^* G_{q+2,n-p}^{(1)}(x',\Delta_*^2)\right)$$
$$+\lambda_i^2\delta_i^{*2}\left(2G_{q+2,n-p}^{(2)}(x',\Delta_*^2) - G_{q+4,n-p}^{(2)}(x',\Delta_*^2)\right)$$
$$-\alpha\lambda_i^2\Delta_i^2 G_{q+2,n-p}^{(2)}(x',\Delta_*^2)\Big]\Big\}. \quad (8.4.25)$$

Hence, a sufficient condition for (8.4.25) to be negative is that $0 < k < k_4^*$, where

$$k_4^* = \min_{1\le i\le p}\Big[\sigma_{\boldsymbol{\varepsilon}}^2\left(\lambda_i - h_{ii}^* G_{q+2,n-p}^{(1)}(x',\Delta_*^2)\right)$$
$$+\lambda_i^2\delta_i^{*2}\left(2G_{q+2,n-p}^{(2)}(x',\Delta_*^2) - G_{q+4,n-p}^{(2)}(x',\Delta_*^2)\right)$$
$$-\alpha_i\lambda_i^2\Delta_i^2 G_{q+2,n-p}^{(2)}(x',\Delta^2)\Big] \Big/ \max_{1\le i\le p}\alpha_i\lambda_i\left(\alpha_i - \Delta_i^2 G_{q+2,n-p}^{(2)}(x',\Delta_*^2)\right). \quad (8.4.26)$$

Thus, $\hat{\boldsymbol{\beta}}_n^{PT}(k)$ has risk value less than the risk of $\hat{\boldsymbol{\beta}}_n^{PT}$ for $0 < k < k_4^*$.

8.4.5.1 Comparison of $\hat{\boldsymbol{\beta}}_n^{PT}(k)$ with $\tilde{\boldsymbol{\beta}}_n(k)$ and $\hat{\boldsymbol{\beta}}_n(k)$

We shall now proceed to compare the performance of $\hat{\boldsymbol{\beta}}_n^{PT}(k)$ against that of $\tilde{\boldsymbol{\beta}}_n(k)$.

Theorem 8.4.6. *A sufficient condition for PTRRE to have risk value less than or equal to that of URRE is that there exists a value of k such that* $0 < k < k_2$ *where* k_2 *is given by* (8.4.28).

Proof Using (8.4.10) and (8.4.24), we obtain

$$R_1(\tilde{\boldsymbol{\beta}}_n(k);\boldsymbol{\beta}) - R_3(\hat{\boldsymbol{\beta}}_n^{PT}(k);\boldsymbol{\beta}) = \sum_{i=1}^{p}\frac{1}{(\lambda_i+k)^2}\Big\{\sigma_{\boldsymbol{\varepsilon}}^2 h_{ii}^* G_{q+2,n-p}^{(1)}(x',\Delta_*^2)$$

$$-\lambda_i^2\delta_i^{*2}\Big[2G^{(2)}_{q+2,n-p}(x',\Delta_*^2) \\ -G^{(2)}_{q+4,n-p}(x',\Delta_*^2)\Big] \\ -2k\alpha_i\lambda_i\Delta_i^2 G^{(2)}_{q+2,n-p}(x',\Delta_*^2)\Big\}. \quad (8.4.27)$$

This difference is nonnegative (≥ 0), whenever

$$k_2 = \frac{\min_{1\leq i\leq p}\left\{\begin{array}{l}\sigma_{\boldsymbol{\varepsilon}}^2 h_{ii}^* G^{(1)}_{q+2,n-p}(x',\Delta_*^2) \\ -\lambda_i^2\delta_i^{*2}\Big[2G^{(2)}_{q+2,n-p}(x',\Delta_*^2) - G^{(2)}_{q+4,n-p}(x',\Delta_*^2)\Big]\end{array}\right\}}{\max_{1\leq i\leq p}\Big[2\alpha_i\lambda_i\Delta_i^2 G^{(2)}_{q+2,n-p}(x',\Delta_*^2)\Big]} \quad (8.4.28)$$

Thus, the theorem is proved.

Similarly, we compare PTRRE to RRRE.

Theorem 8.4.7. *A sufficient condition for PTRRE to have risk value less than or equal to that of RRRE is that there exists a value of k such that $0 < k_3 < k$ where k_3 is given by* (8.4.30).

Proof. Using (8.4.16) and (8.4.24), we find

$$\begin{aligned} &R_3(\hat{\boldsymbol{\beta}}_n^{PT}(k);\boldsymbol{\beta}) - R_1(\hat{\boldsymbol{\beta}}_n(k);\boldsymbol{\beta}) \\ =\ &\sum_{i=1}^p \frac{1}{(\lambda_i+k)^2}\Big\{\sigma_{\boldsymbol{\varepsilon}}^2 h_{ii}^*[1-G^{(1)}_{q+2,n-p}(x',\Delta_*^2)] \\ &-\lambda_i^2\delta^{*2}\Big[1-2G^{(2)}_{q+2,n-p}(x',\Delta_*^2)+G^{(2)}_{q+4,n-p}(x',\Delta_*^2)\Big] \\ &-2k\alpha_i\lambda_i\Delta_i^2[1-G^{(2)}_{q+2,n-p}(x',\Delta_*^2)]\Big\}. \end{aligned} \quad (8.4.29)$$

This difference is negative (< 0) whenever

$$\begin{aligned} k_3^* = \max_{1\leq i\leq p}&\Big\{\sigma_{\boldsymbol{\varepsilon}}^2 h_{ii}^*[1-G^{(1)}_{q+2,n-p}(x',\Delta_*^2)] - \lambda_i^2\delta^{*2}\Big[1-2G^{(2)}_{q+2,n-p}(x',\Delta_*^2) \\ &+G^{(2)}_{q+4,n-p}(x',\Delta_*^2)\Big]\Big\}\Big/\min_{1\leq i\leq p}\Big\{2\alpha_i\lambda_i\Delta_i^2[1-G^{(2)}_{q+2,n-p}(x',\Delta_*^2)]\Big\}. \end{aligned} \quad (8.4.30)$$

Hence, the theorem is proved.

We note that under $H_0 : \boldsymbol{H\beta} = \boldsymbol{h}$, we have

$$R_3(\hat{\boldsymbol{\beta}}_n^{PT}(k);\boldsymbol{\beta}) - R_2(\hat{\boldsymbol{\beta}}_n(k);\boldsymbol{\beta}) = \sigma_{\boldsymbol{\varepsilon}}^2[1-G^{(1)}_{q+2,n-p}(x',0)]\sum_{i=1}^p \frac{h_{ii}^*}{(\lambda_i+k)^2} \geq 0.$$

Since

$$G^{(1)}_{q+2,n-p}(x',0) = F_{q+2,n-p}(\ell^{*(1)}_{\alpha},0) = 1-\alpha, \ell^{*(1)}_{\alpha} = \frac{qF_{\alpha}}{q+2}$$

and $\Delta_i^2 = 0, 1, ..., p$. Thus, $\hat{\boldsymbol{\beta}}_n(k)$ is superior to $\hat{\boldsymbol{\beta}}_n^{PT}(k)$ under $H_0 : \boldsymbol{H\beta} = \boldsymbol{h}$.

8.4.6 Comparison between $\hat{\boldsymbol{\beta}}_n^S$ and $\hat{\boldsymbol{\beta}}_n^S(k)$

Theorem 8.4.8. *There always exists a $k > 0$ in the region $0 < k < k_5^*$ such that SRRE has smaller risk value than SRE under $H_0 : \boldsymbol{H\beta} = \boldsymbol{h}$.*

Proof. Using (8.3.15) and (8.4.2)-(8.4.4), we obtain

$$R_4(\hat{\boldsymbol{\beta}}_n^S(k);\boldsymbol{\beta}) = \sum_{i=1}^{p} \frac{1}{(\lambda_i+k)^2}\{\sigma_{\varepsilon}^2(\lambda_i - dh_{ii}^*) + k^2\alpha_i^2\}. \tag{8.4.31}$$

Differentiating with respect to k, we find

$$\frac{\partial R_4(\hat{\boldsymbol{\beta}}_n^S(k);\boldsymbol{\beta})}{\partial k} = 2\sum_{i=1}^{p} \frac{1}{(\lambda_i+k)^3}\{k\alpha_i^2\lambda_i - \sigma_{\varepsilon}^2(\lambda_i - dh_{ii}^*)\}. \tag{8.4.32}$$

Thus, a sufficient condition for (8.4.32) to be negative is that $0 < k < k_5^*$, where

$$k_5^* = \frac{\sigma_{\varepsilon}^2 \min_{1\le i\le p}(\lambda_i - dh_{ii}^*)}{\max_{1\le i\le p}\alpha_i^2\lambda_i}. \tag{8.4.33}$$

Hence, the proof is complete.

Theorem 8.4.9. *There always exists a $k > 0$ in the region $0 < k < k_6^*$ such that SRRE has smaller risk value than SRE under $H_A : \boldsymbol{H\beta} \neq \boldsymbol{h}$.*

Proof. Using (8.3.15) and (8.4.2)-(8.4.4), the risk expression is given by

$$\begin{aligned} R_4(\hat{\boldsymbol{\beta}}_n^S(k);\boldsymbol{\beta}) &= \sum_{i=1}^{p} \frac{1}{(\lambda_i+k)^2}\Bigg\{\sigma_{\varepsilon}^2\Bigg\{\lambda_i - dqh_{ii}^*\Bigg[(q-2)E^{(1)}(\chi_{q+2}^{-4}(\Delta_*^2)) \\ &+\left(1-\frac{(q+2)\lambda_i^2\delta_i^{*2}}{2\sigma_{\varepsilon}^2\Delta_*^2h_{ii}^*}\right)(2\Delta_*^2)E^{(2)}(\chi_{q+4}^{-4}(\Delta_*^2))\Bigg]\Bigg\} \\ &+2qdk\alpha_i\lambda_i\Delta_i^2E^{(2)}(\chi_{q+2}^{-2}(\Delta_*^2)) + k^2\alpha_i^2\Bigg\}. \end{aligned} \tag{8.4.34}$$

Differentiating with respect to k, we obtain

$$\begin{aligned}\frac{\partial R_4(\hat{\boldsymbol{\beta}}_n^S(k);\boldsymbol{\beta})}{\partial k} &= 2\sum_{i=1}^{p}\frac{1}{(\lambda_i+k)^3}\Bigg\{k\lambda_i\alpha_i\left[\alpha_i - qd\Delta_i^2E^{(2)}(\chi_{q+2}^{-2}(\Delta_*^2))\right]\\ &-\sigma_\varepsilon^2\Bigg\{\lambda_i - dqh_{ii}^*\Bigg[(q-2)E^{(1)}(\chi_{q+2}^{-4}(\Delta_*^2))\\ &+\left(1-\frac{(q+2)\lambda_i^2\delta_i^{*2}}{2\sigma_\varepsilon^2\Delta_*^2h_{ii}^*}\right)(2\Delta_*^2)E^{(2)}(\chi_{q+4}^{-4}(\Delta_*^2))\Bigg]\Bigg\}\\ &-qd\alpha_i\lambda_i^2\Delta_i^2E^{(2)}(\chi_{q+2}^{-2}(\Delta_*^2))\Bigg\}. \end{aligned} \tag{8.4.35}$$

Hence, a sufficient condition for (8.4.35) to be negative is that $0 < k < k_6^*$, where

$$\begin{aligned}k_6^* &= \min_{1\le i\le p}\Bigg\{\sigma_\varepsilon^2\Bigg\{\lambda_i - dqh_{ii}^*\Bigg[(q-2)E^{(1)}(\chi_{q+2}^{-4}(\Delta_*^2))\\ &+\left(1-\frac{(q+2)\lambda_i^2\delta_i^{*2}}{2\sigma_\varepsilon^2\Delta_*^2h_{ii}^*}\right)(2\Delta_*^2)E^{(2)}(\chi_{q+4}^{-4}(\Delta_*^2))\Bigg]\Bigg\}\\ &+qd\alpha_i\lambda_i^2\Delta_i^2E^{(2)}(\chi_{q+2}^{-2}(\Delta_*^2))\Bigg\}\Big/\max_{1\le i\le p}\alpha_i\lambda_i\left[\alpha_i - qd\Delta_i^2E^{(2)}(\chi_{q+2}^{-2}(\Delta_*^2))\right].\end{aligned} \tag{8.4.36}$$

Thus, $\hat{\boldsymbol{\beta}}_n^S(k)$ has risk value less than the risk of $\hat{\boldsymbol{\beta}}_n^S$ for $0 < k < k_6^*$.

8.4.6.1 Comparison of $\hat{\boldsymbol{\beta}}_n^S(k)$ with $\tilde{\boldsymbol{\beta}}_n(k)$, $\hat{\boldsymbol{\beta}}_n(k)$ and $\hat{\boldsymbol{\beta}}_n^{PT}(k)$

Theorem 8.4.10. *A sufficient condition for SRRE to have risk value less than or equal to that of URRE is that there exists a value of k such that* $0 < k < k_4$ *where* k_4 *is given by* (8.4.38).

Proof. Using (8.4.10) and (8.4.34), we find

$$\begin{aligned}&R_1(\tilde{\boldsymbol{\beta}}_n(k);\boldsymbol{\beta}) - R_4(\hat{\boldsymbol{\beta}}_n^S(k);\boldsymbol{\beta})\\ &\quad= \sum_{i=1}^{p}\frac{1}{(\lambda_i+k)^2}\Bigg\{dq\sigma_\varepsilon^2h_{ii}^*\Bigg[(q-2)E^{(1)}(\chi_{q+2}^{-4}(\Delta_*^2))\\ &\quad+\left(1-\frac{(q+2)\lambda_i^2\delta_i^{*2}}{2\sigma_\varepsilon^2\Delta_*^2h_{ii}^*}\right)(2\Delta_*^2)E^{(2)}(\chi_{q+4}^{-4}(\Delta_*^2))\Bigg]\\ &\quad-2qdk\alpha_i\lambda_i\Delta_i^2E^{(i)}(\chi_{q+2}^{-2}(\Delta_*^2))\Bigg\}.\end{aligned} \tag{8.4.37}$$

This difference is non-negative (≥ 0) whenever

$$k_4 = \frac{\sigma_{\varepsilon}^2 \min_{1\leq i\leq p}\left\{h_{ii}^*\left[(q-2)E^{(1)}(\chi_{q+2}^{-4}(\Delta_*^2)) + \left(1 - \frac{(q+2)\lambda_i^2\delta_i^{*2}}{2\sigma_{\varepsilon}^2\Delta_*^2 h_{ii}^*}\right)(2\Delta_*^2)E^{(2)}(\chi_{q+4}^{-4}(\Delta_*^2))\right]\right\}}{max_{1\leq i\leq p}\left[2\alpha_i\lambda_i\Delta_i^2 E^{(2)}(\chi_{q+2}^{-2}(\Delta_*^2))\right]}. \tag{8.4.38}$$

Hence, the theorem is proved.

Theorem 8.4.11. *A sufficient condition for SRRE to have smaller risk value than RRRE is that* $0 < k_5 < k$ *where* k_5 *is given by* (8.4.35).

Proof. By making use of equations (8.4.16) and (8.4.34), we have

$$\begin{aligned} & R_4(\hat{\boldsymbol{\beta}}_n^S(k); \boldsymbol{\beta}) - R_2(\hat{\boldsymbol{\beta}}_n(k); \boldsymbol{\beta}) \\ &= \sum_{i=1}^p \frac{1}{(\lambda_i + k)^2}\left\{\sigma_{\varepsilon}^2 h_{ii}^*\left\{1 - qd\left[(q-2)E^{(1)}(\chi_{q+2}^{-4}(\Delta_*^2))\right.\right.\right. \\ &\quad \left.\left. + \left(1 - \frac{(q+2)\lambda_i^2\delta_i^{*2}}{2\sigma_{\varepsilon}^2\Delta^2 h_{ii}^*}\right)(2\Delta_*^2)E^{(2)}(\chi_{q+4}^{-4}(\Delta_*^2))\right]\right\} \\ &\quad \left. -\lambda_i^2\delta_i^{*2} - 2k\alpha_i\lambda_i\Delta_i^2\left[1 - qdE^{(2)}(\chi_{q+2}^{-2}(\Delta_*^2))\right]\right\}. \end{aligned} \tag{8.4.39}$$

This difference is nonpositive (≤ 0) whenever

$$\begin{aligned} k_5 &= \max_{1\leq i\leq p}\left\{\sigma_{\varepsilon}^2 h_{ii}^*\left\{1 - dq\left[(q-2)E^{(1)}(\chi_{q+2}^{-4}(\Delta_*^2))\right.\right.\right. \\ &\quad \left.\left.\left. + \left(1 - \frac{(q-2)\lambda_i^2\delta_i^{*2}}{2\sigma_{\varepsilon}^2\Delta_*^2 h_{ii}^*}\right)(2\Delta_*^2)E^{(2)}(\chi_{q+4}^{-4}(\Delta_*^2))\right]\right\} - \lambda_i^2\delta_i^{*2}\right\} \\ &\quad \Big/ \min_{1\leq i\leq p}\left\{2\alpha_i\lambda_i\Delta_i^2\left[1 - qdE^{(2)}(\chi_{q+2}^{-2}(\Delta_*^2))\right]\right\}. \end{aligned} \tag{8.4.40}$$

Thus, the theorem is proved.

We note that under $H_0 : \boldsymbol{H\beta} = \boldsymbol{h}$, we have

$$R_4(\hat{\boldsymbol{\beta}}_n^S(k); \boldsymbol{\beta}) - R_2(\hat{\boldsymbol{\beta}}_n(k); \boldsymbol{\beta}) = \sum_{i-1}^p \frac{(1-d)\sigma_{\varepsilon}^2 h_{ii}^*}{(\lambda_i + k)^2} \geq 0.$$

Thus, $\hat{\boldsymbol{\beta}}_n(k)$ is superior to $\hat{\boldsymbol{\beta}}_n^S(k)$ under $H_0 : \boldsymbol{H\beta} = \boldsymbol{h}$.

Theorem 8.4.12. *A sufficient condition for SRRE to have smaller risk value than PTRRE is that* $0 < k_6 < k$ *where* k_6 *is given by* (8.4.42).

Proof. Using (8.4.24) and (8.4.34), we get

$$R_4(\hat{\boldsymbol{\beta}}_n^S(k);\boldsymbol{\beta}) - R_3(\hat{\boldsymbol{\beta}}_n^{PT}(k);\boldsymbol{\beta}) =$$
$$\sum_{i=1}^{p} \frac{1}{(\lambda_i+k)^2}\Bigg\{\sigma_{\varepsilon}^2 h_{ii}^*\Bigg\{G_{q+2,n-p}^{(1)}(x',\Delta_*^2) - dq\Bigg[(q-2)E^{(1)}(\chi_{q+2}^{-4}(\Delta_*^2))$$
$$+\left(1-\frac{(q+2)\lambda_i^2\delta_i^{*2}}{2\sigma_{\varepsilon}^2\Delta_*^2 h_{ii}^*}\right)(2\Delta_*^2)E^{(2)}(\chi_{q+4}^{-4}(\Delta_*^2))\Bigg]\Bigg\} - \lambda_i^2\delta_i^{2*}\Bigg[2G_{q+2,n-p}^{(2)}(x',\Delta_*^2)$$
$$-G_{q+4,n-p}^{(2)}(x',\Delta_*^2)\Bigg] - 2k\alpha_i\Delta_i^2\lambda_i\left[G_{q+2,n-p}^{(2)}(x',\Delta_*^2) - qdE^{(2)}(\chi_{q+2}^{-2}(\Delta_*^2))\right]\Bigg\}. \tag{8.4.41}$$

This difference is nonpositive (≤ 0) whenever

$$\begin{aligned} k_6 \quad &= \max_{1\leq i\leq p}\Bigg\{\sigma_{\varepsilon}^2 h_{ii}^*\Bigg\{G_{q+2,n-p}^{(1)}(x',\Delta_*^2) - dq\Bigg[(q-2)E^{(1)}(\chi_{q+2}^{-4}(\Delta_*^2)) \\ &+\left(1-\frac{(q+2)\lambda_i^2\delta_i^{*2}}{2\sigma_{\varepsilon}^2\Delta_*^2 h_{ii}^*}\right)(2\Delta_*^2)E^{(2)}(\chi_{q+4}^{-4}(\Delta_*^2))\Bigg]\Bigg\} \\ &-\lambda_i^2\delta_i^{2*}\left[2G_{q+2,n-p}^{(2)}(x',\Delta_*^2) - G_{q+4,n-p}^{(2)}(x',\Delta_*^2)\right]\Bigg\} \\ &\Big/ \min_{1\leq i\leq p}\left\{2\alpha_i\lambda_i\Delta_i^2\left[G_{q+2,n-p}^{(2)}(x',\Delta_*^2) - qdE^{(2)}(\chi_{q+2}^{-2}(\Delta_*^2))\right]\right\}. \end{aligned} \tag{8.4.42}$$

Thus, the proof is complete.

We note that under $H_0: \boldsymbol{H\beta} = \boldsymbol{h}$, we have

$$R_4(\hat{\boldsymbol{\beta}}_n^S(k);\boldsymbol{\beta}) - R_3(\hat{\boldsymbol{\beta}}_n^{PT}(k);\boldsymbol{\beta}) = \sum_{i=1}^{p}\frac{\sigma_{\varepsilon}^2[G_{q+2,n-p}^{(1)}(x',0)-d]h_{ii}^*}{(\lambda_i+k)^2},$$

where $G_{q+2,n-p}^{(1)}(x',0) = F_{q+2,n-p}(\ell_\alpha^{*(1)},0) = 1-\alpha$, $\ell_\alpha^{*(1)} = \frac{qF_\alpha}{q+2}$. Thus, the risk of $\hat{\boldsymbol{\beta}}_n^S(k)$ is smaller than that of the risk of $\hat{\boldsymbol{\beta}}_n^{PT}(k)$ when the critical value $\ell_\alpha^{(1)}$ satisfies the inequality

$$\left\{\alpha : \ell_\alpha^{(1)} \leq F_{q+2,n-p}^{-1}(d,0)\right\}, \tag{8.4.43}$$

otherwise $\hat{\boldsymbol{\beta}}_n^{PT}(k)$ is smaller than $\hat{\boldsymbol{\beta}}_n^S(k)$.

8.4.7 Comparison between $\hat{\boldsymbol{\beta}}_n^{S+}$ and $\hat{\boldsymbol{\beta}}_n^{S+}(\boldsymbol{k})$

Theorem 8.4.13. *There always exists a $k > 0$ in the region $0 < k < k_7^*$ such that SRRE has smaller risk value than PRSRE under $H_0: \boldsymbol{H\beta} = \boldsymbol{h}$.*

Proof. Using (8.3.17), (8.4.31), and (8.4.2)–(8.4.4), the risk expression is given by

$$R_4(\hat{\boldsymbol{\beta}}_n^S(k);\boldsymbol{\beta}) = \sum_{i=1}^{p}\frac{1}{(\lambda_i+k)^2}\Bigg\{\sigma_{\varepsilon}^2(\lambda_i - dh_{ii}^*) + k^2\alpha_i^2 - \sigma_{\varepsilon}^2 h_{ii}^* \times E^{(1)}\left[\left(1-\frac{qd}{q+2}F_{q+2,n-p}^{-1}(0)\right)^2 I\left(F_{q+2,n-p}(0)\le\frac{qd}{q+2}\right)\right]\Bigg\}. \tag{8.4.44}$$

Differentiating with respect to k, we obtain

$$\frac{\partial R_4(\hat{\boldsymbol{\beta}}_n^S(k);\boldsymbol{\beta}}{\partial k} = 2\sum_{i=1}^{p}\frac{1}{\lambda_i+k^3}\Bigg\{k\alpha_i^2\lambda_i - \sigma_{\varepsilon}^2\Bigg\{\lambda_i - dh_{ii}^* - h_{ii}^* \times E^{(1)}\left[\left(1-\frac{qd}{q+2}F_{q+2,n-p}^{-1}(0)\right)^2 I\left(F_{q+2,n-p}(0)\le\frac{qd}{q+2}\right)\right]\Bigg\}\Bigg\}. \tag{8.4.45}$$

Thus, a sufficient condition for (8.4.45) to be negative is that $0<k<k_7^*$ where

$$k_7^* = \frac{\sigma_{\varepsilon}^2\min_{1\le i\le p}\left\{\lambda_i - dh_{ii}^*E^{(1)}\left[\left(1-\frac{qd}{q+2}F_{q+2,n-p}^{-1}(0)\right)^2 I\left(F_{q+2,n-p}(0)\le\frac{qd}{q+2}\right)\right]\right\}}{\max_{1\le i\le p}\alpha_i^2\lambda_i}. \tag{8.4.46}$$

Hence, the proof is complete.

Theorem 8.4.14. *There always exists a $k>0$ in the region $0<k<k_8^*$ such that PRSRRE has smaller risk value than PRSRE under $H_A: \boldsymbol{H\beta}\neq\boldsymbol{h}$.*

Proof. Using (8.3.17) and (8.4.2)-(8.4.4), we obtain

$$\begin{aligned}
R_5(\hat{\boldsymbol{\beta}}_n^{S+}(k);\boldsymbol{\beta}) &= R_4(\hat{\boldsymbol{\beta}}_n^S(k);\boldsymbol{\beta}) - \sum_{i=1}^{p}\frac{1}{(\lambda_i+k)^2}\Bigg\{\sigma_{\varepsilon}^2\Bigg\{h_{ii}^*E^{(1)}\left[\left(1-\frac{qd}{q+2}F_{q+2,n-p}^{-1}(\Delta_*^2)\right)^2\right.\\
&\times I\left.\left(F_{q+2,n-p}(\Delta_*^2)\le\frac{qd}{q+2}\right)\right] + \frac{\lambda_i^2\delta_i^{*2}}{\sigma_{\varepsilon}^2}E^{(2)}\left[\left(1-\frac{qd}{q+4}F_{q+4,n-p}^{-1}(\Delta_*^2)\right)^2\right.\\
&\times I\left.\left(F_{q+4,n-p}(\Delta^2)\le\frac{qd}{q+4}\right)\right]\Bigg\} + 2\lambda_i^2\delta_i^{*2}\\
&\times E^{(2)}\left[\left(\frac{qd}{q+2}F_{q+2,n-p}^{-1}(\Delta_*^2)-1\right)I\left(F_{q+2,n-p}(\Delta_*^2)\le\frac{qd}{q+2}\right)\right]\\
&+2k\alpha_i\lambda_i\Delta_i^2\\
&\times E^{(2)}\left[\left(\frac{qd}{q+2}F_{q+2,n-p}^{-1}(\Delta_*^2)-1\right)I\left(F_{q+2,n-p}(\Delta_*^2)\le\frac{qd}{q+2}\right)\right]\Bigg\}.
\end{aligned} \tag{8.4.47}$$

Differentiating with respect to k, we find

$$\begin{aligned}\frac{\partial R_5(\hat{\boldsymbol{\beta}}_n^{S+}(k);\boldsymbol{\beta})}{\partial k} &= \frac{\partial R_4(\hat{\boldsymbol{\beta}}_n^{S}(k);\boldsymbol{\beta})}{\partial k} + 2\sum_{i=1}^{p}\frac{1}{(\lambda_i+k)^3}\Big\{k\alpha_i\lambda_i\Delta_i^2\\ &\times E^{(2)}\left[\left(\frac{qd}{q+2}F_{q+2,n-p}^{-1}(\Delta_*^2)-1\right)I\left(F_{q+2,n-p}(\Delta_*^2)\le\frac{qd}{q+2}\right)\right]\\ &+\sigma_{\varepsilon}^2\Big\{h_{ii}^{*}E^{(1)}\left[\left(1-\frac{qd}{q+2}F_{q+2,n-p}^{-1}(\Delta_*^2)\right)^2 I\left(F_{q+2,n-p}(\Delta_*^2)\le\frac{qd}{q+2}\right)\right]\\ &+\frac{\lambda_i^2\delta_i^{*2}}{\sigma_{\varepsilon}^2}E^{(2)}\left[\left(1-\frac{qd}{q+4}F_{q+4,n-p}^{-1}(\Delta_*^2)\right)^2 I\left(F_{q+4,n-p}(\Delta_*^2)\le\frac{qd}{q+4}\right)\right]\Big\}\\ &-(\alpha_i-2\Delta_i^2)\lambda_i^2\Delta_i^2\\ &\times E^{(2)}\left[\left(\frac{qd}{q+2}F_{q+2,n-p}^{-1}(\Delta_*^2)-1\right)I\left(F_{q+2,n-p}(\Delta^2)\le\frac{qd}{q+2}\right)\right]\Big\},\end{aligned} \tag{8.4.48}$$

where $\frac{\partial R_4(\hat{\boldsymbol{\beta}}_n^{S}(k);\boldsymbol{\beta})}{\partial k}$ is given by (8.4.35). Hence, a sufficient condition for (8.4.43) to be negative is that $0<k<k_8^*$, where

$$k_8^* = \frac{f_1(\alpha,\Delta^2)}{f_2(\alpha,\Delta^2)}, \tag{8.4.49}$$

with

$$\begin{aligned}f_1(\alpha,\Delta^2) = \min_{1\le i\le p}&\Big\{\sigma_{\varepsilon}^2\Big\{\lambda_i - dqh_{ii}^{*}\Big[(q-2)E^{(1)}(\chi_{q+2}^{-4}(\Delta_*^2))\\ &+\left(1-\frac{(q+2)\lambda_i^2\delta_i^{*2}}{2\sigma_{\varepsilon}^2\Delta_*^2h_{ii}^{*}}\right)\Big](2\Delta_*^2)E^{(2)}(\chi_{q+4}^{-4}(\Delta_*^2)) - h_{ii}^{*}\\ &\times E^{(1)}\left[\left(1-\frac{qd}{q+2}F_{q+2,n-p}^{-1}(\Delta_*^2)\right)^2 I\left(F_{q+2,n-p}(\Delta_*^2)\le\frac{qd}{q+2}\right)\right]\\ &-\frac{\lambda_i^2\delta_i^{*2}}{\sigma_{\varepsilon}^2}E^{(2)}\Big[\left(1-\frac{qd}{q+4}F_{q+4,n-p}^{-1}(\Delta_*^2)-1\right)^2\\ &\times I\left(F_{q+4,n-p}(\Delta_*^2)\le\frac{qd}{q+4}\right)\Big]\Big\} + (\alpha_i-2\Delta_i^2)\lambda_i^2\Delta_i^2\\ &\times E^{(2)}\left[\left(\frac{qd}{q+2}F_{q+2,n-p}^{-1}(\Delta_*^2)-1\right)I\left(F_{q+2,n-p}(\Delta_*^2)\le\frac{qd}{q+2}\right)\right]\\ &+dq\alpha_i\lambda_i^2\Delta_i^2E^{(2)}(\chi_{q+2}^{-2}(\Delta_*^2))\Big\},\end{aligned}$$

and

$$\begin{aligned}&f_2(\alpha,\Delta^2) =\\ &\max_{1\le i\le p}\alpha_i\lambda_i\Big\{\alpha_i + \Delta_i^2E^{(2)}\left[\left(\frac{qd}{q+2}F_{q+2,n-p}^{-1}(\Delta_*^2)-1\right)I\left(F_{q+2,n-p}(\Delta^2)\le\frac{qd}{q+2}\right)\right]\\ &-qd\Delta_i^2E^{(2)}(\chi_{q+2}^{-2}(\Delta_*^2))\Big\}.\end{aligned}$$

8.4.7.1 Comparison of $\hat{\beta}_n^{S+}(k)$ with $\hat{\beta}_n^{S}(k)$, $\tilde{\beta}_n(k)$, $\hat{\beta}_n(k)$, and $\hat{\beta}_n^{PT}(k)$

Theorem 8.4.15. *PRSRRE has smaller risk value than SRRE for all $k > 0$.*

Proof. Using equation (8.4.47) and by the fact that

$$\left(0 < F_{q+2,m}(\Delta_*^2) < \frac{qd}{q+2}\right) \Leftrightarrow \left(\frac{qd}{q+2}\, F_{q+2,m}^{-1}(\Delta_*^2) - 1\right) \geq 0, \quad (8.4.50)$$

the result follows.

Theorem 8.4.16. *A sufficient condition for PRSRRE to have smaller risk value than RRRE is that $0 < k_7 < k$, where k_7 is given by* (8.4.52).

Proof. Using (8.4.39) and (8.4.47), we find

$$\begin{aligned}
&R_5(\hat{\boldsymbol{\beta}}_n^{S+}(k);\boldsymbol{\beta}) - R_2(\hat{\boldsymbol{\beta}}_n(k);\boldsymbol{\beta}) = \\
&\sum_{i=1}^{p} \frac{1}{(\lambda_i+k)^2}\Bigg\{\sigma_{\varepsilon}^2 h_{ii}^*\Bigg\{1 - dq\Bigg[(q-2)E^{(1)}(\chi_{q+2}^{-4}(\Delta_*^2)) \\
&+\left(1 - \frac{(q+2)\lambda_i^2\delta_i^{*2}}{2\sigma_{\varepsilon}^2\Delta_*^2 h_{ii}^*}\right)(2\Delta_*^2)E^{(2)}(\chi_{q+4}^{-4}(\Delta_*^2))\Bigg]\Bigg\} - \lambda_i^2\delta_i^{*2} \\
&-\sigma_{\varepsilon}^2\Bigg\{h_{ii}^*E^{(1)}\left[\left(1 - \frac{qd}{q+2}F_{q+2,n-p}^{-1}(\Delta_*^2)\right)^2 I\left(F_{q+2,n-p}(\Delta_*^2) \leq \frac{qd}{q+2}\right)\right] \\
&+\frac{\lambda_i^2\delta_i^{*2}}{\sigma_{\varepsilon}^2}E^{(2)}\left[\left(1 - \frac{qd}{q+4}F_{q+4,n-p}^{-1}(\Delta_*^2)\right)^2 I\left(F_{q+4,n-p}(\Delta_*^2) \leq \frac{qd}{q+4}\right)\right]\Bigg\} \\
&-2\lambda_i^2\delta_i^{*2}E^{(2)}\left[\left(\frac{qd}{q+2}F_{q+2,n-p}^{-1}(\Delta_*^2) - 1\right) I\left(F_{q+2,n-p}(\Delta_*^2) \leq \frac{qd}{q+2}\right)\right] \\
&-2k\alpha_i\lambda_i\Delta_i^2\Bigg\{1 - qdE^{(2)}(\chi_{q+2}^{-2}(\Delta_*^2)) \\
&+E^{(2)}\left[\left(\frac{qd}{q+2}F_{q+2,n-p}^{-1}(\Delta_*^2) - 1\right) I\left(F_{q+2,n-p}(\Delta_*^2) \leq \frac{qd}{q+2}\right)\right]\Bigg\}\Bigg\}.
\end{aligned} \quad (8.4.51)$$

This difference is nonpositive (≤ 0) whenever

$$k_7 = \frac{f_3(\alpha,\Delta^2)}{f_4(\alpha,\Delta^2)} \quad (8.4.52)$$

with

$$\begin{aligned}
f_3(\alpha,\Delta^2) \quad = \quad & \max_{1\leq i\leq p}\Bigg\{\sigma_{\varepsilon}^2 h_{ii}\Bigg\{1 - dq\Bigg[(q-2)E^{(1)}(\chi_{q+2}^{-4}(\Delta_*^2)) \\
&+\left(1 - \frac{(q+2)\lambda_i^2\delta_i^{*2}}{2\sigma_{\varepsilon}^2\Delta_*^2 h_{ii}^*}\right)(2\Delta_*^2)E^{(2)}(\chi_{q+4}^{-4}(\Delta_*^2))\Bigg]\Bigg\} - \lambda_i^2\delta_i^{*2} \\
&-\sigma_{\varepsilon}^2\Bigg\{h_{ii}^*E^{(1)}\left[\left(1 - \frac{qd}{q+2}F_{q+2,n-p}^{-1}(\Delta_*^2)\right)^2 I\left(F_{q+2,n-p}(\Delta_*^2) \leq \frac{qd}{q+2}\right)\right]
\end{aligned}$$

$$+\frac{\lambda_i^2\delta_i^{*2}}{\sigma_\varepsilon^2}E^{(2)}\left[\left(1-\frac{qd}{q+4}F_{q+4,n-p}^{-1}(\Delta_*^2)\right)^2 I\left(F_{q+4,n-p}(\Delta_*^2)\leq\frac{qd}{q+4}\right)\right]\Bigg\}$$
$$-2\lambda_i^2\delta_i^{*2}E^{(2)}\left[\left(\frac{qd}{q+2}F_{q+2,n-p}^{-1}(\Delta_*^2)-1\right)I\left(F_{q+2,n-p}(\Delta_*^2)\leq\frac{qd}{q+2}\right)\right]\Bigg\}$$

and

$$\begin{aligned} f_4(\alpha,\Delta^2) &= \min_{1\leq i\leq p}\Bigg\{2\alpha_i\lambda_i\Delta_i^2\Bigg\{1-qdE^{(2)}(\chi_{q+2}^{-2}(\Delta_*^2)) \\ &\quad +E^{(2)}\left[\left(\frac{qd}{q+2}F_{q+2,n-p}^{-1}(\Delta_*^2)-1\right)I\left(F_{q+2,n-p}(\Delta_*^2)\leq\frac{qd}{q+2}\right)\right]\Bigg\}\Bigg\} \end{aligned}$$

Theorem 8.4.17. *A sufficient condition for PRSRRE to have smaller risk value than PTRRE is that* $0<k_8<k$ *where* k_8 *is given by* (8.4.54).

Proof. By making use of equations (8.4.41) and (8.4.47), we obtain

$$\begin{aligned} & R_5(\hat{\boldsymbol{\beta}}_n^{S+}(k);\boldsymbol{\beta})-R_3(\hat{\boldsymbol{\beta}}_n^{PT}(k);\boldsymbol{\beta}) \\ = & \sum_{i=1}^{p}\frac{1}{(\lambda_i+k)^2}\Bigg\{\sigma_\varepsilon^2 h_{ii}^*\Bigg\{G_{q+2,n-p}^{(1)}(x',\Delta_*^2)-dq\Big[(q-2)E^{(1)}(\chi_{q+2}^{-4}(\Delta_*^2)) \\ & +\left(1-\frac{(q+2)\lambda_i^2\delta_i^{*2}}{2\sigma_\varepsilon^2\Delta_*^2h_{ii}^*}\right)(2\Delta_*^2)E^{(2)}(\chi_{q+4}^{-4}(\Delta_*^2))\Big]\Bigg\} \\ & -\lambda_i^2\delta_i^{*2}\left[2G_{q+2,n-p}^{(2)}(x',\Delta_*^2)-G_{q+4,n-p}^{(2)}(x',\Delta_*^2)\right] \\ & -\sigma_\varepsilon^2\Bigg\{h_{ii}^*E^{(1)}\left[\left(1-\frac{qd}{q+2}F_{q+2,n-p}^{-1}(\Delta_*^2)\right)^2 I\left(F_{q+2,n-p}(\Delta_*^2)\leq\frac{qd}{q+2}\right)\right] \\ & +\frac{\lambda_i^2\delta_i^{*2}}{\sigma_\varepsilon^2}E^{(2)}\left[\left(1-\frac{qd}{q+4}F_{q+4,n-p}^{-1}(\Delta_*^2)\right)^2 I\left(F_{q+4,n-p}(\Delta^2)\leq\frac{qd}{q+4}\right)\right]\Bigg\} \\ & -2\lambda_i^2\delta_i^{*2}E^{(2)}\left[\left(\frac{qd}{q+2}F_{q+2,n-p}^{-1}(\Delta_*^2)-1\right)I\left(F_{q+2,n-p}(\Delta^2)\leq\frac{qd}{q+2}\right)\right] \\ & -2k\alpha_i\lambda_i\Delta_i^2\{G_{q+2,n-p}^{(2)}(x',\Delta_*^2)-qdE^{(2)}(\chi_{q+2}^{-2}(\Delta_*^2)) \\ & +E^{(2)}\left[\left(\frac{qd}{q+2}F_{q+2,n-p}^{-1}(\Delta^2*)-1\right)I\left(F_{q+2,n-p}(\Delta_*^2)\leq\frac{qd}{q+2}\right)\right]\Bigg\}\Bigg\}. \end{aligned} \tag{8.4.53}$$

This difference is nonpositive (≤ 0) whenever

$$k_8=\frac{f_5(\alpha,\Delta^2)}{f_6(\alpha,\Delta^2)} \tag{8.4.54}$$

with

$$\begin{aligned}
f_5(\alpha, \Delta_*^2) &= \max_{1\leq i\leq p} \Bigg\{ \sigma_{\varepsilon}^2 h_{ii}^* \Bigg\{ G_{q+2,n-p}^{(1)}(x', \Delta_*^2) - dq\Bigg[(q-2)E^{(1)}(\chi_{q+2}^{-4}(\Delta_*^2)) \\
&+ \left(1 - \frac{(q+2)\lambda_i^2\delta_i^{*2}}{2\sigma_{\varepsilon}^2\Delta_*^2 h_{ii}^*}\right)(2\Delta_*^2)E^{(2)}(\chi_{q+4}^{-4}(\Delta_*^2))\Bigg]\Bigg\} \\
&-\lambda_i^2\delta_i^{*2}\left[2G_{q+2,n-p}^{(2)}(x', \Delta_*^2) - G_{q+4,n-p}^{(2)}(x', \Delta_*^2)\right] \\
&-\sigma_{\varepsilon}^2\{h_{ii}^* E^{(1)}\left[\left(1 - \frac{qd}{q+2}F_{q+2,n-p}^{-1}(\Delta_*^2)\right)^2 I\left(F_{q+2,n-p}(\Delta_*^2) \leq \frac{qd}{q+2}\right)\right] \\
&+\frac{\lambda_i^2\delta_i^{*2}}{\sigma_{\varepsilon}^2}E^{(2)}\left[\left(1 - \frac{qd}{q+4}F_{q+4,n-p}^{-1}(\Delta_*^2)\right)^2 I\left(F_{q+4,n-p}(\Delta_*^2) \leq \frac{qd}{q+4}\right)\right]\Bigg\} \\
&-2\lambda_i^2\delta_i^{*2}E^{(2)}\left[\left(\frac{qd}{q+2}F_{q+2,n-p}^{-1}(\Delta_*^2) - 1\right) I\left(F_{q+2,n-p}(\Delta_*^2) \leq \frac{qd}{q+2}\right)\right]\Bigg\},
\end{aligned}$$

and

$$\begin{aligned}
f_6(\alpha, \Delta_*^2) &= \min_{1\leq i\leq p} \Bigg\{ 2\alpha_i\lambda_i\Delta_i^2 \Bigg\{ G_{q+2,n-p}^{(2)}(x', \Delta_*^2) - qdE^{(2)}(\chi_{q+2}^{-2}(\Delta_*^2)) \\
&+E^{(2)}\left[\left(\frac{qd}{q+2}F_{q+2,n-p}^{-1}(\Delta_*^2) - 1\right) I\left(F_{q+2,n-p}(\Delta_*^2) \leq \frac{qd}{q+2}\right)\right]\Bigg\}\Bigg\}.
\end{aligned}$$

Remarks. (i) We note that under $H_0 : \boldsymbol{H\beta} = \boldsymbol{h}$, Theorem 8.4.17 simplifies to the following:

$$R_5(\hat{\boldsymbol{\beta}}_n^{S+}(k); \boldsymbol{\beta}) - R_2(\hat{\boldsymbol{\beta}}_n(k); \boldsymbol{\beta}) = \sum_{i=1}^{p} \frac{1}{(\lambda_i + k)^2} \Bigg\{ \sigma_{\varepsilon}^2(1-d)h_{ii}^*$$

$$-\sigma_{\varepsilon}^2 h_{ii}^* E^{(1)}\left[\left(1 - \frac{qd}{q+2}F_{q+2,n-p}^{-1}(0)\right)^2 I\left(F_{q+2,n-p}(0) \leq \frac{qd}{q+2}\right)\right]\Bigg\} \geq 0$$

since

$$\begin{aligned}
& E^{(1)}\left[\left(1 - \frac{qd}{q+2}F_{q+2,n-p}^{-1}(0)\right)^2 I\left(F_{q+2,n-p}(0) \leq \frac{qd}{q+2}\right)\right] \\
\leq\ & E^{(1)}\left[\left(1 - \frac{qd}{q+2}F_{q+2,n-p}^{-1}(0)\right)^2\right] = 1 - d.
\end{aligned}$$

Thus, $\hat{\boldsymbol{\beta}}_n(k)$ is superior to $\hat{\boldsymbol{\beta}}_n^{S+}(k)$ under $H_0 : \boldsymbol{H\beta} = \boldsymbol{h}$.

(ii) We note that under $H_0 : \boldsymbol{H\beta} = \boldsymbol{h}$, Theorem 8.4.20 simplifies to the following:

$$\begin{aligned}
R_5(\hat{\boldsymbol{\beta}}_n^{S+}(k); \boldsymbol{\beta}) - R_3(\hat{\boldsymbol{\beta}}_n^{PT}(k); \boldsymbol{\beta}) &= \sum_{i=1}^{p} \frac{1}{(\lambda_i + k)^2} \left\{ \sigma_{\varepsilon}^2 h_{ii}^* G_{q+2,n-p}^{(1)}(x', 0) - d - \sigma_{\varepsilon}^2 h_{ii}^* \kappa \right\} \\
&\geq 0
\end{aligned}$$

for

$$\left\{\alpha : F_{q+2,m}(l_\alpha^{*(1)}, 0) \geq d + \kappa\right\}. \tag{8.4.55}$$

Thus, the risk of $\hat{\boldsymbol{\beta}}_n^{S+}(k)$ is smaller than that of the risk of $\hat{\boldsymbol{\beta}}_n^{PT}(k)$ when the critical value satisfied the opposite inequality in (8.4.55), otherwise $\hat{\boldsymbol{\beta}}_n^{PT}(k)$ is smaller than $\hat{\boldsymbol{\beta}}_n^{S+}(k)$.

8.5 Choice of Ridge Parameter

In this section, we proceed with some numerical computations as proofs of our assertions. The process is categorized into two setups as follows.

8.5.1 Real Example

Table 8.1 presents a data set concerning the percentage of conversion of n-heptane to acetylene and three explanatory variables (Himmelblau, 1970, Kunugi, Tamura, and Naito, 1961, and Marquardt and Snee, 1975). These are typical chemical process data for which a full quadratic response surface in all three regressors is often considered to be an appropriate tentative model. A plot of contact time versus reactor temperature is shown in Figure 8.1. Because these two regressors are highly correlated, there are potential multicollinearity problems in these data.

As in Montgomery et al. (2001), the full quadratic model for the acetylene data is

$$\begin{aligned} P &= \beta_0 + \beta_1 T + \beta_2 H + \beta_3 C + \beta_{12} TH + \beta_{13} TC + \beta_{23} HC \\ &+ \beta_{11} T^2 + \beta_{22} H^2 + \beta_{33} C^2 + \epsilon. \end{aligned}$$

It can be investigated that there exists a potential multicollinearity between reactor temperature and contact time. Without making any serious decision, the estimators are computed and reported in Table 8.2 and the graphs of risks based on noncentrality parameter Δ^2 and ridge parameter k are displayed in Figure 8.2. Now, in order to overcome the multicollinearity for better performance of the estimators, we follow the algorithm due to Hoerl et al. (1975) to determine the appropriate choice of k, (see Montgomery et al., 2001, pages 358-359). Therefore, the value of $k = 0.00149$ is obtained. Based on this value, for

$$\boldsymbol{H} = \begin{bmatrix} 1 & -1 & -1 & -1 & 0 & 1 & 2 & 0 & 0 \\ 2 & 2 & 1 & 0 & 0 & 1 & 1 & 0 & 0 \\ 3 & 1 & 2 & 0 & 0 & 0 & -2 & 0 & 0 \end{bmatrix}$$

and $\boldsymbol{h} = [1, -1, 1]'$ the estimators are computed and given in Table 8.2. It is seen that the lengths of the estimators are decreased. Further, the risk functions are computed

Table 8.1 Acetylene data

Observation	Conversion of n-Heptane to Acetylene	Reactor Temperature (^0C)	Ratio of H_2 to n-Heptane (mole ratio)	Contact Time (Sec)
1	49.0	1300	7.5	0.0120
2	50.2	1300	9.0	0.0120
3	50.5	1300	11.0	0.0115
4	48.5	1300	13.5	0.0130
5	47.5	1300	17.0	0.0135
6	44.5	1300	23.0	0.0120
7	28.0	1200	5.3	0.0400
8	31.5	1200	7.5	0.0380
9	34.5	1200	11.0	0.0320
10	35.0	1200	13.5	0.0260
11	38.0	1200	17.0	0.0340
12	38.5	1200	23.0	0.0410
13	15.0	1100	5.3	0.0840
14	17.0	1100	7.5	0.0980
15	20.5	1100	11.0	0.0920
16	29.5	1100	17.0	0.0860

and given in Figure 8.3. The given risks are significantly decreased. Finally, as a proof of the assertions given in the theorems of section 8.4, the plot of risks are given in Figure 8.4. It can be realized that the ridge estimators always perform better than their nonridge counterparts.

Table 8.2 Estimated values of ridge estimators

$\tilde{\boldsymbol{\beta}}_n$	$\hat{\boldsymbol{\beta}}_n$	$\hat{\boldsymbol{\beta}}_n^{PT}$	$\hat{\boldsymbol{\beta}}_n^{S}$	$\hat{\boldsymbol{\beta}}_n^{S+}$
0.37452	0.04160	0.37452	0.36044	0.36044
0.23516	0.23597	0.23516	0.23519	0.23519
-0.62467	-0.96631	-0.62467	0.63911	0.63911
-2.10985	-3.45663	-2.10985	-2.16680	-2.16680
-8.55471	-10.96875	-8.55471	-8.65678	-8.65678
-1.24328	-2.83191	-1.24328	-1.31045	-1.31045
-3.94655	-4.57582	-3.94655	-3.97316	-3.97316
-0.33214	-0.43959	-0.33214	-0.33668	-0.33668
-3.74221	-5.32547	-3.74221	-3.80915	-3.80915

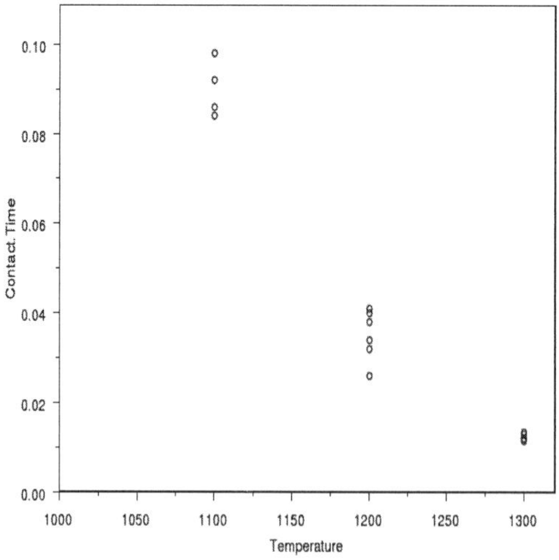

Figure 8.4 Contact time versus reactor temperature

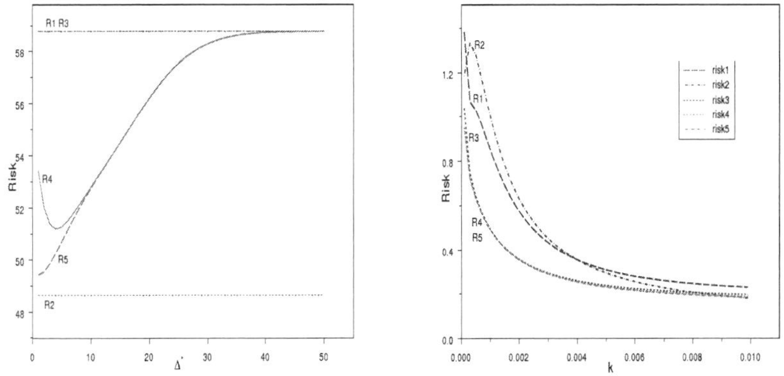

Figure 8.5 Risk performance based on both Δ^2 and k

Table 8.3 Estimated values

$\tilde{\boldsymbol{\beta}}_n$	$\hat{\boldsymbol{\beta}}_n$	$\hat{\boldsymbol{\beta}}_n^{PT}$	$\hat{\boldsymbol{\beta}}_n^{S}$	$\hat{\boldsymbol{\beta}}_n^{S+}$
0.75364	0.55048	0.75364	0.74505	0.74505
0.22655	0.22580	0.22655	0.22652	0.22652
-0.14293	-0.34834	-0.14293	-0.15161	-0.15161
-1.02672	1.41510	-1.02672	-1.04314	-1.04314
-0.24594	-0.30280	-0.24594	-0.24834	-0.24834
0.01682	-0.47517	0.01682	-0.00398	-0.00398
0.38947	0.64410	0.38947	0.40024	0.40024
-0.25045	-0.31433	-0.25045	-0.25315	-0.25315
-0.12995	-0.33026	-0.12995	-0.13842	-0.13842

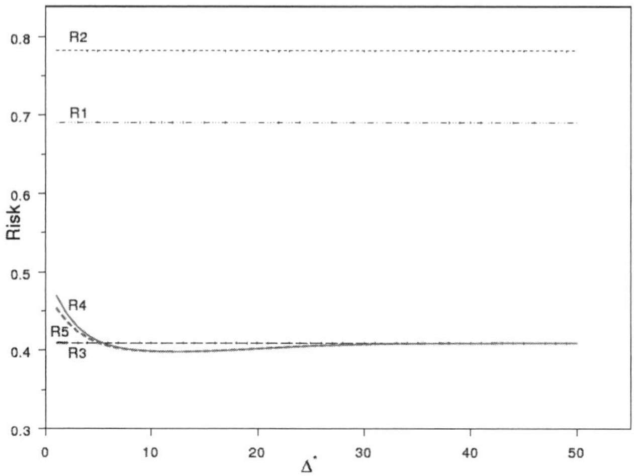

Figure 8.6 Ridge risk performance based on Δ^2

8.5.2 Simulation

The data set consists of four variables with $n = 30$ generated from a multivariate normal model and the following correlation matrix given by Table 8.4. The scatter plots of the variables are displayed in Figure 8.5. As it can be investigated, there exists a potential multicollinearity between x_1 and x_3, and x_2 and x_4. Without making any serious decision, the estimators are computed, reported in Table 8.5, and the graphs of risks based on noncentrality parameter Δ^2 and ridge parameter k are displayed in Figure 8.6. Now, in order to combat the existing multicollinearity for better performance of the estimators, we follow the method of ridge trace due to Hoerl and Kennard (1970). Therefore, for different values of ridge constant k, the estimators are computed and the ridge trace is shown in Tables 8.6 and 8.7 and Figure 8.7. As it is found from the Table 8.6, the ridge parameter is determined as $k = 0.0.015$. Based on this value, for $H = [-1, 0, 1, 1, -1, 1, 2, 0, -1, 1, 2, 0]$ and $h = [1, -1, 0, 1]'$ the estimators are computed. It is seen that the length of the estimators are decreased. Furthermore, the risk functions are computed and given in Figure 8.8. The given risks are significantly decreased. Finally, as a proof of the assertions given in the theorems in this study, the plots of risk are given in Figure 8.9. It can be investigated that the ridge estimators always perform better than others.

To conclude the presented result, it is worthwhile to consider that the risk criterion has been effectively determined a sufficient condition for the superiority of these estimators over traditional estimators. It is subsequently found that there exists a

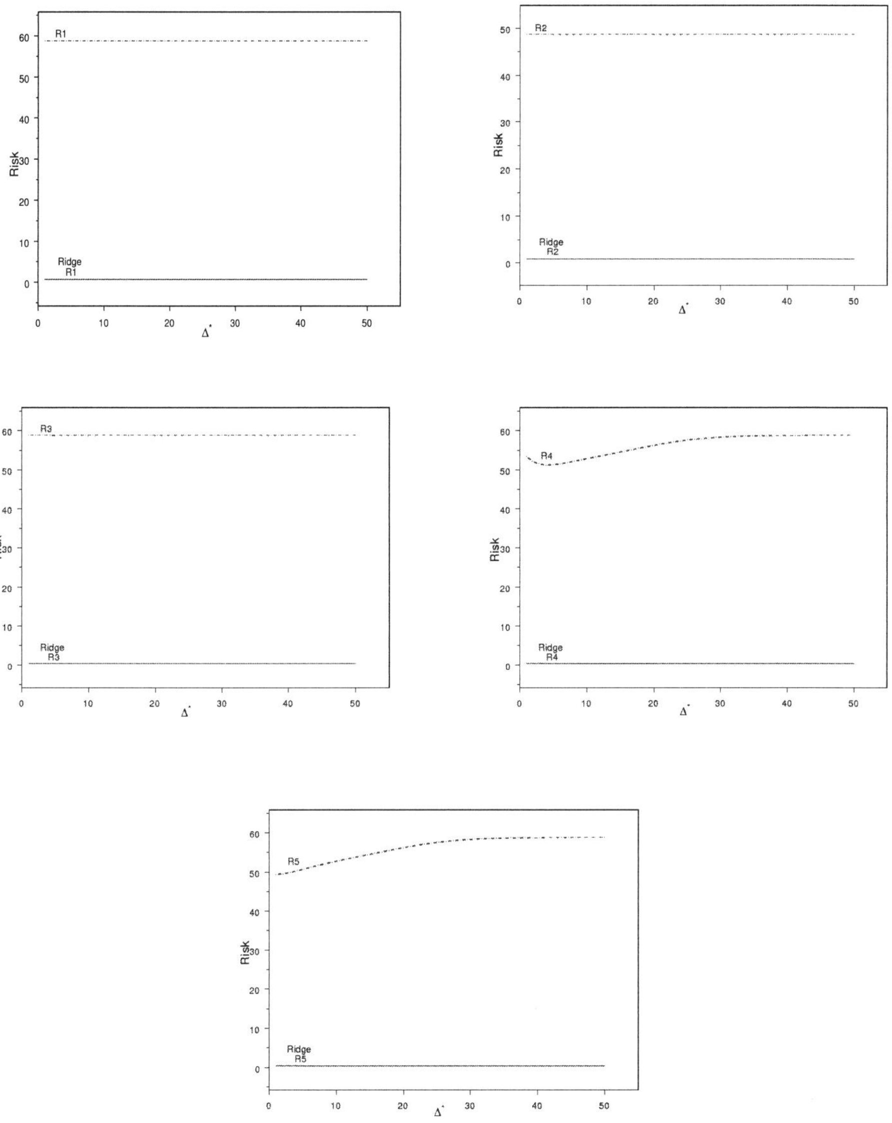

Figure 8.7 Ridge risks performance based on Δ^2

Table 8.4 Correlation matrix

x	x_1	x_2	x_3	x_4
x_1	1.00000	0.18931	0.99443	0.01160
x_2	0.18931	1.00000	0.18592	0.86301
x_3	0.99443	0.18592	1.00000	0.03801
x_4	0.01160	0.86301	0.03801	1.00000

Table 8.5 Estimated values

$\tilde{\boldsymbol{\beta}}_n$	$\hat{\boldsymbol{\beta}}_n$	$\hat{\boldsymbol{\beta}}_n^{PT}$	$\hat{\boldsymbol{\beta}}_n^{S}$	$\hat{\boldsymbol{\beta}}_n^{S+}$
1.97889	0.93059	1.97889	1.83632	1.83632
-0.48620	0.35754	-0.48620	-0.37145	-0.37145
-1.95870	-0.99510	-1.95870	-1.82765	-1.82765
0.46275	-0.07482	0.46275	0.38964	0.38964

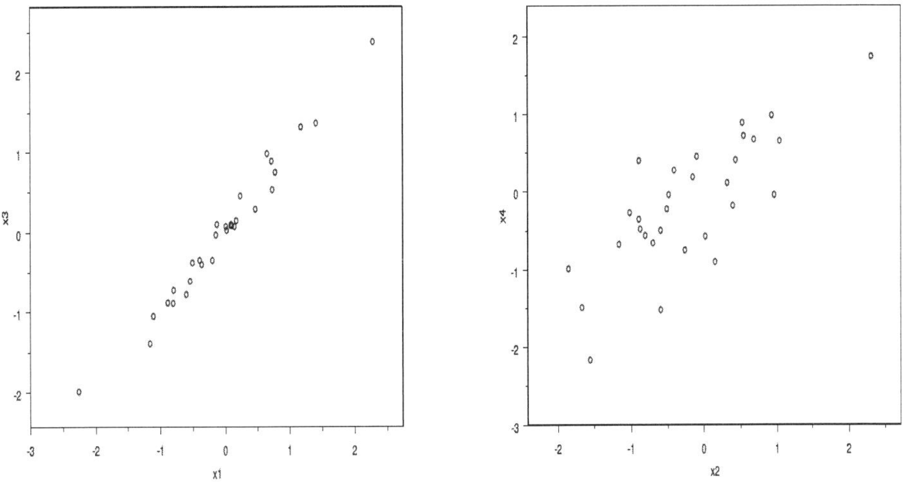

Figure 8.8 Scatter plot of variables versus each other

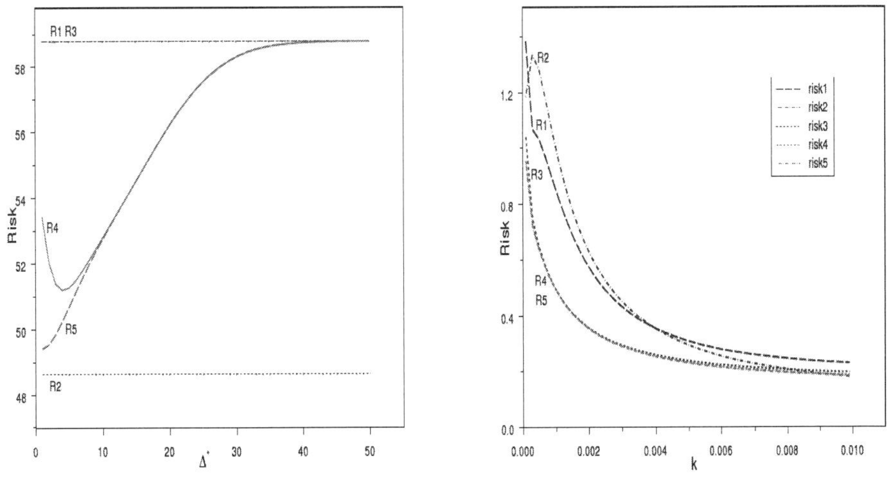

Figure 8.9 Risk performance based on both Δ^2 and k

Table 8.6 Coefficients at various values of k

k	0.001	0.002	0.003	0.004	0.005	0.006	0.007
β_1	1.78951	1.63295	1.50136	1.38921	1.29249	1.20822	1.13415
β_2	-0.46367	-0.44487	-0.42889	-0.41513	-0.40312	-0.39252	-0.38309
β_3	-1.76963	-1.61335	-1.48200	-1.37008	-1.27356	-1.18948	-1.11559
β_4	0.43767	0.41676	0.39903	0.38378	0.37050	0.35880	0.34841
k	0.008	0.009	0.010	0.011	0.012	0.013*	0.014
β_1	1.06853	1.00999	0.95745	0.91003	0.86702	0.82783	0.79198
β_2	-0.37462	-0.36696	-0.35999	-0.35360	-0.34773	-0.34229	-0.33724
β_3	-1.05014	-0.99176	-0.93937	-0.89209	-0.84922	-0.81016	-0.77443
β_4	0.33910	0.33070	0.32306	0.31609	0.30968	0.30377	0.29829
k	0.015	0.016	0.017	0.018	0.019	0.020	0.021
β_1	0.75905	0.72870	0.70065	0.67463	0.65045	0.62790	0.60684
β_2	-0.33253	-0.32812	-0.32397	-0.32007	-0.31638	-0.31288	-0.30956
β_3	-0.74163	-0.71140	-0.68346	-0.65756	-0.63349	-0.61105	-0.59009
β_4	0.29318	0.28842	0.28395	0.27975	0.27579	0.27204	0.26849

Table 8.7 Estimated values using $k = 0.015$

$\tilde{\boldsymbol{\beta}}_n$	$\hat{\boldsymbol{\beta}}_n$	$\hat{\boldsymbol{\beta}}_n^{PT}$	$\hat{\boldsymbol{\beta}}_n^{S}$	$\hat{\boldsymbol{\beta}}_n^{S+}$
0.75905	0.36429	0.75905	0.70536	0.70536
-0.33253	0.39840	-0.33253	-0.23312	-0.23312
-0.74163	-0.42639	-0.74163	-0.69876	-0.69876
0.29318	-0.12715	0.29318	0.23602	0.23602

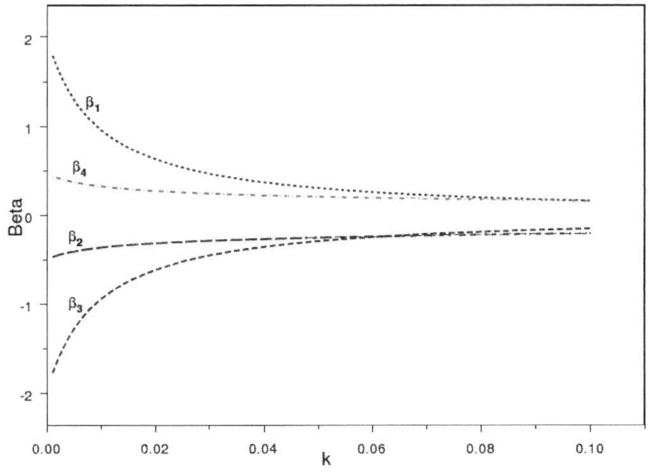

Figure 8.10 Ridge trace for the data using four regressors

k such that the new estimators dominate over the other traditional estimators. The property of the five estimators is robust in the class of multivariate t models. Finally, least and most values of ridge constants in which the ridge estimators dominate each other are given below:

For Real Example

At most k for dominance of URRE over UE is 0.00138.
Under H_0, at most k for dominance of RRRE over RE is ≈ 0.
Under H_0, at most k for dominance of PTRRE over PTE is 0.000000007.
Under H_1, at most k for dominance of PTRRE over PTE is 0.000000007.
Under H_0, at most k for dominance of SRRE over SE is 0.000000005.
Under H_1, at most k for dominance of SRRE over SE is 0.000000007.
Under H_0, at most k for dominance of PRSRRE over PRSE is 0.000000006.
Under H_1, at most k for dominance of PRSRRE over PRSE is 0.000000006.

For Simulation

At most k for dominance of URRE over UE is 0.029723.
Under H_0, at most k for dominance of RRRE over RE is ≈ 0.
Under H_0, at most k for dominance of PTRRE over PTE is 0.016178.
Under H_1, at most k for dominance of PTRRE over PTE is 0.016182.
Under H_0, at most k for dominance of SRRE over SE is 0.011174.
Under H_1, at most k for dominance of SRRE over SE is 0.024111.
Under H_0, at most k for dominance of PRSRRE over PRSE is 0.013036.
Under H_1, at most k for dominance of PRSRRE over PRSE is 0.020398.

Under H_1, at least k for dominance of PTRRE over RRRE is 0.016353.
Under H_1, at least k for dominance of SRRE over RRRE is 0.010854.
Under H_1, at least k for dominance of PRSRRE over RRRE is 0.010493.

It is important to note that, based on the given propositions, we can **always** find a value k for which the ridge estimator performs better than its un-ridge counterpart. But there does not always exist a value k for domination of one ridge estimator over another. However, if it exists, we can specifically determine it through the proposed theorems.

8.6 Problems

1. Verify equation (8.2.1).
2. Verify equations (8.4.2)-(8.4.4).
3. Verify equation (8.4.12).
4. Verify equation (8.4.40).
5. Derive the risk functions of the proposed ridge regression estimators in this chapter, based on the balanced loss function (7.3.1).
6. Define rank-based ridge regression estimators according to the result of section 9.1.2 of Saleh (2006) and then derive the risk functions based on the balanced loss function (7.3.1).

CHAPTER 9

MULTIVARIATE MODELS

Outline

Multivariate statistical analysis of multidimensional data dominates the literature based on multivariate normal distribution like the normal distribution in the case of univariate problems. In this chapter, we consider the statistical theory based on multivariate t-distribution to increase the scope of applications. We consider only two models, namely, (i) the location and (ii) the simple linear regression models and discuss the test of hypothesis and propose several estimators for two models with details of the dominance properties.

9.1 Location Model

Let $\boldsymbol{Y}_1, \boldsymbol{Y}_2, \ldots, \boldsymbol{Y}_N$ be N observation vectors of p-dim satisfying the model

$$\boldsymbol{Y}_\alpha = \boldsymbol{\theta} + \boldsymbol{\varepsilon}_\alpha\,, \qquad \alpha = 1, \ldots, N, \tag{9.1.1}$$

where $\boldsymbol{Y}_\alpha = (Y_{\alpha 1}, \ldots, Y_{\alpha p})'$, $\boldsymbol{\theta} = (\theta_1, \ldots, \theta_p)'$ is the location vector parameter, and $\boldsymbol{\varepsilon}_\alpha = (\varepsilon_{\alpha 1}, , \ldots, \varepsilon_{\alpha p})'$ and $\{\boldsymbol{\varepsilon}_\alpha | \alpha = 1, \ldots, N\}$ are distributed as $M_t^{(p)}(\boldsymbol{0}, \boldsymbol{\Sigma}, \gamma_o)$ for each $\alpha = 1, \ldots, N$.

The unrestricted estimator (UE) of $\boldsymbol{\theta}$ is $\tilde{\boldsymbol{\theta}}_N = \bar{\boldsymbol{Y}}$, and the exact distribution of $\tilde{\boldsymbol{\theta}}_N$ is $M_t^{(p)}(\boldsymbol{\theta}, \frac{1}{N}\boldsymbol{\Sigma}, \gamma_o)$.

Consider a statistic of the form

$$\boldsymbol{A} = \sum_{\alpha=1}^{N} (\boldsymbol{Y}_\alpha - \bar{\boldsymbol{Y}})(\boldsymbol{Y}_\alpha - \bar{\boldsymbol{Y}})'.$$

It is well known that if $\boldsymbol{\varepsilon}_\alpha \sim \mathcal{N}_p(\boldsymbol{0}, \boldsymbol{\Sigma})$ in the model (9.1.1), then $n^{-1}\boldsymbol{A} \sim W_p(\boldsymbol{\Sigma}, n)$, $n = N - 1$ (refer to (2.5.4)). Under the assumption of normal theory, $n^{-1}\boldsymbol{A}|t \sim W_p(t^{-1}\boldsymbol{\Sigma}, n)$. Thus, taking expectation w.r.t. t shows that the unbiased estimator of $\boldsymbol{\Sigma}$ is given by

$$\begin{aligned} \boldsymbol{S}_u &= \frac{\gamma_o - 2}{n\gamma_o}\boldsymbol{A} \\ &= \frac{\gamma_o - 2}{n\gamma_o}\sum_{\alpha=1}^{N} (\boldsymbol{Y}_\alpha - \bar{\boldsymbol{Y}})(\boldsymbol{Y}_\alpha - \bar{\boldsymbol{Y}})', \quad n = N - 1. \end{aligned} \tag{9.1.2}$$

It is not easy to derive the distribution of $\boldsymbol{A}$ by some elementary statistical computations. Sutradhar and Ali (1989) derived the underlying distribution using elliptically contoured models. However, by the representation (2.6.3), we can find its distribution as stated below.

Theorem 9.1.1. *Under the assumptions of this section, the pdf of the elements of* $\boldsymbol{A} = \sum_{\alpha=1}^{N} (\boldsymbol{Y}_\alpha - \bar{\boldsymbol{Y}})(\boldsymbol{Y}_\alpha - \bar{\boldsymbol{Y}})'$ *is given by*

$$C(n,p)|\boldsymbol{\Sigma}|^{-\frac{n}{2}}|\boldsymbol{A}|^{-\frac{n-p-1}{2}}\left(1 + \frac{1}{\gamma_o} tr\boldsymbol{\Sigma}^{-1}\boldsymbol{A}\right)^{-\frac{1}{2}(\gamma_o + np)}, \qquad n = N - 1,$$

where the normalizing constant $C(n,p)$ *is given by*

$$C(n,p) = \frac{\gamma_o^{-\frac{np}{2}}\Gamma\left(\frac{\gamma_o + np}{2}\right)}{\pi^{\frac{p(p-1)}{4}}\Gamma\left(\frac{\gamma_o}{2}\right)\prod_{i=1}^{p}\Gamma\left(\frac{n-i+1}{2}\right)}.$$

Proof: According to the weight representation (2.6.3), conditioning on t, $\boldsymbol{A}$ has the Wishart distribution with parameters n and $t^{-1}\boldsymbol{\Sigma}$. Hence, integrating over t using

(2.6.4) yields

$$\begin{aligned} f(\boldsymbol{A}) &= \int_0^\infty W_o(t) f^*(\boldsymbol{A}|t)dt \\ &= \int_0^\infty W_o(t) \frac{|t^{-1}\boldsymbol{\Sigma}|^{-\frac{n}{2}} |\boldsymbol{A}|^{-\frac{1}{2}(n-p-1)} e^{(-\frac{t}{2} tr \boldsymbol{\Sigma}^{-1} A)}}{2^{\frac{np}{2}} \pi^{\frac{p(p-1)}{4}} \prod_{i=1}^p \Gamma\left(\frac{n-i+1}{2}\right)} dt \\ &= C(n,p) |\boldsymbol{\Sigma}|^{-\frac{n}{2}} |\boldsymbol{A}|^{-\frac{n-p-1}{2}} \left(\gamma_o + tr \boldsymbol{\Sigma}^{-1} \boldsymbol{A}\right)^{-\frac{1}{2}(\gamma_o + np)}. \end{aligned}$$

9.2 Testing of Hypothesis and Several Estimators of Local Parameter

Our purpose is to estimate the location parameter $\boldsymbol{\theta}$ based on a random sample of size N from the model (9.1.1), when it is suspected but not certain that $\boldsymbol{C\theta} = \boldsymbol{0}$, where $\boldsymbol{C}$ is any $q \times p$, $q = p - 1$, matrix of rank q such that $\boldsymbol{C1}_p = \boldsymbol{0}$. The restricted estimator (RE) of $\boldsymbol{\theta}$ is obtained by making use of the hypothesis $H_0 : \boldsymbol{C\theta} = \boldsymbol{0}$

$$\begin{aligned} \hat{\boldsymbol{\theta}}_N &= \frac{1}{p} \boldsymbol{1}_p \boldsymbol{1}_p' \tilde{\boldsymbol{\theta}}_N \\ &= \frac{1}{p} \boldsymbol{1}_p \boldsymbol{1}_p' \bar{\boldsymbol{Y}}. \end{aligned} \tag{9.2.1}$$

In order to test the symmetry hypothesis, i.e., $H_0 : \boldsymbol{C\theta} = \boldsymbol{0}$, one may use the test statistic

$$\begin{aligned} T^2 &= \frac{\gamma_o - 2}{\gamma_o} N \tilde{\boldsymbol{\theta}}_N' \boldsymbol{C}' (\boldsymbol{C} \boldsymbol{S}_u \boldsymbol{C}')^{-1} \boldsymbol{C} \tilde{\boldsymbol{\theta}}_N, \quad n = N - 1 \\ &= N \tilde{\boldsymbol{\theta}}_N' \boldsymbol{C}' (\boldsymbol{CSC}')^{-1} \boldsymbol{C} \tilde{\boldsymbol{\theta}}_N, \quad \boldsymbol{S} = n^{-1} \boldsymbol{A}. \end{aligned}$$

But under the normal theory, i.e., conditioning on t, $\sqrt{t}\boldsymbol{C}\bar{\boldsymbol{Y}} \sim \mathcal{N}_q(\sqrt{t}\boldsymbol{C\theta}, N^{-1}\boldsymbol{C\Sigma C}')$ and $t\boldsymbol{CAC}' \sim W_q(n, \boldsymbol{C\Sigma C}')$ are independent. Thus, $T^2|t$ has a noncentral Hotelling T^2- distribution with n d.f. and noncentrality parameter $\Delta_t^2 = tN\boldsymbol{\theta}'\boldsymbol{C}'(\boldsymbol{C\Sigma C}')^{-1}\boldsymbol{C\theta}$. Now we define

$$\begin{aligned} \mathcal{L}_n &= \frac{m}{qn} T^2 \\ &= \frac{mN}{qn} \tilde{\boldsymbol{\theta}}_N' \boldsymbol{C}' (\boldsymbol{CSC}')^{-1} \boldsymbol{C} \tilde{\boldsymbol{\theta}}_N, \quad m = N - p. \end{aligned} \tag{9.2.2}$$

The test statistic $\mathcal{L}_n$ given by (9.2.2) can be used for testing a more general hypothesis than symmetry, where $\boldsymbol{\theta}$ is suspected to belong to the subspace $\boldsymbol{\theta} = \boldsymbol{B\eta}$, where $\boldsymbol{B}$ is $p \times r$ known matrix of rank r and $\boldsymbol{\eta}$ is an unknown r-vector. In this case, $q = p - r$ and the equality $\boldsymbol{CB} = \boldsymbol{0}$ is satisfied. Note that, in this case, the restricted estimator of $\boldsymbol{\theta}$ is $\boldsymbol{B}(\boldsymbol{B}'\boldsymbol{S}^{-1}\boldsymbol{B})^{-1}\boldsymbol{B}'\boldsymbol{S}^{-1}\tilde{\boldsymbol{\theta}}_N$ (see Srivastava and Khatri, 1979, p. 116 or Srivastava and Saleh, 2005). Saleh and Kibria (2010) have extensively studied the preliminary test, Stein-type, and positive-rule estimators with this assumption and that the error term in the model (9.1.1) has elliptically contoured distribution. See also Khan and Saleh (2005)

Theorem 9.2.1. *Under the assumptions of this section, the LR test statistic given by* (9.2.2) *has the following generalized Hotelling's T^2-distribution:*

$$g^*_{q,m}(\mathcal{L}_n) = \sum_{r\geq 0} \frac{\left(\frac{q}{m}\right)^{\frac{1}{2}(q+2r)} \mathcal{L}_n^{\frac{1}{2}(q+2r-2)} K_r^{(0)}(\Delta^2)}{r!\, B\left(\frac{q+2r}{2}, \frac{m}{2}\right)\left(1+\frac{q}{m}\mathcal{L}_n\right)^{\frac{1}{2}(q+m+2r)}},$$

where $m = N - p$, $\Delta^2 = N\boldsymbol{\theta}'\boldsymbol{C}'(\boldsymbol{C}\boldsymbol{\Sigma}_{\gamma_o}\boldsymbol{C}')^{-1}\boldsymbol{C}\boldsymbol{\theta}$, and $\boldsymbol{\Sigma}_{\gamma_o} = \frac{\gamma_o}{\gamma_o-2}\boldsymbol{\Sigma}$.

Proof: First note that given t, using the representation (2.6.3), and the result of Srivastava and Khatri (1979, p. 113), we have that

$$\frac{n\tilde{\boldsymbol{\theta}}'_N\boldsymbol{C}'(\boldsymbol{C}\boldsymbol{\Sigma}\boldsymbol{C}')^{-1}\boldsymbol{C}\tilde{\boldsymbol{\theta}}_N}{\tilde{\boldsymbol{\theta}}'_N\boldsymbol{C}'(\boldsymbol{C}\boldsymbol{S}_u\boldsymbol{C}')^{-1}\boldsymbol{C}\tilde{\boldsymbol{\theta}}_N} \sim t^{-1}\chi^2_m, \qquad \text{and} \tag{9.2.3}$$

$$N\tilde{\boldsymbol{\theta}}'_N\boldsymbol{C}'(\boldsymbol{C}\boldsymbol{\Sigma}\boldsymbol{C}')^{-1}\boldsymbol{C}\tilde{\boldsymbol{\theta}}_N \sim t^{-1}\chi^2_p(\Delta^2_t), \tag{9.2.4}$$

and from independency of (9.2.3) and (9.3.4), we conclude that

$$\mathcal{L}_n|t = \frac{m}{nq}N\tilde{\boldsymbol{\theta}}'_N\boldsymbol{C}'(\boldsymbol{C}\boldsymbol{S}\boldsymbol{C}')^{-1}\boldsymbol{C}\tilde{\boldsymbol{\theta}}_N|t \sim \frac{m}{q}\frac{\chi^2_q(\Delta^2_t)}{\chi^2_m} \equiv F_{q,m}(\Delta^2_t). \tag{9.2.5}$$

Integrating the respective expressions over t completes the proof.

Note that under normal theory, using (9.2.5), a test rejects H_0 when $\mathcal{L}_n|t > F_{q,m}(\alpha)$, where $F_{q,m}(\alpha)$ is the upper α-level critical value from the central F-distribution. Thus, we reject H_0 when $\mathcal{L}_n > F_{q,m}(\alpha)$ for some constant $F_{q,m}(\alpha)$.

Now, define the preliminary test estimator (PTE) of $\boldsymbol{\theta}$ by

$$\hat{\boldsymbol{\theta}}_N^{PT} = \hat{\boldsymbol{\theta}}_N I(\mathcal{L}_n < F_{q,m}(\alpha)) + \tilde{\boldsymbol{\theta}}_N I(\mathcal{L}_n > F_{q,m}(\alpha)), \tag{9.2.6}$$

The Stein-type estimator (SE) of $\boldsymbol{\theta}$ can be written as

$$\begin{aligned}\hat{\boldsymbol{\theta}}_N^{S} &= \hat{\boldsymbol{\theta}}_N + (1 - d\mathcal{L}_n^{-1})(\tilde{\boldsymbol{\theta}}_N - \hat{\boldsymbol{\theta}}_N)\\ &= \tilde{\boldsymbol{\theta}}_N - d\mathcal{L}_n^{-1}(\tilde{\boldsymbol{\theta}}_N - \hat{\boldsymbol{\theta}}_N),\end{aligned} \tag{9.2.7}$$

where

$$d = \frac{(q-2)m}{q(m+2)}. \tag{9.2.8}$$

Finally, the positive rule Stein-type estimator (PRSE) is given by

$$\begin{aligned}\hat{\boldsymbol{\theta}}_N^{S+} &= \hat{\boldsymbol{\theta}}_N + (1 - d\mathcal{L}_n^{-1})I(\mathcal{L}_n > d)(\tilde{\boldsymbol{\theta}}_N - \hat{\boldsymbol{\theta}}_N)\\ &= \hat{\boldsymbol{\theta}}_N^{S} - (1 - d\mathcal{L}_n^{-1})I(\mathcal{L}_n \leq d)(\tilde{\boldsymbol{\theta}}_N - \hat{\boldsymbol{\theta}}_N).\end{aligned} \tag{9.2.9}$$

9.3 Bias, Quadratic Bias, MSE, and Risk Expressions

Similar to previous chapters, we intend to derive some basic characteristic of the aforementioned estimators. In this regard, we apply some computational techniques provided earlier. We will evaluate bias, quadratic bias, MSE matrices, and risks of the estimators, respectively, by $\boldsymbol{b}(\boldsymbol{\theta}_N^*) = E(\boldsymbol{\theta}_N^* - \boldsymbol{\theta})$, $B(\boldsymbol{\theta}_N^*) = N[\boldsymbol{b}(\boldsymbol{\theta}_N^*)]'\boldsymbol{\Sigma}_{\gamma_o}^{-1}[\boldsymbol{b}(\boldsymbol{\theta}_N^*)]$, $\boldsymbol{M}(\boldsymbol{\theta}_N^*) = E[N(\boldsymbol{\theta}_N^* - \boldsymbol{\theta})(\boldsymbol{\theta}_N^* - \boldsymbol{\theta})']$ and $R(\boldsymbol{\theta}_N^*; \boldsymbol{Q}) = tr[\boldsymbol{Q}\boldsymbol{M}(\boldsymbol{\theta}_N^*)]$, where $\boldsymbol{\theta}_N^*$ stands for any of the proposed estimator in the previous section and $\boldsymbol{\Sigma}_{\gamma_o} = -2\psi'(0)\boldsymbol{\Sigma} = \frac{\gamma_o}{\gamma_o - 2}\boldsymbol{\Sigma}$.

According to Chaudhry (1997), for the calculation of these properties, we choose $\boldsymbol{C}$ as follows:

$$\boldsymbol{C} = \begin{bmatrix} \sqrt{\frac{q}{p}} & -\frac{1}{\sqrt{pq}} & \cdots & -\frac{1}{\sqrt{pq}} & -\frac{1}{\sqrt{pq}} \\ 0 & \sqrt{\frac{p-2}{q}} & \cdots & -\frac{1}{\sqrt{q(p-2)}} & -\frac{1}{\sqrt{q(p-2)}} \\ \vdots & \vdots & \ddots & \vdots & \vdots \\ 0 & 0 & \cdots & \frac{1}{\sqrt{2}} & -\frac{1}{\sqrt{2}} \end{bmatrix},$$

where $\boldsymbol{C}$ is a $q \times p$ matrix of rank q.

For the unrestricted estimator we have

$$\boldsymbol{b}_1(\tilde{\boldsymbol{\theta}}_N) = \boldsymbol{0},\ B_1(\tilde{\boldsymbol{\theta}}_N) = 0,\ \boldsymbol{M}_1(\tilde{\boldsymbol{\theta}}_N) = \boldsymbol{\Sigma}_{\gamma_o},\ \text{and } R_1(\tilde{\boldsymbol{\theta}}_N; \boldsymbol{Q}) = tr(\boldsymbol{Q}\boldsymbol{\Sigma}_{\gamma_o}). \tag{9.3.1}$$

For the restricted estimator

$$\boldsymbol{b}_2(\hat{\boldsymbol{\theta}}_N) = -\boldsymbol{\delta},\ B_2(\hat{\boldsymbol{\theta}}_N) = \boldsymbol{\delta}'\boldsymbol{\delta},\ \boldsymbol{M}_2(\hat{\boldsymbol{\theta}}_N) = \boldsymbol{J}\boldsymbol{\Sigma}_{\gamma_o}\boldsymbol{J} + N\boldsymbol{\delta}\boldsymbol{\delta}',\ \text{and}$$
$$R_2(\hat{\boldsymbol{\theta}}_N; \boldsymbol{Q}) = tr(\boldsymbol{Q}\boldsymbol{J}\boldsymbol{\Sigma}_{\gamma_o}\boldsymbol{J}) + N\boldsymbol{\delta}'\boldsymbol{Q}\boldsymbol{\delta}, \tag{9.3.2}$$

where $\boldsymbol{\delta} = \boldsymbol{H}\boldsymbol{\theta}$, $\boldsymbol{H} = \boldsymbol{C}'\boldsymbol{C} = \boldsymbol{I}_p - \boldsymbol{J}$, $\boldsymbol{J} = \frac{1}{p}\boldsymbol{1}_p\boldsymbol{1}_p'$.

For the PTE, we have

$$\begin{aligned} \boldsymbol{b}_3(\hat{\boldsymbol{\theta}}_N^{PT}) &= -\boldsymbol{\delta}G_{q+2,m}^{(2)}(l_\alpha; \Delta^2),\ B_3(\hat{\boldsymbol{\theta}}_N^{PT}) = \boldsymbol{\delta}'\boldsymbol{\delta}\left[G_{q+2,m}^{(2)}(l_\alpha; \Delta^2)\right]^2 \\ \boldsymbol{M}_3(\hat{\boldsymbol{\theta}}_N^{PT}) &= \boldsymbol{\Sigma}_{\gamma_o} - \boldsymbol{H}\boldsymbol{\Sigma}_{\gamma_o}\boldsymbol{H}G_{q+2,m}^{(1)}(l_\alpha; \Delta^2) \\ &\quad + \boldsymbol{\delta}\boldsymbol{\delta}'\left[2G_{q+2,m}^{(2)}(l_\alpha; \Delta^2) - G_{q+4,m}^{(2)}(l_\alpha; \Delta^2)\right], \\ R_3(\hat{\boldsymbol{\theta}}_N^{PT}; \boldsymbol{Q}) &= tr(\boldsymbol{Q}\boldsymbol{\Sigma}_{\gamma_o}) - tr(\boldsymbol{Q}\boldsymbol{H}\boldsymbol{\Sigma}_{\gamma_o}\boldsymbol{H})G_{q+2,m}^{(1)}(l_\alpha; \Delta^2) \\ &\quad + \boldsymbol{\delta}'\boldsymbol{Q}\boldsymbol{\delta}\left[2G_{q+2,m}^{(2)}(l_\alpha; \Delta^2) - G_{q+4,m}^{(2)}(l_\alpha; \Delta^2)\right], \end{aligned} \tag{9.3.3}$$

where $l_\alpha = \frac{qF_{q,m}(\alpha)}{m+qF_{q,m}(\alpha)}$ and $G_{q,m}^{(j)}(l_\alpha; \Delta^2)$ is given by (5.3.4).

For the respective expressions of SE we have

$$
\begin{aligned}
\boldsymbol{b}_4(\hat{\boldsymbol{\theta}}_N^S) &= -dq\boldsymbol{\delta}E^{(2)}[\chi_{q+2}^{-2}(\Delta^2)],\ B_4(\hat{\boldsymbol{\theta}}_N^S) = d^2q^2\boldsymbol{\delta}'\boldsymbol{\delta}\left\{E^{(2)}[\chi_{q+2}^{-2}(\Delta^2)]\right\}^2, \\
\boldsymbol{M}_4(\hat{\boldsymbol{\theta}}_N^S) &= \boldsymbol{\Sigma}_{\gamma_o} - dq\boldsymbol{H}\boldsymbol{\Sigma}_{\gamma_o}\boldsymbol{H}\Big\{(q-2)E^{(1)}[\chi_{q+2}^{-4}(\Delta^2)] + 2\Delta^2E^{(2)}[\chi_{q+4}^{-4}(\Delta^2)]\Big\} \\
&\quad + dq(q+2)\boldsymbol{\delta}\boldsymbol{\delta}'E^{(2)}[\chi_{q+4}^{-4}(\Delta^2)] \\
R_4(\hat{\boldsymbol{\theta}}_N^S;\boldsymbol{Q}) &= tr(\boldsymbol{Q}\boldsymbol{\Sigma}_{\gamma_o}) \\
&\quad - dq tr(\boldsymbol{Q}\boldsymbol{H}\boldsymbol{\Sigma}_{\gamma_o}\boldsymbol{H})\Big\{(q-2)E^{(1)}[\chi_{q+2}^{-4}(\Delta^2)] + 2\Delta^2E^{(2)}[\chi_{q+4}^{-4}(\Delta^2)]\Big\} \\
&\quad + dq(q+2)\boldsymbol{\delta}'\boldsymbol{Q}\boldsymbol{\delta}E^{(2)}[\chi_{q+4}^{-4}(\Delta^2)], \qquad (9.3.4)
\end{aligned}
$$

where $E^{(j)}[\chi_{q+2}^{-2}(\Delta^2)]$ and $E^{(j)}[\chi_{q+2}^{-4}(\Delta^2)]$ are given by (5.3.6) and (5.3.7), respectively.

Similarly, the expressions for the PRSE, $\hat{\boldsymbol{\theta}}_N^{S+}$, are given by

$$
\begin{aligned}
\boldsymbol{b}_5(\hat{\boldsymbol{\theta}}_N^{S+}) &= \boldsymbol{\delta}\bigg\{\frac{dq}{q+2}E^{(2)}\left[F_{p+2,m}^{-1}(\Delta^2)I\left(F_{p+2,m}(\Delta^2) < \frac{dq}{q+2}\right)\right] \\
&\quad - dqE^{(2)}\left[\chi_{q+2}^{-2}(\Delta^2)\right] - G_{q+2,m}^{(2)}(l,\Delta^2)\bigg\}, \\
B_5(\hat{\boldsymbol{\theta}}_N^{S+}) &= \boldsymbol{\delta}'\boldsymbol{\delta}\bigg\{\frac{dq}{q+2}E^{(2)}\left[F_{p+2,m}^{-1}(\Delta^2)I\left(F_{p+2,m}(\Delta^2) < \frac{dq}{q+2}\right)\right] \\
&\quad - dqE^{(2)}\left[\chi_{q+2}^{-2}(\Delta^2)\right] - G_{q+2,m}^{(2)}(l,\Delta^2)\bigg\}^2, \\
\boldsymbol{M}_5(\hat{\boldsymbol{\theta}}_N^{S+}) &= \boldsymbol{M}_4(\hat{\boldsymbol{\theta}}_N^S) - \boldsymbol{H}\boldsymbol{\Sigma}_{\gamma_o}\boldsymbol{H} \\
&\quad \times E^{(1)}\left[\left(1 - \frac{dq}{q+2}F_{q+2,m}^{-1}(\Delta^2)\right)^2 I\left(F_{q+2,m}(\Delta^2) < \frac{dq}{q+2}\right)\right] \\
&\quad - \boldsymbol{\delta}\boldsymbol{\delta}'\bigg\{E^{(2)}\left[\left(1 - \frac{dq}{q+4}F_{q+4,m}^{-1}(\Delta^2)\right) I\left(F_{q+4,m}(\Delta^2) < \frac{dq}{q+4}\right)\right] \\
&\quad + 2E^{(2)}\left[\left(\frac{dq}{q+2}F_{q+2,m}^{-1}(\Delta^2) - 1\right) I\left(F_{q+2,m}(\Delta^2) < \frac{dq}{q+2}\right)\right]\bigg\} \\
R_5(\hat{\boldsymbol{\theta}}_N^{S+};\boldsymbol{Q}) &= R_4(\hat{\boldsymbol{\theta}}_N^S;\boldsymbol{Q}) - tr(\boldsymbol{Q}\boldsymbol{H}\boldsymbol{\Sigma}_{\gamma_o}\boldsymbol{H}) \\
&\quad \times E^{(1)}\left[\left(1 - \frac{dq}{q+2}F_{q+2,m}^{-1}(\Delta^2)\right)^2 I\left(F_{q+2,m}(\Delta^2) < \frac{dq}{q+2}\right)\right] \\
&\quad - \boldsymbol{\delta}'\boldsymbol{Q}\boldsymbol{\delta}\bigg\{E^{(2)}\left[\left(1 - \frac{dq}{q+4}F_{q+4,m}^{-1}(\Delta^2)\right) I\left(F_{q+4,m}(\Delta^2) < \frac{dq}{q+4}\right)\right] \\
&\quad + 2E^{(2)}\left[\left(\frac{dq}{q+2}F_{q+2,m}^{-1}(\Delta^2) - 1\right) I\left(F_{q+2,m}(\Delta^2) < \frac{dq}{q+2}\right)\right]\bigg\},
\end{aligned}
$$

(9.3.5)

where $l = \frac{dq}{m+dq}$, $E^{(2)}\left[F_{q+2,m}^{-1}(\Delta^2)I\left(F_{q+2,m}(\Delta^2) < \frac{dq}{q+2}\right)\right]$ and $E^{(2)}\left[F_{q+2,m}^{-2}(\Delta^2)I\left(F_{q+2,m}(\Delta^2) < \frac{dq}{q+2}\right)\right]$ are given by (5.3.9) and (5.3.10), respectively.

9.4 Risk Analysis of the Estimators

In this section, we compare the proposed estimators in section 2 using their risk expressions. For the purposes of this study, we follow the approaches in Chaudhry (1997) and Saleh (2006). For the orthogonal matrix $\boldsymbol{\Gamma}$, let

$$\boldsymbol{A} = \boldsymbol{\Gamma}\boldsymbol{\Sigma}_{\gamma_o}^{\frac{1}{2}}\boldsymbol{H}\boldsymbol{Q}\boldsymbol{H}\boldsymbol{\Sigma}_{\gamma_o}^{\frac{1}{2}}\boldsymbol{\Gamma}' = \begin{bmatrix} \boldsymbol{A}_1 & \boldsymbol{A}_2 \\ \boldsymbol{A}_3 & \boldsymbol{A}_4 \end{bmatrix}.$$

Hence,

$$tr(\boldsymbol{Q}\boldsymbol{H}\boldsymbol{\Sigma}_{\gamma_o}\boldsymbol{H}) = tr(\boldsymbol{A}_1), \quad \boldsymbol{\delta}'\boldsymbol{Q}\boldsymbol{\delta} = \boldsymbol{\eta}_1'\boldsymbol{A}_1\boldsymbol{\eta}_1, \text{ and } \boldsymbol{\eta}_1'\boldsymbol{\eta}_1 = \Delta^2. \tag{9.4.1}$$

Now, by Courant's theorem (see Theorem A.2.4 of Anderson, 2003),

$$\begin{aligned} \lambda_p(\boldsymbol{A}_1) \leq & \frac{\boldsymbol{\eta}_1'\boldsymbol{A}_1\boldsymbol{\eta}_1}{\boldsymbol{\eta}_1'\boldsymbol{\eta}_1} \leq \lambda_1(\boldsymbol{A}_1) \\ \Delta^2\lambda_p(\boldsymbol{A}_1) \leq & \boldsymbol{\eta}_1'\boldsymbol{A}_1\boldsymbol{\eta}_1 \leq \Delta^2\lambda_1(\boldsymbol{A}_1), \end{aligned} \tag{9.4.2}$$

where $\lambda_p(\boldsymbol{A}_1)$ and $\lambda_1(\boldsymbol{A}_1)$ are the smallest and largest eigenvalues of $\boldsymbol{A}_1$, respectively.

9.4.1 Comparison between $\tilde{\boldsymbol{\theta}}_N$, $\hat{\boldsymbol{\theta}}_N$, and $\hat{\boldsymbol{\theta}}_N^{PT}$

From (9.3.1), (9.3.2), and (9.4.2) we obtain

$$\begin{aligned} R_1(\tilde{\boldsymbol{\theta}}_N;\boldsymbol{Q}) - tr(\boldsymbol{A}_1) + \Delta^2\lambda_p(\boldsymbol{A}_1) &\leq R_2(\hat{\boldsymbol{\theta}}_N;\boldsymbol{Q}) \\ &\leq R_1(\tilde{\boldsymbol{\theta}}_N;\boldsymbol{Q}) - tr(\boldsymbol{A}_1) + \Delta^2\lambda_1(\boldsymbol{A}_1). \end{aligned} \tag{9.4.3}$$

When $\Delta^2 = 0$, the bounds are equal. In conclusion, using (9.4.3), RE dominates (performs better than) UE, denoted by $\hat{\boldsymbol{\theta}}_N \succ \tilde{\boldsymbol{\theta}}_N$, whenever

$$0 \leq \Delta^2 \leq \frac{tr(\boldsymbol{A}_1)}{\lambda_1(\boldsymbol{A}_1)}.$$

However, $\tilde{\boldsymbol{\theta}}_N \succ \hat{\boldsymbol{\theta}}_N$, whenever

$$\Delta^2 \geq \frac{tr(\boldsymbol{A}_1)}{\lambda_p(\boldsymbol{A}_1)}.$$

From (9.3.1), (9.3.3), and (9.4.1) we have

$$\begin{aligned} R_1(\tilde{\boldsymbol{\theta}}_N;\boldsymbol{Q}) - R_3(\hat{\boldsymbol{\theta}}_N^{PT};\boldsymbol{Q}) &= tr(\boldsymbol{A}_1)G_{q+2,m}^{(1)}(l_\alpha,\Delta^2) \\ &\quad -\boldsymbol{\eta}_1'\boldsymbol{A}_1\boldsymbol{\eta}_1\left[2G_{q+2,m}^{(2)}(l_\alpha,\Delta^2) - G_{q+4,m}^{(2)}(l_\alpha,\Delta^2)\right]. \end{aligned} \tag{9.4.4}$$

Thus, $\hat{\boldsymbol{\theta}}_N^{PT} \succ \tilde{\boldsymbol{\theta}}_N$, whenever

$$0 \leq \Delta^2 \leq \frac{tr(\boldsymbol{A}_1)}{\lambda_1(\boldsymbol{A}_1)} \frac{G_{q+2,m}^{(1)}(l_\alpha,\Delta^2)}{2G_{q+2,m}^{(2)}(l_\alpha,\Delta^2) - G_{q+4,m}^{(2)}(l_\alpha,\Delta^2)}.$$

And $\tilde{\boldsymbol{\theta}}_N \succ \hat{\boldsymbol{\theta}}_N^{PT}$, whenever

$$\Delta^2 \geq \frac{tr(\boldsymbol{A}_1)}{\lambda_p(\boldsymbol{A}_1)} \frac{G_{q+2,m}^{(1)}(l_\alpha,\Delta^2)}{2G_{q+2,m}^{(2)}(l_\alpha,\Delta^2) - G_{q+4,m}^{(2)}(l_\alpha,\Delta^2)}.$$

Under the null-hypothesis H_0, $\hat{\boldsymbol{\theta}}_N^{PT} \succ \tilde{\boldsymbol{\theta}}_N$ because the difference in risk (9.4.4) is positive for all level of significance α. It is also obvious that for $\alpha = 1$,

$$R_3(\hat{\boldsymbol{\theta}}_N^{PT};\boldsymbol{Q}) = R_1(\tilde{\boldsymbol{\theta}}_N;\boldsymbol{Q}).$$

The relative efficiency of $\hat{\boldsymbol{\theta}}_N^{PT}$ can be written, for example, as

$$\text{RE}(\hat{\boldsymbol{\theta}}_N^{PT}:\tilde{\boldsymbol{\theta}}_N) = R_1(\tilde{\boldsymbol{\theta}}_N;\boldsymbol{Q})\left[R_2(\hat{\boldsymbol{\theta}}_N^{PT};\boldsymbol{Q})\right]^{-1} = E[\alpha,\Delta^2], \tag{9.4.5}$$

since it depends on α and Δ^2. As we know, PTE is not uniformly better than $\tilde{\boldsymbol{\theta}}_N$. Nevertheless, we can obtain a PTE with a minimum guaranteed efficiency, say, E_0, by choosing a suitable level of significance, α^*, in solving the equation

$$\max_{\alpha}\min_{\Delta^2} E[\alpha,\Delta^2] \geq E_0. \tag{9.4.6}$$

From (9.3.2), (9.3.3) and (9.4.1) we get

$$\begin{aligned} R_2(\hat{\boldsymbol{\theta}}_N;\boldsymbol{Q}) - R_3(\hat{\boldsymbol{\theta}}_N^{PT};\boldsymbol{Q}) &= -tr(\boldsymbol{A}_1)\left[1 - G_{q+2,m}^{(1)}(l_\alpha,\Delta^2)\right] \\ &\quad +\boldsymbol{\eta}_1'\boldsymbol{A}_1\boldsymbol{\eta}_1\left[1 - 2G_{q+2,m}^{(2)}(l_\alpha,\Delta^2)\right. \\ &\quad \left.+G_{q+4,m}^{(2)}(l_\alpha,\Delta^2)\right]. \end{aligned} \tag{9.4.7}$$

Thus, $\hat{\boldsymbol{\theta}}_N \succ \hat{\boldsymbol{\theta}}_N^{PT}$ whenever

$$0 \leq \Delta^2 \leq \frac{tr(\boldsymbol{A}_1)}{\lambda_1(\boldsymbol{A}_1)} \frac{1 - G_{q+2,m}^{(1)}(l_\alpha,\Delta^2)}{1 - 2G_{q+2,m}^{(2)}(l_\alpha,\Delta^2) + G_{q+4,m}^{(2)}(l_\alpha,\Delta^2)}.$$

And $\hat{\boldsymbol{\theta}}_N^{PT} \succ \hat{\boldsymbol{\theta}}_N$ whenever

$$\Delta^2 \geq \frac{tr(\boldsymbol{A}_1)}{\lambda_p(\boldsymbol{A}_1)} \frac{1 - G_{q+2,m}^{(1)}(l_\alpha, \Delta^2)}{1 - 2G_{q+2,m}^{(2)}(l_\alpha, \Delta^2) + G_{q+4,m}^{(2)}(l_\alpha, \Delta^2)}.$$

It is also obvious that for $\alpha = 0$,

$$R_3(\hat{\boldsymbol{\theta}}_N^{PT}; \boldsymbol{Q}) = R_2(\hat{\boldsymbol{\theta}}_N; \boldsymbol{Q}).$$

Therefore, we propose the following result.

Theorem 9.4.1. *Under H_0, the dominance picture may be given by*

$$\hat{\boldsymbol{\theta}}_N \succ \hat{\boldsymbol{\theta}}_N^{PT} \succ \tilde{\boldsymbol{\theta}}_N.$$

9.4.2 Comparison between $\tilde{\boldsymbol{\theta}}_N$, $\hat{\boldsymbol{\theta}}_N$, $\hat{\boldsymbol{\theta}}_N^{PT}$, and $\hat{\boldsymbol{\theta}}_N^{S}$

In order to compare the UE and the SE estimators using the risk difference, first note that the risk of SE can be rewritten as

$$\begin{aligned} R_4(\hat{\boldsymbol{\theta}}_N^S; \boldsymbol{Q}) &= tr(\boldsymbol{Q}\boldsymbol{\Sigma}_{\gamma_o}) - dqtr(\boldsymbol{Q}\boldsymbol{H}\boldsymbol{\Sigma}_{\gamma_o}\boldsymbol{H})\Big\{(q-2)E^{(1)}[\chi_{p+2}^{-4}(\Delta^2)] \\ &+ \left(1 - \frac{(q+2)\boldsymbol{\delta}'\boldsymbol{Q}\boldsymbol{\delta}}{2\Delta^2 tr(\boldsymbol{Q}\boldsymbol{H}\boldsymbol{\Sigma}_{\gamma_o}\boldsymbol{H})}\right) 2\Delta^2 E^{(2)}[\chi_{q+4}^{-4}(\Delta^2)]\Big\}. \end{aligned} \quad (9.4.8)$$

Thus from (9.3.1), (9.4.1) and (9.4.8), the difference risk between UE and SE is given by

$$\begin{aligned} R_1(\tilde{\boldsymbol{\theta}}_N; \boldsymbol{Q}) - R_4(\hat{\boldsymbol{\theta}}_N^S; \boldsymbol{Q}) &= dqtr(\boldsymbol{A}_1)\Big\{(q-2)E^{(1)}[\chi_{q+2}^{-4}(\Delta^2)] \\ &+ \left(1 - \frac{(q+2)\boldsymbol{\eta}_1'\boldsymbol{A}_1\boldsymbol{\eta}_1}{2\Delta^2 tr(\boldsymbol{A}_1)}\right) 2\Delta^2 E^{(2)}[\chi_{q+4}^{-4}(\Delta^2)]\Big\}. \end{aligned} \quad (9.4.9)$$

The difference in risk (9.4.9) is thus positive for all $\mathcal{A}$ such that

$$\left\{\mathcal{A} : \frac{tr(\boldsymbol{A}_1)}{\lambda_1(\boldsymbol{A}_1)} \geq \frac{q+2}{2}\right\}.$$

Thus, uniformly $\hat{\boldsymbol{\theta}}_N^S \succ \tilde{\boldsymbol{\theta}}_N$. Moreover, as $\Delta^2 \to \infty$, the risk difference (9.4.9) approached to 0 from below.

To compare RE and SE, using (9.3.2), (9.3.4), and (9.4.1) we have that

$$R_4(\hat{\boldsymbol{\theta}}_N^S; \boldsymbol{Q}) = R_2(\hat{\boldsymbol{\theta}}_N; \boldsymbol{Q}) - \boldsymbol{\eta}_1'\boldsymbol{A}_1\boldsymbol{\eta}_1 - dqtr(\boldsymbol{A}_1)\Big\{(q-2)E^{(1)}[\chi_{q+2}^{-4}(\Delta^2)]$$

$$+\left(1-\frac{(q+2)\boldsymbol{\eta}_1'\boldsymbol{A}_1\boldsymbol{\eta}_1}{2\Delta^2 tr(\boldsymbol{A}_1)}\right)2\Delta^2 E^{(2)}[\chi_{q+4}^{-4}(\Delta^2)]\Big\}. \qquad (9.4.10)$$

Under the symmetry assumption, i.e., occurring H_0, we get

$$R_4(\hat{\boldsymbol{\theta}}_N^S;\boldsymbol{Q}) = R_2(\hat{\boldsymbol{\theta}}_N;\boldsymbol{Q})+(1-d)tr(\boldsymbol{A}_1) \geq R_2(\hat{\boldsymbol{\theta}}_N;\boldsymbol{Q}), \quad (9.4.11)$$

which shows $\hat{\boldsymbol{\theta}}_N \succ \hat{\boldsymbol{\theta}}_N^S$. However, as Δ^2 increases, the risk of $\hat{\boldsymbol{\theta}}_N$ becomes unbounded, while the risk of $\hat{\boldsymbol{\theta}}_N^S$ remains below the risk of $\tilde{\boldsymbol{\theta}}_N$ and merges with it as $\Delta^2 \to \infty$. Thus, $\hat{\boldsymbol{\theta}}_N^S \succ \hat{\boldsymbol{\theta}}_N$ outside an interval around the origin.

To compare the PTE and the SE, first note that under H_0, we have $G_{q+2,m}^{(1)}(l_\alpha,0) = F_{q+2,m}(l_\alpha^*,0) = 1-\alpha$, where $l_\alpha^* = \frac{qF_{q,m}(\alpha)}{q+2}$. Thus, from (9.3.3), (9.3.4), and (9.4.1), under H_0, we obtain

$$\begin{aligned} R_4(\hat{\boldsymbol{\theta}}_N^S;\boldsymbol{Q}) &= R_3(\hat{\boldsymbol{\theta}}_N^{PT};\boldsymbol{Q}) + tr(\boldsymbol{A}_1)\left[G_{q+2,m}^{(1)}(l_\alpha,0)-d\right] \\ &\geq R_3(\hat{\boldsymbol{\theta}}_N^{PT};\boldsymbol{Q}), \end{aligned} \qquad (9.4.12)$$

as far as, for some suitable α we have $1-\alpha-d \geq 0$, i.e., for all α such that l_α^* satisfies the following inequality:

$$l_\alpha^* > F_{q+2,m}^{-1}(l_\alpha^*,0). \qquad (9.4.13)$$

The risk of RE is smaller than the risk of PTE when l_α^* satisfies the opposite inequality in (9.4.13). Thus, we can propose the following result.

Theorem 9.4.2. *Under H_0, the dominance picture may be given by*

$$\hat{\boldsymbol{\theta}}_N \succ \hat{\boldsymbol{\theta}}_N^{PT} \succ \hat{\boldsymbol{\theta}}_N^S \succ \tilde{\boldsymbol{\theta}}_N,$$

when α satisfies (9.4.13).

In accordance with Theorem 9.4.2, the dominance picture changes as Δ^2 moves away from 0. As $\Delta^2 \to \infty$, the risks of $\hat{\boldsymbol{\theta}}_N^{PT}$ and $\hat{\boldsymbol{\theta}}_N^S$ converge to that of $\tilde{\boldsymbol{\theta}}_N$.

9.4.3 Comparison between $\tilde{\boldsymbol{\theta}}_N$, $\hat{\boldsymbol{\theta}}_N^S$, and $\hat{\boldsymbol{\theta}}_N^{S+}$

Now we compare the SE and the PRSE. Using (9.3.4), (9.3.5), and (9.4.1), we obtain

$$\begin{aligned} &R_5(\hat{\boldsymbol{\theta}}_N^{S+};\boldsymbol{Q}) - R_4(\hat{\boldsymbol{\theta}}_N^S;\boldsymbol{Q}) \\ &= -\Big\{tr(\boldsymbol{A}_1)E^{(1)}\left[\left(1-\frac{dq}{q+2}F_{q+2,m}^{-1}(\Delta^2)\right)^2 I\left(F_{q+2,m}(\Delta^2)<\frac{dq}{q+2}\right)\right] \end{aligned}$$

$$+\boldsymbol{\eta}_1'\boldsymbol{A}_1\boldsymbol{\eta}_1 E^{(2)}\left[\left(1-\frac{dq}{q+4}F_{q+4,m}^{-1}(\Delta^2)\right)^2 I\left(F_{q+4,m}(\Delta^2)<\frac{dq}{q+4}\right)\right]$$
$$-2\boldsymbol{\eta}_1'\boldsymbol{A}_1\boldsymbol{\eta}_1 E^{(2)}\left[\left(\frac{dq}{q+2}F_{q+2,m}^{-1}(\Delta^2)-1\right) I\left(F_{q+2,m}(\Delta^2)<\frac{dq}{q+2}\right)\right]\Bigg\}. \tag{9.4.14}$$

Because the expectation of a positive random variable is positive, the R.H.S. of (9.4.14) is negative and, therefore, $\hat{\boldsymbol{\theta}}_N^{S+} \succ \hat{\boldsymbol{\theta}}_N^{S+}$.

Further, becuase for all Δ^2, we have $\hat{\boldsymbol{\theta}}_N^{S} \succ \tilde{\boldsymbol{\theta}}_N$. Thus, we conclude that for Δ^2, $\hat{\boldsymbol{\theta}}_N^{S+} \succ \tilde{\boldsymbol{\theta}}_N$. In this regard, we have the following result.

Theorem 9.4.3. *The dominance picture may be given by*

$$\hat{\boldsymbol{\theta}}_N^{S+} \succ \hat{\boldsymbol{\theta}}_N^{S} \succ \tilde{\boldsymbol{\theta}}_N.$$

The above scenario can be similarly repeated for MSE analysis. The superiority results will be the same. See Chaudhry (1997) for more details.

9.5 Simple Multivariate Linear Model

In this section, we take the simple multivariate linear model into account, and in the same fashion as in section 9.2, we propose the estimators under study for the parameters of interest in the model, as well as their respective properties.

Consider the simple linear model

$$\boldsymbol{Y}_\alpha = \boldsymbol{\theta} + \boldsymbol{\beta}\boldsymbol{x}_\alpha + \boldsymbol{\varepsilon}_\alpha\ ; \qquad \alpha = 1,\ldots,N, \tag{9.5.1}$$

where $\boldsymbol{Y}_\alpha = (Y_{\alpha 1},\ldots,Y_{\alpha p})'$, $\boldsymbol{\theta} = (\theta_1,\ldots,\theta_p)'$ is the *intercept* vector, $\boldsymbol{\beta} = (\beta_1,\ldots,\beta_p)'$ is *slope* vector, $\boldsymbol{x}_\alpha = (x_1,\ldots,x_p)'$ is the p-vector of fixed constants for every α, and $\boldsymbol{\varepsilon}_\alpha = (\varepsilon_{\alpha 1},,\ldots,\varepsilon_{\alpha p})'$ is the vector of errors, having the distribution $M_t^{(p)}(\boldsymbol{0},\boldsymbol{\Sigma},\gamma_o)$ for each $\alpha = 1,\ldots,N$.

The point unrestricted estimators of $\boldsymbol{\theta}$ and $\boldsymbol{\beta}$ are given by

$$\begin{aligned}\tilde{\boldsymbol{\theta}}_N &= \bar{\boldsymbol{Y}}_N - \tilde{\boldsymbol{\beta}}_N\bar{x}_N,\\ \tilde{\boldsymbol{\beta}}_N &= \frac{1}{Q_N}\left\{\boldsymbol{x}'\begin{pmatrix}\boldsymbol{Y}_1\\ \vdots\\ \boldsymbol{Y}_N\end{pmatrix} - \frac{1}{N}(\boldsymbol{1}_N'\boldsymbol{x})\left(\boldsymbol{1}_N'\begin{pmatrix}\boldsymbol{Y}_1\\ \vdots\\ \boldsymbol{Y}_N\end{pmatrix}\right)\right\},\end{aligned} \tag{9.5.2}$$

with $Q_N = \boldsymbol{x}'\boldsymbol{x} - \frac{1}{N}(\boldsymbol{1}_N'\boldsymbol{x})$, $\boldsymbol{x} = (\boldsymbol{x}_1,\ldots,\boldsymbol{x}_N)'$, $\bar{x}_N = \frac{1}{N}\boldsymbol{1}_N'\boldsymbol{x}$, and $\bar{\boldsymbol{Y}}_N = \frac{1}{N}\boldsymbol{1}_N'\begin{pmatrix}\boldsymbol{Y}_1\\ \vdots\\ \boldsymbol{Y}_N\end{pmatrix}$.

The point estimator of $\mathbf{\Sigma}$ is given by

$$\boldsymbol{S}_u = (N-2)^{-1}\sum_{\alpha=1}^{N}\boldsymbol{Z}\boldsymbol{Z}', \tag{9.5.3}$$

where

$$\boldsymbol{Z} = (\boldsymbol{Y}_\alpha - \bar{\boldsymbol{Y}}_N) - \tilde{\boldsymbol{\beta}}_N(\boldsymbol{x}_\alpha - \bar{\boldsymbol{x}}_N).$$

For the test of null hypothesis $H_0 : \boldsymbol{\beta} = \mathbf{0}$, the LR-test is given by

$$\mathcal{L}_n = \frac{m}{p}Q_N\tilde{\boldsymbol{\beta}}_N'\boldsymbol{S}_u^{-1}\tilde{\boldsymbol{\beta}}_N, \tag{9.5.4}$$

which follows the central F-distribution with (p, m) d.f. under H_0.

9.5.1 More Estimators for $\boldsymbol{\beta}$ and $\boldsymbol{\theta}$

Now, we propose the expressions for preliminary test, Stein-type, and its positive part estimators. Let $F_{p,m}(\alpha)$ be the α-level critical value of the $\mathcal{L}_n$-statistic under H_0. Then we have

$$\hat{\boldsymbol{\beta}}_N^{PT} = \tilde{\boldsymbol{\beta}}_N - \tilde{\boldsymbol{\beta}}_N I(\mathcal{L}_n < F_{p,m}(\alpha)), \tag{9.5.5}$$

$$\hat{\boldsymbol{\beta}}_N^{S} = \tilde{\boldsymbol{\beta}}_N - d\tilde{\boldsymbol{\beta}}_N\mathcal{L}_n^{-1}, \quad d = \frac{(p-2)m}{p(m+2)}, \tag{9.5.6}$$

$$\hat{\boldsymbol{\beta}}_N^{S+} = \hat{\boldsymbol{\beta}}_N^{S} - \tilde{\boldsymbol{\beta}}_N(1 - d\mathcal{L}_n^{-1})I(\mathcal{L}_n < d). \tag{9.5.7}$$

Similarly, by knowing that the restricted estimator of $\boldsymbol{\theta}$ is $\hat{\boldsymbol{\theta}}_N = \bar{\boldsymbol{Y}}$, the corresponding estimators for $\boldsymbol{\theta}$ are given by

$$\begin{aligned}\hat{\boldsymbol{\theta}}_N^{PT} &= \bar{\boldsymbol{Y}}I(\mathcal{L}_n < F_{p,m}(\alpha)) + (\bar{\boldsymbol{Y}} - \tilde{\boldsymbol{\beta}}_N\bar{\boldsymbol{x}}_N)\\ &= \tilde{\boldsymbol{\theta}}_N + \tilde{\boldsymbol{\beta}}_N\bar{\boldsymbol{x}}_N I(\mathcal{L}_n < F_{p,m}(\alpha)),\end{aligned} \tag{9.5.8}$$

$$\hat{\boldsymbol{\theta}}_N^{S} = \tilde{\boldsymbol{\theta}}_N + d\tilde{\boldsymbol{\beta}}_N\bar{\boldsymbol{x}}_N\mathcal{L}_n^{-1}, \tag{9.5.9}$$

and

$$\begin{aligned}\hat{\boldsymbol{\theta}}_N^{S+} &= \hat{\boldsymbol{\theta}}_N + (\tilde{\boldsymbol{\theta}}_N - \hat{\boldsymbol{\theta}}_N)(1 - d\mathcal{L}_n^{-1})I(\mathcal{L}_n > d)\\ &= \hat{\boldsymbol{\theta}}_N^{S} + \tilde{\boldsymbol{\theta}}_N\bar{\boldsymbol{x}}_N\left[1 - (1 - d\mathcal{L}_n^{-1})I(\mathcal{L}_n < d)\right].\end{aligned} \tag{9.5.10}$$

9.5.2 Bias, Quadratic Bias, and MSE Expressions

In this section, like as section 9.3, we propose some expressions for the characteristics of the given estimators. To save space, we state that the risk of any estimator is the trace of MSE expression, thus ignoring it. Note that, in this part, for convenience, we assume $B(\boldsymbol{\theta}_N^*) = Q_N[\boldsymbol{b}(\boldsymbol{\theta}_N^*)]'\boldsymbol{\Sigma}_{\gamma_o}^{-1}[\boldsymbol{b}(\boldsymbol{\theta}_N^*)]$, i.e., we replace N by Q_N in the definition of quadratic bias as it was in section 9.3.

We categorize the results into five different cases as follows:

(i) Unrestricted Estimator
Directly we have $\boldsymbol{b}_1(\tilde{\boldsymbol{\beta}}_N) = \mathbf{0}$, and $B_1(\tilde{\boldsymbol{\beta}}_N) = 0$.
Also

$$\boldsymbol{M}_1(\tilde{\boldsymbol{\beta}}_N) = \frac{1}{Q_N}\boldsymbol{\Sigma}_{\gamma_o}, \tag{9.5.11}$$

where $\boldsymbol{\Sigma}_{\gamma_o} = -2\psi'(0)\boldsymbol{\Sigma} = \frac{\gamma_o}{\gamma_o-2}\boldsymbol{\Sigma}$.

In the same fashion, $\boldsymbol{b}_1(\tilde{\boldsymbol{\theta}}_N) = \mathbf{0}$ and $B_1(\tilde{\boldsymbol{\theta}}_N) = 0$. And

$$\boldsymbol{M}_1(\tilde{\boldsymbol{\theta}}_N) = \left(\frac{1}{N} + \frac{\bar{\boldsymbol{x}}_N^2}{Q_N}\right)\boldsymbol{\Sigma}_{\gamma_o}. \tag{9.5.12}$$

(ii) Restricted Estimator

$$\boldsymbol{b}_2(\hat{\boldsymbol{\beta}}_N) = -\boldsymbol{\beta}, \quad B_2(\hat{\boldsymbol{\beta}}_N) = \Delta_*^2, \quad \text{and} \quad \boldsymbol{M}_2(\hat{\boldsymbol{\beta}}_N) = \boldsymbol{\beta}\boldsymbol{\beta}', \tag{9.5.13}$$

where

$$\Delta_*^2 = Q_N\boldsymbol{\beta}'\boldsymbol{\Sigma}_{\gamma_o}^{-1}\boldsymbol{\beta}. \tag{9.5.14}$$

Note that Δ_*^2 in (9.5.14) can be derived from the non-null distribution of $\mathcal{L}_n$.

$$\begin{aligned}
\boldsymbol{b}_2(\hat{\boldsymbol{\theta}}_N) &= \boldsymbol{\beta}\bar{\boldsymbol{x}}_N, \quad B_2(\hat{\boldsymbol{\theta}}_N) = \bar{\boldsymbol{x}}_N^2\Delta_*^2, \quad \text{and} \\
\boldsymbol{M}_2(\hat{\boldsymbol{\theta}}_N) &= \left(\frac{-2}{N} + \frac{\bar{\boldsymbol{x}}_N^2}{Q_N}\Delta_*^2\right)\boldsymbol{\Sigma}_{\gamma_o}.
\end{aligned} \tag{9.5.15}$$

(iii) PTE
For the preliminary test estimator we have

$$\boldsymbol{b}_3(\hat{\boldsymbol{\beta}}_N^{PTE}) = -\boldsymbol{\beta}G_{p+2,m}^{(2)}(l_\alpha;\Delta_*^2),$$

and

$$B_3(\hat{\boldsymbol{\beta}}_N^{PTE}) = \Delta_*^2\left[G_{p+2,m}^{(2)}(l_\alpha;\Delta_*^2)\right]^2$$

$$\begin{aligned}
\boldsymbol{M}_3(\hat{\boldsymbol{\beta}}_N^{PTE}) &= \left[\frac{\boldsymbol{\Sigma}_{\gamma_o}}{Q_N}\left(1 - G_{p+2,m}^{(1)}(l_\alpha;\Delta_*^2)\right)\right] \\
&\quad +\boldsymbol{\beta}\boldsymbol{\beta}'\left[2G_{p+2,m}^{(2)}(l_\alpha;\Delta_*^2) - G_{p+4,m}^{(2)}(l_\alpha;\Delta_*^2)\right],
\end{aligned} \tag{9.5.16}$$

$$\boldsymbol{b}_3(\hat{\boldsymbol{\theta}}_N^{PTE}) = \boldsymbol{\beta}\bar{\boldsymbol{x}}_N G_{p+2,m}^{(1)}(l_\alpha; \Delta_*^2)$$

and

$$B_3(\hat{\boldsymbol{\theta}}_N^{PTE}) = \bar{\boldsymbol{x}}_N^2 \Delta_*^2 \left[G_{p+2,m}^{(1)}(l_\alpha; \Delta_*^2)\right]^2$$

$$\begin{aligned} \boldsymbol{M}_3(\hat{\boldsymbol{\theta}}_N^{PTE}) = \boldsymbol{M}_2(\hat{\boldsymbol{\theta}}_N) - \frac{\bar{\boldsymbol{x}}_N^2}{Q_N} \boldsymbol{\Sigma}_{\gamma_o} G_{p+2,m}^{(1)}(l_\alpha; \Delta_*^2) \\ + \bar{\boldsymbol{x}}_N^2 \boldsymbol{\beta}\boldsymbol{\beta}' \left[2G_{p+2,m}^{(2)}(l_\alpha; \Delta_*^2) - G_{p+4,m}^{(2)}(l_\alpha; \Delta_*^2)\right] \end{aligned} \quad (9.5.17)$$

(iv) SE

Similar expressions can be proposed for the Stein-type estimator as given by

$$\boldsymbol{b}_4(\hat{\boldsymbol{\beta}}_N^S) = -dp\boldsymbol{\beta}E^{(2)}[\chi_{p+2}^{-2}(\Delta_*^2)] \quad (9.5.18)$$

and

$$B_4(\hat{\boldsymbol{\beta}}_N^S) = d^2p^2\Delta_*^2 \left[E^{(2)}[\chi_{p+2}^{-2}(\Delta_*^2)]\right]^2 \quad (9.5.19)$$

$$\begin{aligned} \boldsymbol{M}_4(\hat{\boldsymbol{\beta}}_N^S) &= \left[\frac{dp}{Q_N}\left(\frac{1}{dp} - 2E^{(2)}[\chi_{p+2}^{-2}(\Delta_*^2)] + (p-2)E^{(2)}[\chi_{p+2}^{-4}(\Delta_*^2)]\right)\right] \boldsymbol{\Sigma}_{\gamma_o} \\ &\quad + dp(p+2)\boldsymbol{\beta}\boldsymbol{\beta}' E^{(2)}[\chi_{p+4}^{-4}(\Delta_*^2)] \end{aligned}$$

$$\boldsymbol{b}_4(\hat{\boldsymbol{\theta}}_N^S) = dp\boldsymbol{\beta}\bar{\boldsymbol{x}}_N E^{(2)}[\chi_{p+2}^{-2}(\Delta_*^2)]$$

and

$$B_4(\hat{\boldsymbol{\theta}}_N^S) = d^2p^2\bar{\boldsymbol{x}_N}^2\Delta_*^2 \left[E^{(2)}[\chi_{p+2}^{-2}(\Delta_*^2)]\right]^2,$$

$$\begin{aligned} \boldsymbol{M}_4(\hat{\boldsymbol{\theta}}_N^S) &= \boldsymbol{M}_2(\hat{\boldsymbol{\theta}}_N^S) - \frac{dp\bar{\boldsymbol{x}}_N^2}{Q_N} \boldsymbol{\Sigma}_{\gamma_o} \\ &\quad \times \left(2E^{(1)}[\chi_{p+2}^{-2}(\Delta_*^2)] - (p-2)E^{(1)}[\chi_{p+2}^{-4}(\Delta_*^2)]\right) \\ &\quad + dp(p+2)\bar{\boldsymbol{x}}_N^2\boldsymbol{\beta}\boldsymbol{\beta}' E^{(2)}[\chi_{p+4}^{-4}(\Delta_*^2)]. \end{aligned} \quad (9.5.20)$$

(v) PRSE

Finally, for the positive rule Stein-type estimator we have

$$\begin{aligned}
\boldsymbol{b}_5(\hat{\boldsymbol{\beta}}_N^{S+}) &= \boldsymbol{\beta}\Big\{d_1 E^{(2)}[F_{p+2,m}^{-1}(\Delta_*^2)] \\
&\quad + E^{(2)}[(1 - d_1 F_{p+2,m}^{-1}(\Delta_*^2))I(F_{p+2,m}(\Delta_*^2) < d_1)]\Big\} \\
B_5(\hat{\boldsymbol{\beta}}_N^{S+}) &= \Delta_*^2\Big\{d_1 E^{(2)}[F_{p+2,m}^{-1}(\Delta_*^2)] \\
&\quad + E^{(2)}[(1 - d_1 F_{p+2,m}^{-1}(\Delta_*^2))I(F_{p+2,m}(\Delta_*^2) < d_1)]\Big\}^2 \\
\boldsymbol{M}_5(\hat{\boldsymbol{\beta}}_N^{S+}) &= \boldsymbol{M}_4(\hat{\boldsymbol{\beta}}_N^{S}) \\
&\quad - \frac{\boldsymbol{\Sigma}_{\gamma_o}}{Q_N} E^{(1)}[(1 - d_1 F_{p+2,m}^{-1}(\Delta_*^2))^2 I(F_{p+2,m}(\Delta_*^2) < d_1)] \\
&\quad + \boldsymbol{\beta}\boldsymbol{\beta}'\Big\{2E^{(2)}[(1 - d_1 F_{p+2,m}^{-1}(\Delta_*^2))I(F_{p+2,m}(\Delta_*^2) < d_1)] \\
&\quad - E^{(2)}[(1 - d_2 F_{p+4,m}^{-1}(\Delta_*^2))^2 I(F_{p+4,m}(\Delta_*^2) < d_2)]\Big\}.
\end{aligned} \tag{9.5.21}$$

$$\begin{aligned}
\boldsymbol{b}_5(\hat{\boldsymbol{\theta}}_N^{S+}) &= \boldsymbol{b}_4(\hat{\boldsymbol{\theta}}_N^{S}) \\
&\quad + \boldsymbol{\beta}\bar{\boldsymbol{x}}_N E^{(1)}[(1 - d_1 F_{p+2,m}^{-1}(\Delta_*^2))I(F_{p+2,m}(\Delta_*^2) < d_1)] \\
B_5(\hat{\boldsymbol{\theta}}_N^{S+}) &= \bar{\boldsymbol{x}}_N^2 B_5(\hat{\boldsymbol{\beta}}_N^{S+}) \\
\boldsymbol{M}_5(\hat{\boldsymbol{\theta}}_N^{S+}) &= \boldsymbol{M}_4(\hat{\boldsymbol{\theta}}_N^{S}) - \frac{\bar{\boldsymbol{x}}_N^2}{Q_N}\boldsymbol{\Sigma}_{\gamma_o} \\
&\quad \times E^{(1)}[(1 - d_1 F_{p+2,m}^{-1}(\Delta_*^2))I(F_{p+2,m}(\Delta_*^2) < d_1)] \\
&\quad + \bar{\boldsymbol{x}}_N^2\boldsymbol{\beta}\boldsymbol{\beta}'\Big\{2E^{(2)}[(1 - d_1 F_{p+2,m}^{-1}(\Delta_*^2))I(F_{p+2,m}(\Delta_*^2) < d_1)] \\
&\quad - E^{(2)}[(1 - d_2 F_{p+4,m}^{-1}(\Delta_*^2))^2 I(F_{p+4,m}(\Delta_*^2) < d_2)]\Big\},
\end{aligned} \tag{9.5.22}$$

where $d_i = \frac{dp}{p+2i}$, $i = 1, 2$.

For the comparison of the proposed estimators, the reader is referred to the problems.

9.6 Problems

1. Prove that the following equalities hold:

$$E^{(2)}[\chi_q^{-2}(\Delta^2)] - E^{(2)}[\chi_{q+2}^{-2}(\Delta^2)] \quad = \quad 2E^{(2)}[\chi_{q+2}^{-4}(\Delta^2)],$$

$$E^{(1)}[\chi_{q+2}^{-2}(\Delta^2)] - (q-2)E^{(1)}[\chi_{q+2}^{-4}(\Delta^2)] = \Delta^2 E^{(2)}[\chi_{q+4}^{-4}(\Delta^2)].$$

(Hint: Use (2.2.13d) and (2.2.13e) of Saleh, 2006).

2. Specify the expressions (9.3.1)-(9.3.5).

3. Provide tables for maximum and minimum efficiencies of SE and efficiency of PTE at Δ_0 using (4.4.10).

4. According to the result of section 9.4, do the comparison between $\hat{\boldsymbol{\theta}}_N$, $\hat{\boldsymbol{\theta}}_N^S$ and $\hat{\boldsymbol{\theta}}_N^{S+}$.

5. Do the MSE analysis for the estimators in section 9.2.

6. Prove (9.5.4).

7. Compare the proposed estimators of the intercept parameter based on the MSE criterion in section 9.5.

8. Consider model (9.5.1). Set $\boldsymbol{x} = (0, \cdots, 0, 1, \cdots, 1)'$ containing N_1 zero's and N_2 one's. Further, let $\mu_1 = \theta$ and $\mu_2 = \theta + \beta$. Show that the test statistic for testing $H_o : \mu_1 = \mu_2$ against $H_1 : \mu_1 \neq \mu_2$ is Hotelling's T^2-test and derive its non-null distribution.

9. Referring to problem 8, define unrestricted, restricted, PT, S, and PRS estimators of μ_1 and their respective bias and MSE expressions.

CHAPTER 10

BAYESIAN ANALYSIS

Outline

Bayesian analysis has become an influential topic in modern statistics. In this chapter we discuss the Bayesian analysis when the error distribution is the multivariate t-model.

10.1 Introduction (Zellner's Model)

Zellner (1976) was first to initiate the use of multivariate t-error in from a Bayesian analysis of regression models. In his seminal paper, he considered linear multivariate

Statistical Inference for Models with Multivariate t-Distributed Errors, First Edition.

A. K. Md. Ehsanes Saleh, M. Arashi, S.M.M. Tabatabaey.

t-regression models under Bayesian viewpoint. We consider the multiple regression model (7.1.1) as

$$\boldsymbol{y} = \boldsymbol{X}\boldsymbol{\beta} + \boldsymbol{\varepsilon}, \tag{10.1.1}$$

where $\boldsymbol{y} = (y_1, \ldots, y_n)'$ is an $(n \times 1)$ vector of observations, $\boldsymbol{X} = (\boldsymbol{x}'_1, \ldots, \boldsymbol{x}'_n)'$ is a nonstochastic $(n \times p)$ matrix of full rank p, $\boldsymbol{\beta}$ is a $(p \times 1)$ vector of unknown regression coefficients, and $\boldsymbol{\varepsilon}$ is an $(n \times 1)$ random error-vector distributed as $M_t^{(n)}(\boldsymbol{0}, \sigma^2\boldsymbol{I}_n, \gamma_o)$ (in this case). With this error assumption, we get into the theory for uncorrelated but dependent errors, making the analysis more applicable in real-life situations.

To begin with, we assume that the prior knowledge for $\boldsymbol{\beta}$ and σ^2 is a diffuse (noninformative/vague/flat) prior with the pdf

$$p(\boldsymbol{\beta}, \sigma^2) \propto \frac{1}{\sigma^2}, \tag{10.1.2}$$

where the elements of $\boldsymbol{\beta}$ and $\log\sigma^2$ are uniformly and independently distributed. Then, the posterior pdf for the parameters is given by

$$\begin{aligned}
p(\boldsymbol{\beta}, \sigma^2|\boldsymbol{y}) &\propto (\sigma^2)^{-(n+2)}\left\{\gamma_o + \frac{1}{\sigma^2}(\boldsymbol{y} - \boldsymbol{X}\boldsymbol{\beta})'(\boldsymbol{y} - \boldsymbol{X}\boldsymbol{\beta})\right\}^{-\frac{1}{2}(\gamma_o+\gamma_1)} \\
&\propto \frac{(\sigma^2)^{\frac{\gamma_o}{2}-1}}{(\sigma_*^2)^{\frac{n+\gamma_o}{2}}}\left\{\gamma_1 + \frac{1}{\sigma_*^2}(\boldsymbol{\beta} - \hat{\boldsymbol{\beta}})'\boldsymbol{X}'\boldsymbol{X}(\boldsymbol{\beta} - \hat{\boldsymbol{\beta}})\right\}^{-\frac{1}{2}(k+\gamma_1)} \\
&\propto [A(\boldsymbol{\beta})]^{-\frac{n}{2}}\left\{\frac{[B(\boldsymbol{\beta})]^{\frac{\gamma_o}{2}-1}}{A(\boldsymbol{\beta})[1 + B(\boldsymbol{\beta})]^{\frac{1}{2}(n+\gamma_o)}}\right\},
\end{aligned} \tag{10.1.3}$$

where

$$\begin{aligned}
&\gamma_1 = \gamma_o + \gamma, \quad \gamma = n - p, \\
&\sigma_*^2 = (\gamma_o\sigma^2 + \gamma s^2)/\gamma_1, \quad \gamma s^2 = (\boldsymbol{y} - \boldsymbol{X}\hat{\boldsymbol{\beta}})'(\boldsymbol{y} - \boldsymbol{X}\hat{\boldsymbol{\beta}}), \quad \hat{\boldsymbol{\beta}} = (\boldsymbol{X}'\boldsymbol{X})^{-1}\boldsymbol{X}'\boldsymbol{y}, \\
&B(\boldsymbol{\beta}) = \gamma_o\sigma^2/A(\boldsymbol{\beta}), \quad A(\boldsymbol{\beta}) = (\boldsymbol{y} - \boldsymbol{X}\boldsymbol{\beta})'(\boldsymbol{y} - \boldsymbol{X}\boldsymbol{\beta}).
\end{aligned}$$

It can be concluded that the conditional posterior pdf of $\boldsymbol{\beta}$ given σ^2 is of multivariate t-pdf with mean $\hat{\boldsymbol{\beta}}$ and covariance (see Zellner, 1976):

$$\begin{aligned}
Cov(\boldsymbol{\beta}|\sigma^2, \boldsymbol{y}) &= \frac{\gamma_1\sigma_*^2}{\gamma_1 - 2}(\boldsymbol{X}'\boldsymbol{X})^{-1} \\
&= \frac{\gamma_o\sigma^2 + \gamma s^2}{\gamma_o + \gamma - 2}(\boldsymbol{X}'\boldsymbol{X})^{-1}, \quad \gamma_1 > 2.
\end{aligned} \tag{10.1.4}$$

It can be shown that the marginal posterior distribution is again a multivariate t with the pdf (see Zellner, 1976):

$$\begin{aligned}
p(\boldsymbol{\beta}|\boldsymbol{y}) &\propto [A(\boldsymbol{\beta})]^{-\frac{n}{2}} \\
&\propto \left\{\gamma s^2 + (\boldsymbol{\beta} - \hat{\boldsymbol{\beta}})'\boldsymbol{X}'\boldsymbol{X}(\boldsymbol{\beta} - \hat{\boldsymbol{\beta}})\right\}^{-\frac{1}{2}(\gamma+p)}.
\end{aligned} \tag{10.1.5}$$

Further, Zellner (1976) showed that the marginal posterior distribution of σ^2 is the central F-distribution with (γ_o, γ) d.f.

Clearly, one may note that the posterior distribution of $\boldsymbol{\beta}$ under the t-model is the same as for the normal linear model. However, the posterior distribution of the scale parameter is affected by departures from normality.

With the above background, the idea of a natural conjugate prior (NCP) for the multivariate t-model is the product of a marginal F-pdf for σ^2 times a conditional p-dim multivariate t-pdf for $\boldsymbol{\beta}$ given σ^2, i.e.,

$$\pi(\boldsymbol{\beta}, \sigma^2) = p_F(\sigma^2) p_S(\boldsymbol{\beta}|\sigma^2), \tag{10.1.6}$$

where $p_F(\sigma^2)$ denotes an F pdf and $p_S(\boldsymbol{\beta}|\sigma^2)$ a conditional multivariate t-pdf. Recently, Arashi, Iranmanesh, Norouzirad, and Salarzadeh Jenatabadi (2013) developed the above methodology for matrix elliptically contoured models.

Although Zellner is credited with the Bayesian analysis with t-errors, many others have used non-normal errors. In fact, it was Box and Tiao (1973) who took power exponential distribution as the error term to motivate the use of non-normal error. Later, under the Bayesian perspective, many other systematic studies for regression models with non-normal errors came into the literature. The reader is refered to the works of Singh, Misra, and Pandey (1995); Jammalamadaka, Tiwari, and Chib(1987); Chib, Osiewalski, and Steel (1991); Osiewalski and Steel (1993); and, more recently, Fang and Li (1999), Arellano-Valle, Galea-Rojas, and Iglesias (2000); Ng (2002, 2010); Arashi (2010); Vidal and Arellano-Valle (2010); and Tsukuma (2010), to mention a few.

10.2 Conditional Bayesian Inference

In this section, we consider the Bayesian inference for the multivariate t-regression model in a more detailed analysis. The material is well developed for a broader class of distributions in Arellano-Valle, Galea-Rojas, and Iglesias (2000).

Again, consider the regression model (10.1.1), where

$$\boldsymbol{\varepsilon} \sim M_t^{(n)}(\mathbf{0}, \phi^{-1}\boldsymbol{I}_n, \gamma_o), \quad \phi = \frac{1}{\sigma^2}. \tag{10.2.1}$$

We recall that $\boldsymbol{y}$ can be expressed as (using the weight representation (2.6.3))

$$\int \mathcal{N}_n(\boldsymbol{X}\boldsymbol{\beta}, t^{-1}\phi^{-1}\boldsymbol{I}_n) W_o(t)dt,$$

where

$$\boldsymbol{y}|t \sim \mathcal{N}_n(\boldsymbol{X}\boldsymbol{\beta}, t^{-1}\phi^{-1}\boldsymbol{I}_n), \quad \text{and} \quad t \sim W_o. \tag{10.2.2}$$

Thus, for the Bayesian analysis with the t-model, one may use the normal theory due to the representation (10.2.2). In this regard, one may consider the following conditional NCP guess given by

$$\boldsymbol{\beta}|\phi, t \sim \mathcal{N}_p(\boldsymbol{b}_o, \phi^{-1}\boldsymbol{B}_o), \quad \phi|t \sim G\left(\frac{d_o}{2}, \frac{d_o c_o}{2}\right), \tag{10.2.3}$$

where G is the gamma distribution. This gives rise to the following structure, conditioning on t (see Arellano-Valle, Galea-Rojas, and Iglesias, 2000):

$$\pi^*(\boldsymbol{\beta}, \phi|t) \propto \phi^{\frac{d_o+p}{2}-1} \exp\left\{-\left(\frac{\phi}{2}\right)\left[d_o c_o + (\boldsymbol{\beta} - \boldsymbol{b}_o)'\boldsymbol{B}_o^{-1}(\boldsymbol{\beta} - \boldsymbol{b}_o)\right]\right\}, \tag{10.2.4}$$

for the known required hyper parameters.

Thus, we may use the following structure:

$$\pi^*(\boldsymbol{\beta}, \phi) = \int W_o(t)\pi^*(\boldsymbol{\beta}, \phi|t)dt, \tag{10.2.5}$$

as one plausible suggestion for a NCP distribution.

The NCP distribution as in (10.2.5) is the basis of the studies of Ng (2010, 2012) and Arashi, Iranmanesh, Norouzirad, and Salarzadeh Jenatabadi (2013), which will be considered in the next section.

To conclude the result of this section, by some algebra, the conditional posterior distribution is given by

$$\pi(\boldsymbol{\beta}, \phi|\boldsymbol{y}, t) \propto \phi^{\frac{d+p}{2}-1} \exp\left\{-\left(\frac{\phi t}{2}\right)\left[d\nu(t) + (\boldsymbol{\beta} - \boldsymbol{b}(t))'\left(t^{-1}\boldsymbol{B}(t)\right)^{-1}(\boldsymbol{\beta} - \boldsymbol{b}(t))\right]\right\},$$

where

$$\begin{aligned} d &= n + d_o \\ \boldsymbol{b}(t) &= \boldsymbol{B}(t)\left(t^{-1}\boldsymbol{B}_o^{-1}\boldsymbol{b}_o + \boldsymbol{X}'\boldsymbol{y}\right), \quad \boldsymbol{B}(t) = \left(t^{-1}\boldsymbol{B}_o^{-1} + \boldsymbol{X}'\boldsymbol{X}\right)^{-1}, \\ \nu(t) &= td^{-1}\left[c_o t^{-1} + (\boldsymbol{y} - \boldsymbol{X}\boldsymbol{b}(t))'\boldsymbol{y} + (\boldsymbol{b}_o - \boldsymbol{b}(t))'\boldsymbol{B}(t)^{-1}\boldsymbol{b}_o\right]. \end{aligned}$$

We should not forget that

$$\boldsymbol{\beta}|\boldsymbol{y}, t \sim M_t^{(p)}\left(\boldsymbol{b}(t), t^{-1}\nu(t)\boldsymbol{B}(t), d\right). \tag{10.2.6}$$

Thus incorporating the NCP distribution (10.2.5), the posterior distribution of $\boldsymbol{\beta}$ is given by

$$\boldsymbol{\beta}|\boldsymbol{y} \sim \int M_t^{(p)}\left(\boldsymbol{b}(t), t^{-1}\nu(t)\boldsymbol{B}(t), d\right) W_o(t)dt. \tag{10.2.7}$$

Further important characteristics of $\boldsymbol{\beta}|\boldsymbol{y}$ are

$$\begin{aligned} E(\boldsymbol{\beta}|\boldsymbol{y}) &= E(t^{-1}\boldsymbol{B}(t)|\boldsymbol{y})\boldsymbol{B}_o^{-1}\boldsymbol{b}_o + E(\boldsymbol{B}(t)|\boldsymbol{y})\boldsymbol{X}'\boldsymbol{y} \\ V(\boldsymbol{\beta}|\boldsymbol{y}) &= V(\boldsymbol{b}(t)|y) + (d-2)^{-1}E(t^{-1}\nu(t)\boldsymbol{B}(t)|\boldsymbol{y}). \end{aligned}$$

Note that taking $t = 1$, the result reduces to that of under the normal theory.

10.3 Matrix Variate t-Distribution

Before proceeding to the Bayesian analysis for the matrix case, we bring some necessary tools, such as the definition of matrix variate/valued t-distribution in this section.

Let $\boldsymbol{X}$ be an $n \times p$ random matrix, which can be expressed in terms of its elements, column, and rows as

$$\boldsymbol{X} = (x_{ij}) = (\boldsymbol{x}_1, \cdots, \boldsymbol{x}_p) = (\boldsymbol{x}_{(1)}, \cdots, \boldsymbol{x}_{(n)})'. \tag{10.3.1}$$

Here $\boldsymbol{x}_{(1)}, \cdots, \boldsymbol{x}_{(n)}$ can be regarded as a sample of size n from a p-dim population.

In our setup, we need to start with the matrix normal distribution, denoted by $\mathcal{N}_{n,p}(\boldsymbol{M}, \boldsymbol{\Sigma}, \boldsymbol{\Omega})$ or $\mathcal{N}_{n,p}(\boldsymbol{M}, \boldsymbol{\Sigma} \otimes \boldsymbol{\Omega})$, where $\otimes$ is the Kronecker product, with the following pdf:

$$f(\boldsymbol{X}|\boldsymbol{M}, \boldsymbol{\Sigma}, \boldsymbol{\Omega}) = \frac{|\boldsymbol{\Omega}|^{-\frac{n}{2}}|\boldsymbol{\Sigma}|^{-\frac{p}{2}}}{(2\pi)^{\frac{np}{2}}} e^{\left\{-\frac{1}{2}\operatorname{tr}\left[\boldsymbol{\Omega}^{-1}(\boldsymbol{X}-\boldsymbol{M})'\boldsymbol{\Sigma}^{-1}(\boldsymbol{X}-\boldsymbol{M})\right]\right\}}, \tag{10.3.2}$$

where $\boldsymbol{X} = (x_{ij})$, is in $\mathbb{R}^{n\times p}$, $\boldsymbol{M} \in \mathbb{R}^{n\times p}$ and $\boldsymbol{\Sigma} = (\sigma_{ik})$ and $\boldsymbol{\Omega} = (\omega_{jl})$ are $p \times p$ and $n \times n$ positive definite matrices respectively. Then we have

$$E(\boldsymbol{X}) = \boldsymbol{M}, \quad Cov(\boldsymbol{X}) = \boldsymbol{\Sigma} \otimes \boldsymbol{\Omega}, \text{ i.e., } Cov(x_{ij}, x_{kl}) = \omega_{ik}\sigma_{jl}. \tag{10.3.3}$$

Note that using the vectorial operator (see Kollo and von Rosen, 2005) we have

$$\operatorname{vec}\boldsymbol{X} \sim \mathcal{N}_{np}(\operatorname{vec}\boldsymbol{M}, \boldsymbol{\Sigma} \otimes \boldsymbol{\Omega}).$$

The matrix variate t (MT) distribution with mean $\boldsymbol{M}$, row and column scale matrices $\boldsymbol{\Omega}$ and $\boldsymbol{\Sigma}$, respectively, and γ_o d.f. denoted by $M_t^{(n,p)}(\boldsymbol{M}, \boldsymbol{\Sigma}, \boldsymbol{\Omega}, \gamma_o)$, has the pdf

$$f(\boldsymbol{Y}|\boldsymbol{\Sigma}, \boldsymbol{\Omega}, \gamma_o) = \frac{|\boldsymbol{\Omega}|^{-\frac{n}{2}}|\boldsymbol{\Sigma}|^{-\frac{p}{2}}}{g_{n,p}} \left[1 + \frac{1}{\gamma_o}\boldsymbol{\Omega}^{-1}(\boldsymbol{X}-\boldsymbol{M})'\boldsymbol{\Sigma}^{-1}(\boldsymbol{X}-\boldsymbol{M})\right]^{-\frac{\gamma_o+np}{2}}, \tag{10.3.4}$$

where $\boldsymbol{Y} \in \mathbb{R}^{n\times p}$ and the normalizing constant is given by

$$g_{n,p} = \frac{(\gamma_o\pi)^{\frac{np}{2}}\Gamma_p\left(\frac{\gamma_o}{2}\right)}{\Gamma_p\left(\frac{\gamma_o+np}{2}\right)},$$

where

$$\begin{aligned}\Gamma_p(a) &= \int_{\boldsymbol{A}>\boldsymbol{0}} |\boldsymbol{A}|^{a-\frac{1}{2}(p+1)} e^{tr(-\boldsymbol{A})} d\boldsymbol{A} \\ &= \pi^{\frac{1}{4}p(p-1)} \prod_{i=1}^{p} \Gamma\left(a - \frac{i-1}{2}\right)\end{aligned}$$

is the multivariate gamma function.

Thus, the MT distribution can be written as

$$f(\boldsymbol{Y}|\boldsymbol{\Sigma},\boldsymbol{\Omega},\gamma_o) = \int W_o(t)\mathcal{N}_{n,p}(\boldsymbol{M},t^{-1}\boldsymbol{\Sigma}\otimes\boldsymbol{\Omega})dt, \tag{10.3.5}$$

where $W_o(t)$ is given by (2.6.4). This representation was used by Gupta and Varga (1995), as well as by Chu (1973).

We present some necessary tools for our pursuit to follow as given below:

A partition λ of n is a sequence $\lambda = (\lambda_1, \cdots, \lambda_l)$, where the $\lambda_j \geq 0$ are weakly decreasing and $\sum_j \lambda_j = n$. We denote this by $\lambda \vdash j$. The number of nonzero parts of λ is called the length of λ, denoted $l(\lambda)$.

Let $\boldsymbol{Y}$ be an $m \times m$ symmetric matrix with latent roots $y_1, \cdots, y_m$ and let $\kappa = (k_1, \cdots, k_m)$ be a partition of k into not more than m parts. The zonal polynomial of $\boldsymbol{Y}$ corresponding to κ, denoted by $C_\kappa(\boldsymbol{Y})$, is a symmetric, homogeneous polynomial of degree k in the latent roots $y_1, \cdots, y_m$ such that:

(i) The term of highest weight in $C_\kappa(\boldsymbol{Y})$ is $y_1^{k_1}, \cdots y_m^{k_m}$; that is,
(1) $C_\kappa(\boldsymbol{Y}) = d_k y_1^{k_1}, \cdots y_m^{k_m}$+terms of lower weight,
where d_k is a constant.

(ii) $C_\kappa(\boldsymbol{Y})$ is an eigenfunction of the differential operator $\triangle_{\boldsymbol{Y}}$ given by
(2) $\triangle_{\boldsymbol{Y}} = \sum_{i=1}^m y_i^2 \frac{\partial^2}{\partial y_i^2} + \sum_{i=1}^m \sum_{j=1, j\neq i}^m \frac{y_i^2}{y_i - y_j}\frac{\partial}{\partial y_i}$.

(iii) As κ varies over all partitions of k, the zonal polynomial has unit coefficients in the expansion of $(tr\boldsymbol{Y})^k$; that is,
(3) $(tr\boldsymbol{Y})^k = (y_1 + \cdots + y_m)^k = \sum_\kappa C_\kappa(\boldsymbol{Y})$.

See section 7.2 of Muirhead (2005) for more details.

Lemma 10.3.1. *(Teng, Fang and Deng, 1989) Let $\boldsymbol{Z}$ be a complex symmetric $p \times p$ matrix with $Re(\boldsymbol{Z}) > 0$ and let $\boldsymbol{Y}$ be a symmetric $p \times p$ matrix, if k is a non-negative integer and κ is a partition of k. Then, for $Re(a) > (p-1)/2$*

$$\int_{\boldsymbol{X}>0} g(tr\boldsymbol{X}\boldsymbol{Z})|\boldsymbol{X}|^{\frac{a-(p+1)}{2}} C_\kappa(\boldsymbol{X}\boldsymbol{Y})d\boldsymbol{X} = \frac{|\boldsymbol{Z}|^{-a}(a)_\kappa\Gamma_p(a)C_\kappa(\boldsymbol{Y}\boldsymbol{Z}^{-1})}{\Gamma(pa+k)} S,$$

where,

$$S = \int_0^\infty g(w)w^{pa+k-1}dw < \infty,$$

$(a)_\kappa$ is the generalized hypergeometric coefficient (generalized Pochhammer symbol of weight κ) given by

$$(a)_\kappa = \prod_{i=1}^p \left(a - \frac{i-1}{2}\right)_{k_i} = \frac{\pi^{\frac{1}{4}p(p-1)}}{\Gamma_p(a)} \prod_{i=1}^p \Gamma\left(a + k_i - \frac{i-1}{2}\right),$$

and $(a)_i = a(a+1)\dots(a+i-1)$.

As a direct consequence of Lemma 10.3.1, putting $\kappa = 0$ yields $(a)_\kappa = 1$ and $C_\kappa(\boldsymbol{Y}\boldsymbol{Z}^{-1}) = 1$, thus we have the following result.

Lemma 10.3.2. *Let* $\boldsymbol{Z}$ *be a complex symmetric* $p \times p$ *matrix with* $Re(\boldsymbol{Z}) > 0$ *and let* $\boldsymbol{Y}$ *be a symmetric* $p \times p$ *matrix, if* k *is a non-negative integer and* κ *is a partition of* k. *Then, for* $Re(a) > (p-1)/2$

$$\int_{\boldsymbol{X}>0} g(tr\boldsymbol{X}\boldsymbol{Z})|\boldsymbol{X}|^{\frac{a-(p+1)}{2}} d\boldsymbol{X} = \frac{|\boldsymbol{Z}|^{-a}\Gamma_p(a)}{\Gamma(pa)} S^*,$$

where,

$$S^* = \int_0^\infty g(w)w^{pa-1}dw < \infty.$$

The following is the generalization of Sverdrup's (1947) lemma.

Lemma 10.3.3. *(Kabe, 1965) Let* $\boldsymbol{G}$ *and* $\boldsymbol{\Sigma}$ *be positive definite symmetric* $p \times p$ *matrices and* $\boldsymbol{\delta} \in \mathbb{R}^p$, *then*

$$\int_{\boldsymbol{G}>\boldsymbol{0}} |\boldsymbol{G}|^{\frac{1}{2}(N-p-1)}(\boldsymbol{\delta}'\boldsymbol{G}\boldsymbol{\delta})^r \exp\left[-\frac{1}{2}tr(\boldsymbol{\Sigma}^{-1}\boldsymbol{G})\right] d\boldsymbol{G}$$
$$= \frac{(2\pi)^{\frac{1}{2}pN}2^{p+r}\Gamma(\frac{1}{2}N+r)}{\prod_{i=1}^p C(N-p+i)\Gamma\left(\frac{1}{2}N\right)} |\boldsymbol{\Sigma}|^{\frac{1}{2}N}(\boldsymbol{\delta}'\boldsymbol{\Sigma}\boldsymbol{\delta})^r,$$

where $C(n)$ *represents the surface area of a unit* n*-dim sphere.*

10.4 Bayesian Analysis in Multivariate Regression Model

In this section, we consider a multivariate regression model with MT errors. For a precise setup consider the following regression model:

$$\boldsymbol{Y} = \boldsymbol{B}\boldsymbol{X} + \boldsymbol{E}, \tag{10.4.1}$$

where the n columns of the $p \times n$ response matrix $\boldsymbol{Y}$ can be regarded as a sample of size n from a p-dimensional population, $\boldsymbol{X}$ is the $k \times n$ design matrix of known values of rank k, $\boldsymbol{B}$ is the $p \times k$ matrix of unknown regression parameters, and $n > p + k$. The $p \times n$ error matrix $\boldsymbol{E}$ is assumed to have a MT distribution, $M_t^{(p,n)}(\boldsymbol{0}, \boldsymbol{\Phi}^{-1} \otimes \boldsymbol{I}_n, \gamma_o)$.

Now, following Ng (2010, 2012), we consider the weight representation (10.3.5) and let $\mathbf{\Sigma} = t\mathbf{\Phi}$ with the Jacobian of transformation $J(\mathbf{\Sigma} \to \mathbf{\Phi}) = t^{\frac{p(p+1)}{2}}$. Further, adopt a normal-Wishart prior for $(\boldsymbol{B}, \mathbf{\Sigma})$ as

$$\pi(\boldsymbol{B}|\mathbf{\Sigma}) \propto |\mathbf{\Sigma}|^{\frac{k}{2}} etr\left\{-\frac{1}{2}\,\mathbf{\Sigma}(\boldsymbol{B}-\boldsymbol{B}^*)\mathbf{\Upsilon}(\boldsymbol{B}-\boldsymbol{B}^*)'\right\}, \tag{10.4.2}$$

$$\pi(\mathbf{\Sigma}) \propto |\mathbf{\Sigma}|^{\nu-\frac{p+1}{2}} etr\left\{-\frac{1}{2}\,\mathbf{\Omega}\mathbf{\Sigma}\right\}, \tag{10.4.3}$$

where ν, $p \times k$ matrix $\boldsymbol{B}^*$, $k \times k$ matrix $\mathbf{\Upsilon}$, and $p \times p$ matrix $\mathbf{\Omega}$ are all known hyperparameters.

Suppose

$$\begin{aligned}\pi(\boldsymbol{B},\mathbf{\Phi}|t) &\propto \pi(\boldsymbol{B}|\mathbf{\Sigma})\pi(\mathbf{\Sigma})J(\mathbf{\Sigma}\to\mathbf{\Phi})\\ &\propto t^{\frac{p(\nu+k)}{2}}|\mathbf{\Phi}|^{\frac{\nu+k}{2}-\frac{p+1}{2}} etr\left\{-\frac{1}{2}\,t\mathbf{\Phi}\left[(\boldsymbol{B}-\boldsymbol{B}^*)\mathbf{\Upsilon}(\boldsymbol{B}-\boldsymbol{B}^*)'+\mathbf{\Omega}\right]\right\}.\end{aligned} \tag{10.4.4}$$

Then, similar to (10.2.5), the NCP distribution for the MT distribution can be obtained as

$$\pi(\boldsymbol{B},\mathbf{\Phi}) \propto \int_0^\infty \pi(\boldsymbol{B},\mathbf{\Phi}|t)W_o(t)dt. \tag{10.4.5}$$

Theorem 10.4.1. *Using the conjugate prior on $(\boldsymbol{B}, \mathbf{\Phi})$ given by* (10.4.5)*, the posterior distribution of the regression parameter $\boldsymbol{B}$ for the multivariate regression model* (10.4.1) *is MT with the following pdf:*

$$\begin{aligned}f(\boldsymbol{B}|\boldsymbol{Y}) &= c(p,k,\nu)|\boldsymbol{X}\boldsymbol{X}'+\mathbf{\Upsilon}|^{\frac{p}{2}}|\mathbf{\Psi}|^{\frac{k}{2}}\\ &\quad\times|\boldsymbol{I}_p+\mathbf{\Psi}^{-1}(\boldsymbol{B}-\hat{\boldsymbol{B}})'(\boldsymbol{X}\boldsymbol{X}'+\mathbf{\Upsilon})(\boldsymbol{B}-\hat{\boldsymbol{B}})|^{-\frac{1}{2}(n+\nu+k)},\end{aligned} \tag{10.4.6}$$

where

$$\begin{aligned}c(p,k,\nu)^{-1} &= \left[\Gamma\left(\frac{1}{2}\right)\right]^{pk}\frac{\Gamma_p\left[\frac{1}{2}(n+\nu)\right]}{\Gamma_p\left[\frac{1}{2}(n+\nu+k)\right]},\\ \hat{\boldsymbol{B}} &= (\boldsymbol{Y}\boldsymbol{X}'+\boldsymbol{B}^*\mathbf{\Upsilon})(\boldsymbol{X}\boldsymbol{X}'+\mathbf{\Upsilon})^{-1},\\ \mathbf{\Psi} &= \boldsymbol{Y}\boldsymbol{Y}'+\boldsymbol{B}^*\mathbf{\Upsilon}\boldsymbol{B}^{*\prime}-\hat{\boldsymbol{B}}(\boldsymbol{Y}\boldsymbol{X}'+\boldsymbol{B}^*\mathbf{\Upsilon})'+\mathbf{\Omega}.\end{aligned} \tag{10.4.7}$$

Proof: Let $f(\boldsymbol{Y}|\boldsymbol{B}, \mathbf{\Phi}, t)$ be the density function of $\boldsymbol{Y}$ under the normality assumption; then, by definition we have

$$\begin{aligned}f(\boldsymbol{B}|\boldsymbol{Y}) &\propto \int\int_{\mathbf{\Phi}>\mathbf{0}} f(\boldsymbol{Y}|\boldsymbol{B},\mathbf{\Phi},t)\pi(\boldsymbol{B},\mathbf{\Phi}|t)d\mathbf{\Phi}W_o(t)dt\\ &\propto \int f(\boldsymbol{B}|Y,t)W_o(t)dt,\end{aligned}$$

where

$$\begin{aligned}
f(\boldsymbol{B}|Y,t) &= \int_{\boldsymbol{\Phi}>\boldsymbol{0}} f(\boldsymbol{Y}|\boldsymbol{B},\boldsymbol{\Phi},t)\pi(\boldsymbol{B},\boldsymbol{\Phi}|t)d\boldsymbol{\Phi} \\
&\propto \int t^{\frac{p(n+\nu+k)}{2}} |\boldsymbol{\Phi}|^{\frac{n+\nu+k}{2}-\frac{p+1}{2}} \\
&\quad \times etr\left\{-\frac{1}{2}\, t\boldsymbol{\Phi}(\boldsymbol{Y}-\boldsymbol{B}\boldsymbol{X})(\boldsymbol{Y}-\boldsymbol{B}\boldsymbol{X})'\right\} \\
&\quad \times etr\left\{-\frac{1}{2}\, t\boldsymbol{\Phi}\left[(\boldsymbol{B}-\boldsymbol{B}^*)\boldsymbol{\Upsilon}(\boldsymbol{B}-\boldsymbol{B}^*)'+\boldsymbol{\Omega}\right]\right\} d\boldsymbol{\Phi}. \quad (10.4.8)
\end{aligned}$$

Using the fact that

$$\begin{aligned}
&(\boldsymbol{Y}-\boldsymbol{X}\boldsymbol{B})(\boldsymbol{Y}-\boldsymbol{X}\boldsymbol{B})'+(\boldsymbol{B}-\boldsymbol{B}^*)\boldsymbol{\Upsilon}(\boldsymbol{B}-\boldsymbol{B}^*)'+\boldsymbol{\Omega} \\
&\qquad = (\boldsymbol{B}-\hat{\boldsymbol{B}})(\boldsymbol{X}\boldsymbol{X}'+\boldsymbol{\Upsilon})(\boldsymbol{B}-\hat{\boldsymbol{B}})'+\boldsymbol{\Psi} \quad (10.4.9)
\end{aligned}$$

the expression in (10.4.8) simplifies to

$$\begin{aligned}
f(\boldsymbol{B}|\boldsymbol{Y},t) &\propto \int_{\boldsymbol{\Phi}>\boldsymbol{0}} t^{\frac{1}{2}p(n+\nu+k)} |\boldsymbol{\Phi}|^{\frac{n+\nu+k}{2}-\frac{p+1}{2}} \\
&\quad \times etr\left\{-\frac{t}{2}\left[(\boldsymbol{B}-\hat{\boldsymbol{B}})(\boldsymbol{X}\boldsymbol{X}'+\boldsymbol{\Upsilon})(\boldsymbol{B}-\hat{\boldsymbol{B}})'+\boldsymbol{\Psi}\right]\boldsymbol{\Phi}\right\} d\boldsymbol{\Phi}.
\end{aligned}$$

Now make the transformation

$$\boldsymbol{Z} = t\left[(\boldsymbol{B}-\hat{\boldsymbol{B}})(\boldsymbol{X}\boldsymbol{X}'+\boldsymbol{\Upsilon})(\boldsymbol{B}-\hat{\boldsymbol{B}})'+\boldsymbol{\Psi}\right]\boldsymbol{\Phi}$$

with the Jacobian of the transformation $|t\left[(\boldsymbol{B}-\hat{\boldsymbol{B}})(\boldsymbol{X}\boldsymbol{X}'+\boldsymbol{\Upsilon})(\boldsymbol{B}-\hat{\boldsymbol{B}})'+\boldsymbol{\Psi}\right]|^{-\frac{p+1}{2}}$ to get

$$\begin{aligned}
f(\boldsymbol{B}|\boldsymbol{Y},t) &\propto \int_{\boldsymbol{Z}>\boldsymbol{0}} t^{\frac{1}{2}p(n+\nu+k)} \\
&\quad \times \left| \left[(\boldsymbol{B}-\hat{\boldsymbol{B}})(\boldsymbol{X}\boldsymbol{X}'+\boldsymbol{\Upsilon})(\boldsymbol{B}-\hat{\boldsymbol{B}})'+\boldsymbol{\Psi}\right]^{-1} t^{-1}\boldsymbol{Z}\right|^{\frac{n+\nu+k}{2}-\frac{p+1}{2}} \\
&\quad \times \left| t\left[(\boldsymbol{B}-\hat{\boldsymbol{B}})(\boldsymbol{X}\boldsymbol{X}'+\boldsymbol{\Upsilon})(\boldsymbol{B}-\hat{\boldsymbol{B}})'+\boldsymbol{\Psi}\right]\right|^{-\frac{p+1}{2}} etr\left(-\frac{1}{2}\,\boldsymbol{Z}\right) d\boldsymbol{Z} \\
&\propto \left|\left[(\boldsymbol{B}-\hat{\boldsymbol{B}})(\boldsymbol{X}\boldsymbol{X}'+\boldsymbol{\Upsilon})(\boldsymbol{B}-\hat{\boldsymbol{B}})'+\boldsymbol{\Psi}\right]\right|^{-\frac{n+\nu+k}{2}} \\
&\quad \times \int_{\boldsymbol{Z}>\boldsymbol{0}} |\boldsymbol{Z}|^{\frac{n+\nu+k}{2}-\frac{p+1}{2}} etr\left(-\frac{1}{2}\,\boldsymbol{Z}\right) d\boldsymbol{Z} \\
&\propto \left|\left[(\boldsymbol{B}-\hat{\boldsymbol{B}})(\boldsymbol{X}\boldsymbol{X}'+\boldsymbol{\Upsilon})(\boldsymbol{B}-\hat{\boldsymbol{B}})'+\boldsymbol{\Psi}\right]\right|^{-\frac{n+\nu+k}{2}}.
\end{aligned}$$

Thus, we conclude that

$$f(\boldsymbol{B}|\boldsymbol{Y}) \propto \left|\left[(\boldsymbol{B}-\hat{\boldsymbol{B}})(\boldsymbol{X}\boldsymbol{X}'+\boldsymbol{\Upsilon})(\boldsymbol{B}-\hat{\boldsymbol{B}})'+\boldsymbol{\Psi}\right]\right|^{-\frac{n+\nu+k}{2}} \int W_o(t)dt$$

$$= \left| \left[(\boldsymbol{B} - \hat{\boldsymbol{B}})(\boldsymbol{X}\boldsymbol{X}' + \boldsymbol{\Upsilon})(\boldsymbol{B} - \hat{\boldsymbol{B}})' + \boldsymbol{\Psi} \right] \right|^{-\frac{n+\nu+k}{2}}.$$

Remark 10.4.1. *The posterior distribution of $\boldsymbol{B}$ under matrix normal responses is identical to the MT distribution above (see Broemeling 1985, p.379). Thus one can carry out the inference for $\boldsymbol{B}$ and find the highest posterior density (HPD) region of $\boldsymbol{B}$ similar to the result of section 8.4 of Box and Tiao (1992).*

Ng (2010) showed that when random responses in a multivariate regression model are assumed to have multivariate scale mixtures of normal distributions, the Bayesian analysis using a prior in the conjugate family yields a posterior distribution of the regression parameters identical to those obtained under independently distributed normal responses. The marginal distribution of the regression parameters is therefore invariant to a wider class of distributions of the responses in a Bayesian analysis using an informative prior.

Theorem 10.4.2. *Using the conjugate prior on $(\boldsymbol{B}, \boldsymbol{\Phi})$ given by* (10.4.6), *the posterior distribution of the precision matrix $\boldsymbol{\Phi}$ for the multivariate regression model* (10.4.1) *is the generalized Wishart denoted by $\boldsymbol{\Phi}|\boldsymbol{Y} \sim GW_p(\boldsymbol{\Psi}^{-1}, n+\nu)$ with the following density:*

$$f(\boldsymbol{\Phi}|\boldsymbol{Y}) = \frac{\Gamma\left(\frac{p(n+\nu)+\gamma_o}{2}\right)}{\Gamma_p\left(\frac{n+\nu}{2}\right)\Gamma\left(\frac{\gamma_o}{2}\right)\gamma_o^{\frac{1}{2}p(n+\nu)}} |\boldsymbol{\Psi}|^{\frac{n+\nu}{2}} \times |\boldsymbol{\Phi}|^{\frac{n+\nu}{2}-\frac{p+1}{2}} \left(1 + \frac{1}{\gamma_o} tr\boldsymbol{\Psi}\boldsymbol{\Phi}\right)^{-\frac{1}{2}[p(n+\nu)+\gamma_o]}.$$

Proof: Using the notation defined in Theorem 11.4.1, we have

$$\begin{aligned}
f(\boldsymbol{\Phi}|\boldsymbol{Y}) &\propto \int\int_{\boldsymbol{B}\in\mathbb{R}^{p\times k}} f(\boldsymbol{Y}|\boldsymbol{B}, \boldsymbol{\Phi}, t)\pi(\boldsymbol{B}, \boldsymbol{\Phi}|t)W_o(t)d\boldsymbol{B}dt \\
&\propto \int\int_{\boldsymbol{B}\in\mathbb{R}^{p\times k}} t^{\frac{1}{2}p(n+\nu+k)}|\boldsymbol{\Phi}|^{\frac{n+\nu+k}{2}-\frac{p+1}{2}} \\
&\quad \times etr\left\{-\frac{t}{2}\left[(\boldsymbol{B} - \hat{\boldsymbol{B}})(\boldsymbol{X}\boldsymbol{X}' + \boldsymbol{\Upsilon})(\boldsymbol{B} - \hat{\boldsymbol{B}})' + \boldsymbol{\Psi}\right]\boldsymbol{\Phi}\right\} W_o(t)d\boldsymbol{B}dt \\
&= |\boldsymbol{\Phi}|^{\frac{n+\nu+k}{2}-\frac{p+1}{2}} \int t^{\frac{1}{2}p(n+\nu+k)} etr\left(-\frac{t}{2}\boldsymbol{\Psi}\boldsymbol{\Phi}\right) \\
&\quad \times \left(\int_{\boldsymbol{B}\in\mathbb{R}^{p\times k}} etr\left\{-\frac{t}{2}(\boldsymbol{B} - \hat{\boldsymbol{B}})(\boldsymbol{X}\boldsymbol{X}' + \boldsymbol{\Upsilon})(\boldsymbol{B} - \hat{\boldsymbol{B}})'\boldsymbol{\Phi}\right\} dB\right) \\
&\quad W_o(t)dt.
\end{aligned} \tag{10.4.10}$$

By the fact that the integral of matrix normal density is equal to one, we have

$$\int_{\boldsymbol{B}\in\mathbb{R}^{p\times k}} etr\left\{-\frac{t}{2}(\boldsymbol{B} - \hat{\boldsymbol{B}})(\boldsymbol{X}\boldsymbol{X}' + \boldsymbol{\Upsilon})(\boldsymbol{B} - \hat{\boldsymbol{B}})'\boldsymbol{\Phi}\right\} dB$$

$$= (2\pi)^{\frac{kp}{2}} t^{-\frac{kp}{2}} |\mathbf{\Phi}|^{-\frac{k}{2}} |\mathbf{X}\mathbf{X}' + \mathbf{\Upsilon}|^{-\frac{p}{2}}. \qquad (10.4.11)$$

Substituting (10.4.11) in (10.4.10) yields

$$\begin{aligned} f(\mathbf{\Phi}|\mathbf{Y}) &= w_{\mathbf{\Phi}} (2\pi)^{\frac{kp}{2}} |\mathbf{\Phi}|^{\frac{n+\nu}{2} - \frac{p+1}{2}} |\mathbf{X}\mathbf{X}' + \mathbf{\Upsilon}|^{-\frac{p}{2}} \\ &\times \int t^{\frac{1}{2}p(n+\nu)} etr\left(-\frac{t}{2}\mathbf{\Psi}\mathbf{\Phi}\right) W_o(t)dt, \end{aligned}$$

where $w_{\mathbf{\Phi}}$ is the normalizing constant. Because $\int_{\mathbf{\Phi}>\mathbf{0}} f(\mathbf{\Phi}|\mathbf{Y})d\mathbf{\Phi} = 1$, from the Wishart integral one can find

$$\begin{aligned} w_{\mathbf{\Phi}}^{-1} &= \int_{\mathbf{\Phi}>\mathbf{0}} (2\pi)^{\frac{kp}{2}} |\mathbf{\Phi}|^{\frac{n+\nu}{2} - \frac{p+1}{2}} |\mathbf{X}\mathbf{X}' + \mathbf{\Upsilon}|^{-\frac{p}{2}} \\ &\times \int t^{\frac{1}{2}p(n+\nu)} etr\left(-\frac{t}{2}\mathbf{\Psi}\mathbf{\Phi}\right) W_o(t)dtd\mathbf{\Phi} \\ &= (2\pi)^{\frac{kp}{2}} |\mathbf{X}\mathbf{X}' + \mathbf{\Upsilon}|^{-\frac{p}{2}} \\ &\times \int t^{\frac{1}{2}p(n+\nu)} \int_{\mathbf{\Phi}>\mathbf{0}} |\mathbf{\Phi}|^{\frac{n+\nu}{2} - \frac{p+1}{2}} etr\left(-\frac{t}{2}\mathbf{\Psi}\mathbf{\Phi}\right) d\mathbf{\Phi} W_o(t)dt \\ &= (2\pi)^{\frac{kp}{2}} |\mathbf{X}\mathbf{X}' + \mathbf{\Upsilon}|^{-\frac{p}{2}} \\ &\times \int t^{\frac{1}{2}p(n+\nu)} \Gamma_p\left(\frac{n+\nu}{2}\right) |\frac{t\mathbf{\Psi}}{2}|^{-\frac{n+\nu}{2}} W_o(t)dt \\ &= 2^{\frac{1}{2}p(n+\nu+k)} \pi^{\frac{kp}{2}} \Gamma_p\left(\frac{n+\nu}{2}\right) |\mathbf{\Psi}|^{-\frac{n+\nu}{2}} |\mathbf{X}\mathbf{X}' + \mathbf{\Upsilon}|^{-\frac{p}{2}}, \end{aligned}$$

inasmuch as $\int W_o(t)dt = 1$. Proof will be completed by knowing the fact that

$$\begin{aligned} \mathcal{I} &= \int t^{\frac{1}{2}p(n+\nu)} etr\left(-\frac{t}{2}\mathbf{\Psi}\mathbf{\Phi}\right) W_o(t)dt \\ &= \int t^{\frac{1}{2}p(n+\nu)} etr\left(-\frac{t}{2}\mathbf{\Psi}\mathbf{\Phi}\right) \frac{1}{\Gamma\left(\frac{\gamma_o}{2}\right)} \left(\frac{\gamma_o t}{2}\right)^{\frac{\gamma_o}{2}} e^{-\frac{\gamma_o t}{2}} t^{-1} dt \\ &= \left(\frac{2}{\gamma_o}\right)^{\frac{1}{2}p(n+\nu)} \frac{\Gamma\left(\frac{p(n+\nu)+\gamma_o}{2}\right)}{\Gamma\left(\frac{\gamma_o}{2}\right)} \left(1 + \frac{1}{\gamma_o} tr\mathbf{\Psi}\mathbf{\Phi}\right)^{-\frac{1}{2}[p(n+\nu)+\gamma_o]}. \end{aligned}$$

Remark 10.4.2. *Taking a better look at Theorem 10.4.2, it is easy to realize that the generalized Wishart (GW) distribution is nothing more than the distribution of* $t^{-1}\mathbf{W}$, *where* $\mathbf{W} \sim W_p(\mathbf{\Psi}^{-1}, n+\nu)$. *Thus, one can do the inference on* $\mathbf{\Phi}$ *or* $\mathbf{\Sigma}$ *similar to the result of section 8.5 of Box and Tiao (1992).*

10.4.1 Properties of B and Φ

Now, based on the results of the Theorems 10.4.1 and 10.4.2, we explore the posterior mean and covariance of $\mathbf{B}$ and linear transformation of $\mathbf{\Phi}$ given $\mathbf{Y}$. It can be directly concluded that

$$\begin{aligned} E(\boldsymbol{B}) &= \hat{\boldsymbol{B}}, \quad \text{and} \\ Cov(\boldsymbol{\mathcal{B}}) &= E(\boldsymbol{\mathcal{B}} - \hat{\boldsymbol{\mathcal{B}}})(\boldsymbol{\mathcal{B}} - \hat{\boldsymbol{\mathcal{B}}})' = \frac{1}{n+\nu-(p+1)} \boldsymbol{\Psi} \otimes (\boldsymbol{X}\boldsymbol{X}' + \boldsymbol{\Upsilon})^{-1}, \end{aligned} \tag{10.4.12}$$

where $\boldsymbol{\mathcal{B}}' = (\boldsymbol{B}_1', \cdots, \boldsymbol{B}_p')$ and $\hat{\boldsymbol{\mathcal{B}}}' = (\hat{\boldsymbol{B}}_1', \cdots, \hat{\boldsymbol{B}}_p')$ are $kp \times 1$ vectors.

Theorem 10.4.3. *Under the assumptions of Theorem 10.4.2 we have*

(i) $E(\boldsymbol{\Phi}) = \frac{(n+\nu)\gamma_o}{\gamma_o - 2} \boldsymbol{\Psi}^{-1}$,

(ii) $E(\boldsymbol{\Phi}^{-1}) = \frac{2\Gamma_p\left(\frac{\gamma_o}{2}+1\right)}{\gamma_o(n+\nu-p-1)\Gamma_p\left(\frac{\gamma_o}{2}\right)} \boldsymbol{\Psi}, \ n+\nu-p-1 > 0$;

(iii)

$$\begin{aligned} E(\boldsymbol{\Phi}^{-1} \otimes \boldsymbol{\Phi}^{-1}) &= \frac{4\Gamma_p\left(\frac{\gamma_o}{2}+2\right)}{\gamma_o^2 \Gamma_p\left(\frac{\gamma_o}{2}\right)(n+\nu-p)(n+\nu-p-1)(n+\nu-p-3)} \\ &\times \Big[(n+\nu-p-2)\boldsymbol{\Psi} \otimes \boldsymbol{\Psi} \\ &+ (\text{vec}\,\boldsymbol{\Psi}\,\text{vec}'\,\boldsymbol{\Psi} + \boldsymbol{K}_{p,p}(\boldsymbol{\Psi} \otimes \boldsymbol{\Psi}))\Big], \ n+\nu-p-3 > 0, \end{aligned}$$

(iv)

$$\begin{aligned} E(tr(\boldsymbol{\Phi}^{-1})\boldsymbol{\Phi}^{-1}) &= \frac{4\Gamma_p\left(\frac{\gamma_o}{2}+2\right)}{\gamma_o^2 \Gamma_p\left(\frac{\gamma_o}{2}\right)(n+\nu-p)(n+\nu-p-1)(n+\nu-p-3)} \\ &\times \left[2\boldsymbol{\Psi}^2 + (n+\nu-p-2)\boldsymbol{\Psi} tr(\boldsymbol{\Psi})\right], \ n+\nu-p-3 > 0, \end{aligned}$$

(v) $E(|\boldsymbol{\Phi}|^r) = \left(\frac{4}{\gamma_o}\right)^{rp} \frac{\Gamma_p\left(\frac{\gamma_o}{2}+rp\right)\Gamma_p\left(\frac{n+\nu}{2}+r\right)}{\Gamma_p\left(\frac{\gamma_o}{2}\right)\Gamma_p\left(\frac{n+\nu}{2}\right)} |\boldsymbol{\Psi}^{-1}|^r$,

(vi)

$$E\left[(\boldsymbol{\delta}'\boldsymbol{\Phi}\boldsymbol{\delta})^r\right] = \frac{\pi^{\frac{1}{2}\,p(n+\nu)} 2^{p+2r} \Gamma_p\left(\frac{\gamma_o}{2}+r\right)\Gamma_p\left(\frac{n+\nu}{2}+r\right)}{\gamma_o^r \prod_{i=1}^p C(n+\nu-p+i)} \left(\boldsymbol{\delta}'\boldsymbol{\Psi}^{-1}\boldsymbol{\delta}\right)^r,$$

(vii)

$$E\left[C_\kappa(\boldsymbol{A}\boldsymbol{\Phi})\right] = \frac{\gamma_o^k \left(\frac{n+\nu}{2}\right)_\kappa \Gamma\left(\frac{\gamma_o}{2}-k\right)}{\Gamma\left(\frac{\gamma_o}{2}\right)} C_\kappa\left(\boldsymbol{A}\boldsymbol{\Psi}^{-1}\right),$$

where $\boldsymbol{K}_{p,p}$ *is the commutation matrix (see Magnus and Neudecker, 1999),* $C(n)$ *represents the surface area of a unit* n *dimensional sphere, and* $\boldsymbol{A} \in \mathbb{R}^{p\times p}$ *is a known symmetric matrix.*

Proof: (i) Using Remark 10.2.2 we have

$$\begin{aligned} E(\boldsymbol{\Phi}) &= (n+\gamma_o)\int t^{-1}\boldsymbol{\Psi}^{-1}W_o(t)dt \\ &= \boldsymbol{\Psi}^{-1}E(t^{-1}) \\ &= \frac{(n+\gamma_o)\gamma_o}{\gamma_o-2}\boldsymbol{\Psi}^{-1}. \end{aligned}$$

In the same fashion, one can directly deduce (ii)-(v) using the result of Muirhead (2005) and Kollo and von Rosen (2005) and the fact that

$$E(t^i) = \left(\frac{\gamma_o}{2}\right)^{-i}\left(\frac{\Gamma_p\left(\frac{\gamma_o}{2}+i\right)}{\Gamma_p\left(\frac{\gamma_o}{2}\right)}\right).$$

(vi) From Lemma 10.3.3 and Remark 10.4.2 for the r-th moment of linear transformation $\boldsymbol{\delta}'\boldsymbol{\Phi}\boldsymbol{\delta}$, for any $\boldsymbol{\delta}\in\mathbb{R}^p$, we have

$$\begin{aligned} E\left[(\boldsymbol{\delta}'\boldsymbol{\Phi}\boldsymbol{\delta})^r\right] &= \int E\left[(\boldsymbol{\delta}'t^{-1}\boldsymbol{W}\boldsymbol{\delta})^r\right]W_o(t)dt \\ &= \frac{|\boldsymbol{\Psi}|^{\frac{n+\nu}{2}}}{2^{\frac{1}{2}p(n+\nu)}\Gamma_p\left(\frac{n+\nu}{2}\right)}E\left(t^{-r}\right) \\ &\quad\times\int_{\boldsymbol{W}>\boldsymbol{0}}|\boldsymbol{W}|^{\frac{n+\nu}{2}-\frac{p+1}{2}}\left(\boldsymbol{\delta}'\boldsymbol{W}\boldsymbol{\delta}\right)^r\exp\left[-\frac{1}{2}\,tr(\boldsymbol{\Psi}\boldsymbol{W})\right]d\boldsymbol{W} \\ &= \frac{|\boldsymbol{\Psi}|^{\frac{n+\nu}{2}}}{2^{\frac{1}{2}p(n+\nu)}\Gamma_p\left(\frac{n+\nu}{2}\right)}\left(\frac{2}{\gamma_o}\right)^r\frac{\Gamma_p\left(\frac{\gamma_o}{2}+r\right)}{\Gamma_p\left(\frac{\gamma_o}{2}\right)} \\ &\quad\times\frac{(2\pi)^{\frac{1}{2}p(n+\nu)}2^{p+r}\Gamma_p\left(\frac{n+\nu}{2}+r\right)}{\prod_{i=1}^p C(n+\nu-p+i)\Gamma_p\left(\frac{n+\nu}{2}\right)}|\boldsymbol{\Psi}|^{-\frac{n+\nu}{2}}\left(\boldsymbol{\delta}'\boldsymbol{\Psi}^{-1}\boldsymbol{\delta}\right)^r \\ &= \frac{\pi^{\frac{1}{2}p(n+\nu)}2^{p+r}\Gamma_p\left(\frac{n+\nu}{2}+r\right)}{\prod_{i=1}^p C(n+\nu-p+i)}\left(\frac{2}{\gamma_o}\right)^r\frac{\Gamma_p\left(\frac{\gamma_o}{2}+r\right)}{\Gamma_p\left(\frac{\gamma_o}{2}\right)}\left(\boldsymbol{\delta}'\boldsymbol{\Psi}^{-1}\boldsymbol{\delta}\right)^r. \end{aligned}$$

(vii) The result follows by taking $g(x) = (1+x/\gamma_o)^{-[p(n+\nu)+\gamma_o]/2}$ in Lemma 10.3.1 and the fact that, by using the integral of beta type II, we have

$$\begin{aligned} S &= \int_0^\infty w^{\frac{1}{2}p(n+\nu)+k-1}g(w)dw \\ &= \int_0^\infty w^{\frac{1}{2}p(n+\nu)+k-1}\left(1+\frac{1}{\gamma_o}w\right)^{-\frac{1}{2}[p(n+\nu)+\gamma_o]}dw \\ &= \gamma_o^{\frac{1}{2}p(n+\nu)+k}B\left(\frac{1}{2}p(n+\nu)+k,\frac{\gamma_o}{2}-k\right). \end{aligned}$$

10.5 Problems

1. Let $\boldsymbol{W} \sim GW_p(\boldsymbol{\Sigma}, n)$ and $\boldsymbol{A} \in \mathbb{R}^{p\times q}$. Then prove

$$\boldsymbol{A}(\boldsymbol{A}'\boldsymbol{W}^{-1}\boldsymbol{A})^{-}\boldsymbol{A}' \sim GW_p(\boldsymbol{A}(\boldsymbol{A}'\boldsymbol{\Sigma}^{-1}\boldsymbol{A})^{-}\boldsymbol{A}', n-p+r(\boldsymbol{A})),$$

where $(\boldsymbol{A}'\boldsymbol{W}^{-1}\boldsymbol{A})^{-}$ is the generalized inverse of $(\boldsymbol{A}'\boldsymbol{W}^{-1}\boldsymbol{A})$. (Hint: See Theorem 2.4.13 of Kollo and Rosen, 2005.)

2. Let $\boldsymbol{W} \sim GW_p(\boldsymbol{\Sigma}, n)$ and consider the following partitions:

$$\boldsymbol{W} = \begin{pmatrix} \boldsymbol{W}_{11} & \boldsymbol{W}_{12} \\ \boldsymbol{W}_{21} & \boldsymbol{W}_{22} \end{pmatrix} \quad \begin{pmatrix} r\times r & r\times(p-r) \\ (p-r)\times r & (p-r)\times(p-r) \end{pmatrix},$$

where on the R.H.S. the sizes of the matrices are indicated, and

$$\boldsymbol{\Sigma} = \begin{pmatrix} \boldsymbol{\Sigma}_{11} & \boldsymbol{\Sigma}_{12} \\ \boldsymbol{\Sigma}_{21} & \boldsymbol{\Sigma}_{22} \end{pmatrix} \quad \begin{pmatrix} r\times r & r\times(p-r) \\ (p-r)\times r & (p-r)\times(p-r) \end{pmatrix}.$$

Further, let $\boldsymbol{W}_{1.2} = \boldsymbol{W}_{11} - \boldsymbol{W}_{12}\boldsymbol{W}_{22}^{-1}\boldsymbol{W}_{21}$ and $\boldsymbol{\Sigma}_{1.2} = \boldsymbol{\Sigma}_{11} - \boldsymbol{\Sigma}_{12}\boldsymbol{\Sigma}_{22}^{-1}\boldsymbol{\Sigma}_{21}$. Then prove $\boldsymbol{W}_{1.2} \sim GW_r(\boldsymbol{\Sigma}_{1.2}, n-p+r)$.

3. Let $\boldsymbol{W}_1 \sim GW_p(\boldsymbol{I}_p, n)$, $p \le n$, and $\boldsymbol{W}_2 \sim GW_p(\boldsymbol{I}_p, m)$, $p \le m$, be independently distributed. Find the distributions of the following transformations:

$$\begin{aligned} \boldsymbol{F}_1 &= \boldsymbol{W}_2^{-\frac{1}{2}}\boldsymbol{W}_1\boldsymbol{W}_2^{-\frac{1}{2}} \\ \boldsymbol{F}_2 &= (\boldsymbol{W}_1 + \boldsymbol{W}_2)^{-\frac{1}{2}}\boldsymbol{W}_2(\boldsymbol{W}_1 + \boldsymbol{W}_2)^{-\frac{1}{2}}. \end{aligned}$$

4. Under the assumptions of Section 10.2, derive the MLE of $(\boldsymbol{\mu}, \boldsymbol{\Sigma})$.

5. Under the assumptions of Section 10.2, suppose that $\boldsymbol{X} \sim M_t^{(n,p)}(\boldsymbol{\mu}, \boldsymbol{I}_n, \boldsymbol{\Sigma}, \gamma_o)$. Show that under the entropy loss function given by (2.5.8), the best Bayesian estimator of $\boldsymbol{\Sigma}$ is given by

$$\hat{\boldsymbol{\Sigma}} = \frac{(n-p)(\gamma_o - 2)}{n(n-p-1)\gamma_o}\boldsymbol{S}.$$

6. Prove that in Theorem 10.4.2, the posterior distribution of $\boldsymbol{\Phi}$ is the matrix beta type II.

7. Using the result of item (vii) of Theorem 10.4.3, find the moment generating function of $\boldsymbol{\Phi}$.

CHAPTER 11

LINEAR PREDICTION MODELS

Outline

The predictive inference had been the oldest form of statistical inference used in real life. In general, predictive inference is directed towards inference involving the observable rather than the parameters. However, recently, Khan (2002b, 2004, 2006b) proposed the prediction distribution for the future regression vector and residual sum of squares. Predictive inference for a set of future responses of the model, conditional on the realized responses from the same model, has been derived by many authors, including Aitchison and Sculthorpe (1965), Fraser and Haq (1969),

A. K. Md. Ehsanes Saleh, M. Arashi, S.M.M. Tabatabaey.

Guttman (1970), Haq and Rinco (1973), Aitchison and Dunsmore (1975), Geisser (1993), Khan and Haq (1994), Khan (2002b, 2006b), and Ng (2010). Khan and Haq (1994), Anderson and Fang (1990), and Khan (2002a) provide predictive analyses of linear models with multivariate t and spherical errors. The contribution of Prof. Shahjahan Khan to this field should be acknowledged.

11.1 Model and Preliminaries

Consider the regression model

$$\boldsymbol{y} = \boldsymbol{\beta X} + \sigma \boldsymbol{e}, \tag{11.1.1}$$

where the n-dim row vector $\boldsymbol{y}$ is the vector of the response variable; $\boldsymbol{X}$ is the $p \times n$ dimensional matrix of the values of the p regressors; $\boldsymbol{e}$ is the $1 \times n$ row vector of the error component associated with the response vector $\boldsymbol{y}$; and the regression vector $\boldsymbol{\beta}$ and scale parameter $\sigma > 0$. Assume the error vector follows the multivariate t-distribution with mean $\boldsymbol{0}$, a vector of n-tuple of zeros, and covariance matrix, $\kappa(1)\boldsymbol{I}_n = \frac{\gamma_o}{\gamma_o - 2}\boldsymbol{I}_n$. Therefore, the joint density function of the vector of errors becomes

$$g(\boldsymbol{e}) = \frac{\Gamma\left(\frac{\gamma_o+n}{2}\right)}{(\pi\gamma_o)^{\frac{n}{2}}\Gamma\left(\frac{\gamma_o}{2}\right)}\left(1 + \frac{1}{\gamma_o}\boldsymbol{ee}'\right)^{-\frac{\gamma_o+n}{2}}. \tag{11.1.2}$$

Consequently, the response vector follows the multivariate t-distribution with mean vector $\boldsymbol{\beta X}$ and Cov matrix $\frac{\gamma_o \sigma^2}{\gamma_o - 2}\boldsymbol{I}_n$. Thus, the joint density function of the response vector becomes

$$g(\boldsymbol{y}; \boldsymbol{\beta}, \sigma^2) = \frac{\Gamma\left(\frac{\gamma_o+n}{2}\right)}{(\pi\gamma_o\sigma^2)^{\frac{n}{2}}\Gamma\left(\frac{\gamma_o}{2}\right)}\left(1 + \frac{1}{\gamma_o\sigma^2}(\boldsymbol{y} - \boldsymbol{\beta X})(\boldsymbol{y} - \boldsymbol{\beta X})'\right)^{-\frac{\gamma_o+n}{2}}. \tag{11.1.3}$$

In this chapter, we call the above multiple regression model the realized model of the responses from the performed experiment. The above joint density becomes the likelihood function of $\boldsymbol{\beta}$ and σ^2 when treated as a function of the parameters rather than the sample response.

Some useful notations are introduced here to facilitate the derivation of the results in the forthcoming sections. First, we denote the sum of regression vector (SRV) of $\boldsymbol{e}$ on $\boldsymbol{X}$ by $\boldsymbol{b}(\boldsymbol{e})$ and the residual sum of square (RSS) of the error vector by $s^2(\boldsymbol{e})$. Then we have

$$\boldsymbol{b}(\boldsymbol{e}) = \boldsymbol{eX}'(\boldsymbol{XX}')^{-1} \text{ and } s^2(\boldsymbol{e}) = [\mathrm{e} - \boldsymbol{b}(\boldsymbol{e})\boldsymbol{X}][\mathrm{e} - \boldsymbol{b}(\boldsymbol{e})\boldsymbol{X}]'. \tag{11.1.4}$$

Let $s(\boldsymbol{e})$ be the positive square root of the residual sum of squares based on the error regression, $s^2(\boldsymbol{e})$, and $\boldsymbol{d}(\boldsymbol{e}) = s^{-1}(\boldsymbol{e})[\mathrm{e} - \boldsymbol{b}(\boldsymbol{e})\boldsymbol{X}]$ be the "standardized" residual vector of the error regression.

Now write the error vector, $\boldsymbol{e}$, as a function of $\boldsymbol{b}(\boldsymbol{e})$ and $s(\boldsymbol{e})$ in the following way:

$$\boldsymbol{e} = \boldsymbol{b}(\boldsymbol{e})\boldsymbol{X} + s(\boldsymbol{e})\boldsymbol{d}(\boldsymbol{e}) \text{ and hence we get } \boldsymbol{ee}' = \boldsymbol{b}(\boldsymbol{e})\boldsymbol{XX}'\boldsymbol{b}'(\boldsymbol{e}) + s^2(\boldsymbol{e}) \tag{11.1.5}$$

since $\boldsymbol{d}(\boldsymbol{e})\boldsymbol{d}'(\boldsymbol{e}) = 1$, inner product of two orthonormal vectors, and $\boldsymbol{X}\boldsymbol{d}'(\boldsymbol{e}) = 0$, inasmuch as $\boldsymbol{X}$ and $\boldsymbol{d}(\boldsymbol{e})$ are orthogonal. From (11.1.4) and (11.1.5), the following relations (cf. Fraser, 1968, p.127) can easily be established:

$$\boldsymbol{b}(\boldsymbol{e}) = \sigma^{-1}\{\boldsymbol{b}(\boldsymbol{y}) - \boldsymbol{\beta}\}, \text{ and } s^2(\boldsymbol{e}) = \sigma^{-2}s^2(\boldsymbol{y}), \tag{11.1.6}$$

where

$$\boldsymbol{b}(\boldsymbol{y}) = \boldsymbol{y}\boldsymbol{X}'(\boldsymbol{X}\boldsymbol{X}')^{-1} \text{ and } s^2(\boldsymbol{y}) = [\boldsymbol{y} - \boldsymbol{b}(\boldsymbol{y})\boldsymbol{X}][\boldsymbol{y} - \boldsymbol{b}(\boldsymbol{y})\boldsymbol{X}]' \tag{11.1.7}$$

are the SRV of $\boldsymbol{y}$ on $\boldsymbol{X}$, and the RSS of the regression based on the realized responses, respectively. It may be mentioned here that both $s^2(\boldsymbol{e})$ and $s^2(\boldsymbol{y})$ have the same structure because the definitions of $s^2(\boldsymbol{e})$ in (11.1.6) and that of $s^2(\boldsymbol{y})$ in (11.1.7) ensure the same format of the two residual statistics of errors and realized responses, respectively. Haq (1982) called the relation in (11.1.7) the structural relation. It can easily be shown that $\boldsymbol{d}(\boldsymbol{e}) = s^{-1}(\boldsymbol{y})[\boldsymbol{y} - \boldsymbol{b}(\boldsymbol{y})\boldsymbol{X}] = \boldsymbol{d}(\boldsymbol{y})$. From the above results, the density of the error vector in (2.3) can be written as a function of $\boldsymbol{b}(\boldsymbol{e})$ and $s(\boldsymbol{e})$ as follows:

$$g(\boldsymbol{e}) = \psi \times \left[1 + \frac{1}{\gamma_o}\left(\boldsymbol{b}(\boldsymbol{e})\boldsymbol{X}\boldsymbol{X}'\boldsymbol{b}'(\boldsymbol{e}) + s^2(\boldsymbol{e})\right)\right]^{-\frac{\gamma_o+n}{2}}, \tag{11.1.8}$$

where ψ is an appropriate normalizing constant.

11.2 Distribution of SRV and RSS

From the probability density of $\boldsymbol{e}$ in (11.1.2) and the relation (11.1.6), the joint probability density of $\boldsymbol{b}(\boldsymbol{e})$ and $s^2(\boldsymbol{e})$, conditional on the $\boldsymbol{d}(\boldsymbol{e})$, is obtained using the invariant differentials (see Eaton, 1983, p.194-206 or Fraser, 1968, p.30) as follows:

$$g\left(\boldsymbol{b}(\boldsymbol{e}),\, s^2(\boldsymbol{e})|\boldsymbol{d}(\boldsymbol{e})\right) = K_1(\boldsymbol{d})\,[s^2(\boldsymbol{e})]^{\frac{n-p-2}{2}}\left[1 + \frac{1}{\gamma_o}\left(\boldsymbol{b}(\boldsymbol{e})\boldsymbol{X}\boldsymbol{X}'\boldsymbol{b}'(\boldsymbol{e}) + s^2(\boldsymbol{e})\right)\right]^{-\frac{\gamma_o+n}{2}}, \tag{11.2.1}$$

where

$$K_1(\boldsymbol{d}) = \frac{\Gamma\left(\frac{\gamma_o+n}{2}\right)|\boldsymbol{X}\boldsymbol{X}'|^{\frac{1}{2}}}{\pi^{\frac{p}{2}}\gamma_o^{\frac{n-p}{2}}\Gamma\left(\frac{\gamma_o}{2}\right)\Gamma\left(\frac{n-p}{2}\right)}.$$

It can be shown that the above density does not depend on $\boldsymbol{d}(\boldsymbol{e})$ (cf. Fraser, 1979, p.113).

Applying the representation (2.6.3), the joint density in (11.2.1) can be rewritten as

$$\begin{aligned} g\left(\boldsymbol{b}(\boldsymbol{e}),\, s^2(\boldsymbol{e})\right) &= K_1(\boldsymbol{d})\,[s^2(\boldsymbol{e})]^{\frac{n-p-2}{2}} \\ &\times \int_0^\infty W_o(t)e^{-\frac{t}{2}\{\boldsymbol{b}(\boldsymbol{e})\boldsymbol{X}\boldsymbol{X}'\boldsymbol{b}'(\boldsymbol{e})+s^2(\boldsymbol{e})\}}dt. \end{aligned} \tag{11.2.2}$$

Clearly, the joint distribution does not factor, and hence the marginal distributions are not independent of each other. Appropriate integrations lead to

$$\begin{aligned} g_1\left(\boldsymbol{b}(\boldsymbol{e})\right) &= K_1(\boldsymbol{d})\int_0^\infty W_o(t)e^{-\frac{t}{2}\{\boldsymbol{b}(\boldsymbol{e})\boldsymbol{X}\boldsymbol{X}'\boldsymbol{b}'(\boldsymbol{e})\}} \\ &\times \left[\int_0^\infty [s^2(\boldsymbol{e})]^{\frac{n-p-2}{2}} e^{-\frac{t}{2}\{s^2(\boldsymbol{e})\}} ds^2(\boldsymbol{e})\right] dt \\ &= \frac{\Gamma\left(\frac{\gamma_o+n}{2}\right)|\boldsymbol{X}\boldsymbol{X}'|^{\frac{1}{2}}}{\pi^{\frac{p}{2}}\Gamma\left(\frac{\gamma_o}{2}\right)}\left[1+\frac{1}{\gamma_o}\boldsymbol{b}(\boldsymbol{e})\boldsymbol{X}\boldsymbol{X}'\boldsymbol{b}'(\boldsymbol{e})\right]^{-\frac{\gamma_o+n}{2}} \end{aligned} \tag{11.2.3}$$

and

$$\begin{aligned} g_2\left(s^2(\boldsymbol{e})\right) &= K_1(\boldsymbol{d})[s^2(\boldsymbol{e})]^{\frac{n-p-2}{2}}\int_0^\infty W_o(t)e^{-\frac{t}{2}\{s^2(\boldsymbol{e})\}} \\ &\times \left[\int_{\mathbb{R}^p} e^{-\frac{t}{2}\{\boldsymbol{b}(\boldsymbol{e})\boldsymbol{X}\boldsymbol{X}'\boldsymbol{b}'(\boldsymbol{e})\}} d\boldsymbol{b}'(\boldsymbol{e})\right] dt \\ &= \frac{\Gamma\left(\frac{\gamma_o+n-p}{2}\right)}{\gamma_o^{\frac{n-p}{2}}\Gamma\left(\frac{\gamma_o}{2}\right)\Gamma\left(\frac{n-p}{2}\right)}[s^2(\boldsymbol{e})]^{\frac{n-p-2}{2}}\left[1+\frac{1}{\gamma_o}s^2(\boldsymbol{e})\right]^{-\frac{\gamma_o+n-p}{2}} \end{aligned} \tag{11.2.4}$$

To summarize the presented result from Khan (2002b, 2004), the SRV based on the error regression follows a p-dim multivariate t-distribution with mean $\mathbf{0}$ and scale matrix $[\boldsymbol{X}\boldsymbol{X}']^{-1}$. That is,

$$\boldsymbol{b}(\boldsymbol{e}) \sim M_t^{(p)}\left(\mathbf{0}, \frac{\gamma_o+n-p}{\gamma_o+n-p-2}\sigma^2(\boldsymbol{X}\boldsymbol{X}')^{-1}, \gamma_o+n-p\right).$$

The RSS of the error regression, $s^2(\boldsymbol{e})$, follows a scaled beta distribution with arguments $\frac{n-p}{2}$ and $\frac{\gamma_o}{2}$.

Remark 11.2.1. *Taking* $W_o(.)$ *in the representation* (2.6.3) *to be the Dirac (cf. Dirac, 1958) delta function, the result matches that of the normal distribution. For the normal model with independent errors, Khan (2004) found that the joint density of the above two statistics factorizes and hence they are independent.*

To find the distributions of the SRV of the response regression, $\boldsymbol{b}(\boldsymbol{y})$, and the RSS of the response regression, $s^2(\boldsymbol{y})$, we use the following relations:

$$\boldsymbol{b}(\boldsymbol{e}) = \sigma^{-1}[\boldsymbol{b}(\boldsymbol{y}) - \boldsymbol{\beta}] \text{ and } s^2(\boldsymbol{e}) = \sigma^{-2}s^2(\boldsymbol{y}). \tag{11.2.5}$$

So the associated differentials can be expressed as

$$d\boldsymbol{b}(\boldsymbol{e}) = \sigma^{-p}d\boldsymbol{b}(\boldsymbol{y}) \text{ and } ds^2(\boldsymbol{e}) = \sigma^{-2}ds^2(\boldsymbol{y}). \tag{11.2.6}$$

Therefore, the density function of $\boldsymbol{b}(\boldsymbol{y})$ is written as

$$g\left(\boldsymbol{b}(\boldsymbol{y})\right) = \frac{\Gamma\left(\frac{\gamma_o+n}{2}\right)|\boldsymbol{X}\boldsymbol{X}'|^{\frac{1}{2}}}{(\pi\gamma_o\sigma^2)^{\frac{p}{2}}\Gamma\left(\frac{\gamma_o}{2}\right)}\left[1+\frac{1}{\gamma_o\sigma^2}(\boldsymbol{b}(\boldsymbol{y})-\boldsymbol{\beta})\boldsymbol{X}\boldsymbol{X}'(\boldsymbol{b}(\boldsymbol{y})-\boldsymbol{\beta})\right]^{-\frac{\gamma_o+p}{2}}, \tag{11.2.7}$$

and that of $s^2(\boldsymbol{y})$ is given by

$$g\left(s^2(\boldsymbol{y})\right) = \frac{\Gamma\left(\frac{\gamma_o+n-p}{2}\right)}{(\gamma_o\sigma^2)^{\frac{n-p}{2}}\Gamma\left(\frac{\gamma_o}{2}\right)\Gamma\left(\frac{n-p}{2}\right)}[s^2(\boldsymbol{e})]^{\frac{n-p-2}{2}}\left[1+\frac{1}{\gamma_o\sigma^2}s^2(\boldsymbol{e})\right]^{-\frac{\gamma_o+n-p}{2}}. \tag{11.2.8}$$

The above result clearly demonstrates that for the multivariate t-model with γ_o d.f., the SRV of the response regression follows a multivariate t-distribution with mean vector $\boldsymbol{\beta}$ and covariance matrix $\frac{\gamma_o}{\gamma_o-2}\,\sigma^2(\boldsymbol{X}\boldsymbol{X}')^{-1}$, and the residual sum of squares of the response regression, $s^2(\boldsymbol{y})$, is distributed as a scaled beta variable of the second kind (type II). Unlike the multiple regression model with independent normal errors, the SRV and RSS of the response regression of the t-model are not independently distributed. Moreover, the distributions of SRV and RSS, for both errors and responses, depend on the unknown shape parameter γ_o.

11.3 Regression Model for Future Responses

Now consider a set of $n_f \geq p$ future unobserved responses, $\boldsymbol{y}_f = (y_{f1}, \ldots, y_{fn_f})$, from the multiple regression model as given in (11.1.1) with the same regression and scale parameters as

$$\boldsymbol{y}_f = \boldsymbol{\beta}\boldsymbol{X}_f + \sigma\boldsymbol{e}_f, \tag{11.3.1}$$

where $\boldsymbol{X}_f$ is the $p \times n_f$ matrix of the values of the regressors that generate the future response vector $\boldsymbol{y}_f$, and $\boldsymbol{e}_f$ is the n_f-dim row vector of future error terms. We assume noninformative prior distribution of the above parameters. Our objective here is to find the distributions of the future regression vector (FRV) and the residual sum of squares of the future regression model, conditional on the realized responses.

Following the same process as in section 11.2, define:

$$\boldsymbol{b}_f(\boldsymbol{e}_f) = \boldsymbol{e}_f\boldsymbol{X}_f'(\boldsymbol{X}_f\boldsymbol{X}_f')^{-1},\ s_f^2(\boldsymbol{e}_f) = [\boldsymbol{e}_f - \boldsymbol{b}_f(\boldsymbol{e}_f)\boldsymbol{X}_f][\boldsymbol{e}_f - \boldsymbol{b}_f(\boldsymbol{e}_f)\boldsymbol{X}_f]', \tag{11.3.2}$$

in which $\boldsymbol{b}_f(\boldsymbol{e}_f)$ is the FRV and $s_f^2(\boldsymbol{e}_f)$ is the RSS of the future error of the future model. Then we write

$$\boldsymbol{e}_f = \boldsymbol{b}_f(\boldsymbol{e}_f)\boldsymbol{X}_f + s_f(\boldsymbol{e}_f)\boldsymbol{d}_f(\boldsymbol{e}_f), \tag{11.3.3}$$

where $s_f(\boldsymbol{e}_f)$ is the positive square root of $s_f^2(\boldsymbol{e}_f)$, and hence we get

$$\boldsymbol{e}_f\boldsymbol{e}_f' = \boldsymbol{b}_f(\boldsymbol{e}_f)\boldsymbol{X}_f\boldsymbol{X}_f'\boldsymbol{b}_f'(\boldsymbol{e}_f) + s_f^2(\boldsymbol{e}_f), \tag{11.3.4}$$

since $\boldsymbol{X}_f$ and $\boldsymbol{d}(\boldsymbol{e}_f)$ are orthogonal and $\boldsymbol{d}_f(\boldsymbol{e}_f)$ is orthonormal. Moreover, the following relations can easily be observed:

$$\boldsymbol{b}_f(\boldsymbol{e}_f) = \sigma^{-1}\{\boldsymbol{b}_f(\boldsymbol{y}_f) - \boldsymbol{\beta}\}, \text{ and } s_f^2(\boldsymbol{e}_f) = \sigma^{-2}s_f^2(\boldsymbol{y}_f), \qquad (11.3.5)$$

where

$$\boldsymbol{b}_f(\boldsymbol{y}_f) = \boldsymbol{y}_f\boldsymbol{X}'_f(\boldsymbol{X}_f\boldsymbol{X}'_f)^{-1} \text{ and } s_f^2(\boldsymbol{y}) = [\boldsymbol{y}_f - \boldsymbol{b}_f(\boldsymbol{y}_f)\boldsymbol{X}_f][\boldsymbol{y}_f - \boldsymbol{b}_f(\boldsymbol{y}_f)\boldsymbol{X}_f]' \qquad (11.3.6)$$

in which $\boldsymbol{b}_f(\boldsymbol{y}_f)$ is the FRV of the future responses and $s_f^2(\boldsymbol{y}_f)$ is the RSS of future responses respectively. Following the same argument as in section 11.2, the density function of $\boldsymbol{e}_f$ becomes

$$g(\boldsymbol{e}_f) = \frac{\Gamma\left(\frac{\gamma_o+n_f}{2}\right)}{(\pi\gamma_o)^{\frac{n_f}{2}}\Gamma\left(\frac{\gamma_o}{2}\right)}\left[1 + \frac{1}{\gamma_o}\boldsymbol{e}_f\boldsymbol{e}'_f\right]^{-\frac{\gamma_o+n_f}{2}}. \qquad (11.3.7)$$

Hence, by using the invariant differentials, we get the joint distribution of $\boldsymbol{b}_f(\boldsymbol{e}_f)$ and $s_f^2(\boldsymbol{e}_f)$ as follows:

$$\begin{aligned} g\left(\boldsymbol{b}_f(\boldsymbol{e}_f), s_f^2(\boldsymbol{e}_f)\right) &= K_2 \times [s_f^2(\boldsymbol{e}_f)]^{\frac{n_f-p-2}{2}} \\ &\left[1 + \frac{1}{\gamma_o}\left\{\boldsymbol{b}_f(\boldsymbol{e}_f)\boldsymbol{X}_f\boldsymbol{X}'_f\boldsymbol{b}'_f(\boldsymbol{e}_f) + s_f^2(\boldsymbol{e}_f)\right\}\right]^{-\frac{\gamma_o+n_f}{2}}, \end{aligned} \qquad (11.3.8)$$

where K_2 is the normalizing constant. Now the join density function of the combined error vector, $(\boldsymbol{e},\ \boldsymbol{e}_f)$, can be expressed as

$$g(\boldsymbol{e},\ \boldsymbol{e}_f) = \frac{\Gamma\left(\frac{\gamma_o+n+n_f}{2}\right)}{(\pi\gamma_o)^{\frac{n+n_f}{2}}\Gamma\left(\frac{\gamma_o}{2}\right)}\left[1 + \frac{1}{\gamma_o}\left(\boldsymbol{e}\boldsymbol{e}' + \boldsymbol{e}_f\boldsymbol{e}'_f\right)\right]^{-\frac{\gamma_o+n+n_f}{2}}. \qquad (11.3.9)$$

11.4 Predictive Distributions of FRV and FRSS

In this section we derive the predictive distributions of the FRV and future residual sum of squares (FRSS) for the future multiple regression model, conditional on the realized responses. In the absence of any knowledge about the parameters, we consider noninformative prior distribution for the parameters as follows:

$$p(\boldsymbol{\beta}) \propto \text{ constant, and } p(\sigma^2) \propto \sigma^{-2}. \qquad (11.4.1)$$

This prior distribution is used to derive the predictive distributions of $\boldsymbol{b}(\boldsymbol{y}_f)$ and $s^2(\boldsymbol{y}_f)$ from the joint distribution of $\boldsymbol{\beta}$, σ^2, $\boldsymbol{b}(\boldsymbol{y}_f)$ and $s^2(\boldsymbol{y}_f)$. Justification for the use of such a noninformative prior is given by Geisser (1993, p.60,192), Box and Tiao (1992, p.21), Press (1989, p. 132) and Meng (1994), among any others. It is worth noting that no prior distribution is required in the structural approach (cf. Fraser, 1978) as the structural distribution, similar to the Bayes posterior distribution, can be obtained from the structural relation of the model without involving any prior distribution. Fraser and Haq (1969) discussed that, for the noninformative prior, the Bayes posterior density is the same as the structural density.

11.4.1 Distribution of the FRV

The joint density function of the error statistics $\boldsymbol{b}(\boldsymbol{e})$, $s^2(\boldsymbol{e})$, $\boldsymbol{b}_f(\boldsymbol{e}_f)$, and $s_f^2(\boldsymbol{e}_f)$, for given $\boldsymbol{d}(\cdot)$, is derived from the joint density in (11.3.9) by applying the properties of invariant differentials as follows:

$$\begin{aligned} p\Big(\boldsymbol{b}(\boldsymbol{e}), s^2(\boldsymbol{e}), \boldsymbol{b}_f(\boldsymbol{e}_f), s_f^2(\boldsymbol{e}_f)|\boldsymbol{d}(\cdot)\Big) &= \Psi_{11}(\cdot)[s^2(\boldsymbol{e})]^{\frac{n-p-2}{2}}[s_f^2(\boldsymbol{e}_f)]^{\frac{n_f-p-2}{2}} \\ &\times \left[1+\frac{1}{\gamma_o}\Big(h_1\Big(\boldsymbol{b},\boldsymbol{X}\Big)+h_2\Big(\boldsymbol{b}_f,\boldsymbol{X}_f\Big)\Big)\right]^{\frac{\gamma_o+n+n_f}{2}}, \end{aligned} \quad (11.4.2)$$

where

$$\begin{aligned} h_1\Big(\boldsymbol{b},\boldsymbol{X}\Big) &= \boldsymbol{b}(\boldsymbol{e})\boldsymbol{X}\boldsymbol{X}'\boldsymbol{b}'(\boldsymbol{e}), \\ h_2\Big(\boldsymbol{b}_f,\boldsymbol{X}_f\Big) &= \boldsymbol{b}_f(\boldsymbol{e}_f)\boldsymbol{X}_f\boldsymbol{X}'_f\boldsymbol{b}'_f(\boldsymbol{e}_f), \\ \Psi_{11} &= \frac{\Gamma\left(\frac{\gamma_o+n+n_f}{2}\right)}{\pi^p\gamma_o^{\frac{n+n_f}{2}}\Gamma\left(\frac{n-p}{2}\right)\Gamma\left(\frac{n_f-p}{2}\right)\Gamma\left(\frac{\gamma_o}{2}\right)}|\boldsymbol{X}\boldsymbol{X}'|^{\frac{1}{2}}|\boldsymbol{X}_f\boldsymbol{X}'_f|^{\frac{1}{2}}. \end{aligned}$$

Because the above density does not depend on $\boldsymbol{d}(\cdot)$, the conditioning in (11.4.2) can be disregarded, as we need to find the joint density of $\boldsymbol{b}(\boldsymbol{y})$, $s^2(\boldsymbol{y})$, $\boldsymbol{b}_f(\boldsymbol{y}_f)$, and $s_f^2(\boldsymbol{y}_f)$ from the above joint density. The structural relation of the model yields

$$\boldsymbol{b}(\boldsymbol{e}) = \sigma^{-1}[\boldsymbol{b}(\boldsymbol{y}) - \boldsymbol{\beta}] \text{ and } s^2(\boldsymbol{e}) = \sigma^{-2}s^2(\boldsymbol{y}). \quad (11.4.3)$$

The joint distribution of $\boldsymbol{b}(\boldsymbol{y})$, $s^2(\boldsymbol{y})$, $\boldsymbol{b}_f(\boldsymbol{e}_f)$, and $s_f^2(\boldsymbol{e}_f)$ is then obtained by using the Jacobian of the transformation,

$$J\left(\boldsymbol{b}(\boldsymbol{e}), s^2(\boldsymbol{e}) \to \boldsymbol{b}(\boldsymbol{y}), s^2(\boldsymbol{y})\right) = \left(\sigma^2\right)^{-\frac{p+2}{2}}, \quad (11.4.4)$$

as follows:

$$\begin{aligned} p\Big(\boldsymbol{b}(\boldsymbol{y}), s^2(\boldsymbol{y}), \boldsymbol{b}_f(\boldsymbol{e}_f), s_f^2(\boldsymbol{e}_f)\Big) &= \Psi_2 \times [s^2]^{\frac{n-p}{2}}[s_f^2(\boldsymbol{y}_f)]^{\frac{n_f-p-2}{2}}[\sigma^2]^{-\frac{n+n_f-p}{2}} \\ &\times \left[1+\frac{1}{\gamma_o\sigma^2}\left(\xi_1(\boldsymbol{b},\boldsymbol{\beta})+s^2+\xi_2(\boldsymbol{b}_f(\boldsymbol{e}_f))+s_f^2(\boldsymbol{e}_f)\right)\right]^{-\frac{\gamma_o+n+n_f}{2}}, \end{aligned} \quad (11.4.5)$$

where

$$\begin{aligned}\xi_1(\boldsymbol{b},\boldsymbol{\beta}) &= (\boldsymbol{b}-\boldsymbol{\beta})\boldsymbol{X}\boldsymbol{X}'(\boldsymbol{b}-\boldsymbol{\beta})',\\ \xi_2(\boldsymbol{b}_f(\boldsymbol{e}_f)) &= \boldsymbol{b}_f(\boldsymbol{e}_f)\boldsymbol{X}_f\boldsymbol{X}_f'\boldsymbol{b}_f'(\boldsymbol{e}_f),\\ \boldsymbol{b} &= \boldsymbol{b}(\boldsymbol{y}), \quad \text{and} \quad s^2 = s^2(\boldsymbol{y}).\end{aligned}$$

The normalizing constant Ψ_2 can be obtained by integrating the right-hand side of the above function over the appropriate domains of the underlying variables. As we are interested in the distributions of $\boldsymbol{b}_f(\boldsymbol{y}_f)$ and $s_f^2(\boldsymbol{y}_f)$, the FRV and RSS for the future regression, respectively, conditional on the realized responses, we don't pursue the matter any further here.

To derive the joint distribution of $\boldsymbol{\beta}$, σ^2, $\boldsymbol{b}_f(\boldsymbol{y}_f)$, and $s_f^2(\boldsymbol{y}_f)$ from the above joint density, note that from the structure of the future regression equation we have

$$\boldsymbol{b}_f(\boldsymbol{e}_f) = \sigma^{-1}[\boldsymbol{b}_f(\boldsymbol{y}_f) - \boldsymbol{\beta}] \text{ and } s^2(\boldsymbol{e}_f) = \sigma^{-2}s^2(\boldsymbol{y}_f), \tag{11.4.6}$$

where

$$\boldsymbol{b}_f(\boldsymbol{y}_f) = \boldsymbol{y}\boldsymbol{X}_f'(\boldsymbol{X}_f\boldsymbol{X}_f')^{-1}, \;\; s^2(\boldsymbol{y}_f) = [\boldsymbol{y}_f - \boldsymbol{b}_f(\boldsymbol{y}_f)\boldsymbol{X}_f'][\boldsymbol{y}_f - \boldsymbol{b}_f(\boldsymbol{y}_f)\boldsymbol{X}_f]'. \tag{11.4.7}$$

Therefore, the Jacobian of the transformation is found to be

$$J\Big\{[\boldsymbol{b}_f(\boldsymbol{e}_f), s_f^2(\boldsymbol{e}_f)] \to [\boldsymbol{b}_f(\boldsymbol{y}_f), s^2(\boldsymbol{y}_f)]\Big\} = [\sigma^2]^{-\frac{p+2}{2}}. \tag{11.4.8}$$

Now, the joint density of $\boldsymbol{b}(\boldsymbol{y})$, $s^2(\boldsymbol{y})$, $\boldsymbol{b}_f(\boldsymbol{y}_f)$, and $s_f^2(\boldsymbol{y}_f)$ is obtained as

$$\begin{aligned}p\Big(\boldsymbol{b}, s^2, \boldsymbol{b}_f, s_f^2\Big) &= \Psi_3(\cdot) \times [s^2]^{\frac{n-p-2}{2}}[s_f^2(\boldsymbol{y}_f)]^{\frac{n_f-p-2}{2}}[\sigma^2]^{-\frac{n+n_f-p}{2}}\\ &\times\Big[1 + \frac{1}{\gamma_o\sigma^2}\Big\{\Big(\boldsymbol{b}-\boldsymbol{\beta}\Big)\boldsymbol{X}\boldsymbol{X}'\Big(\boldsymbol{b}-\boldsymbol{\beta}\Big)'\\ &+s^2 + \Big(\boldsymbol{b}_f-\boldsymbol{\beta}\Big)\boldsymbol{X}_f\boldsymbol{X}_f'\Big(\boldsymbol{b}_f-\boldsymbol{\beta}\Big)' + s_f^2\Big\}\Big]^{-\frac{\gamma_o+n+n_f}{2}},\end{aligned} \tag{11.4.9}$$

where $\boldsymbol{b}_f = \boldsymbol{b}_f(\boldsymbol{y}_f)$ and $s_f^2 = s_f^2(\boldsymbol{y}_f)$ for notational convenience.

From the noninformative prior distribution of the parameters of the model and the density in (11.4.5), we find the following joint density of $\boldsymbol{\beta}$, σ^2, $\boldsymbol{b}_f(\boldsymbol{y}_f)$, and $s_f^2(\boldsymbol{y}_f)$,

$$\begin{aligned}p\Big(\boldsymbol{\beta}, \sigma^2, \boldsymbol{b}_f, s_f^2|\boldsymbol{d}\Big) &= \Psi_3(\cdot) \times [s^2]^{\frac{n-p-2}{2}}[s_f^2(\boldsymbol{y}_f)]^{\frac{n_f-p-2}{2}}[\sigma^2]^{-\frac{n+n_f-p}{2}}\\ &\times\Big[1 + \frac{1}{\gamma_o\sigma^2}\Big\{\Big(\boldsymbol{b}-\boldsymbol{\beta}\Big)\boldsymbol{X}\boldsymbol{X}'\Big(\boldsymbol{b}-\boldsymbol{\beta}\Big)'\\ &+s^2 + \Big(\boldsymbol{b}_f-\boldsymbol{\beta}\Big)\boldsymbol{X}_f\boldsymbol{X}_f'\Big(\boldsymbol{b}_f-\boldsymbol{\beta}\Big)' + s_f^2\Big\}\Big]^{-\frac{\gamma_o+n+n_f}{2}}.\end{aligned} \tag{11.4.10}$$

Such results can also be obtained by using the Bayesian approach. In particular, the Bayes posterior density can be obtained by assuming uniform prior for the regression and scale parameters of the model. However, the final results here will be the same as that obtained by the Bayesian approach under uniform prior. Interested readers may refer to Fraser and Haq (1969) for details.

To evaluate the normalizing constant $\Psi_3(\cdot)$, let

$$\begin{aligned} I_{\sigma^2} &= \int_{\sigma^2} p\Big(\boldsymbol{\beta}, \sigma^2, \boldsymbol{b}_f,\ s_f \big| \boldsymbol{d}\Big) d\sigma^2 \\ &= [s_f^2]^{\frac{n_f-p-2}{2}} \int_{\sigma^2} [\sigma^2]^{-\frac{n+n_f+2}{2}} \left[\frac{1}{\gamma_o \sigma^2}\, \boldsymbol{Q}\right]^{-\frac{\gamma_o+n+n_f}{2}} d\sigma^2, \end{aligned} \tag{11.4.11}$$

where

$$\boldsymbol{Q} = \Big(\boldsymbol{b}-\boldsymbol{\beta}\Big)\boldsymbol{X}\boldsymbol{X}'\Big(\boldsymbol{b}-\boldsymbol{\beta}\Big)' + s^2 + \Big(\boldsymbol{b}_f-\boldsymbol{\beta}\Big)\boldsymbol{X}_f\boldsymbol{X}_f'\Big(\boldsymbol{b}_f-\boldsymbol{\beta}\Big)' + s_f^2.$$

Therefore,

$$I_{\sigma^2} = \gamma_o^{\frac{n+n_f}{2}} B\left(\frac{\gamma_o}{2}, \frac{n+n_f}{2}\right) [s_f^2]^{\frac{n_f-p-2}{2}} \boldsymbol{Q}^{-\frac{n+n_f}{2}}. \tag{11.4.12}$$

To facilitate the further integrations, the terms involving the regression vector $\boldsymbol{\beta}$ in $\boldsymbol{Q}$ can be expressed as follows:

$$\begin{aligned} &\Big(\boldsymbol{b}-\boldsymbol{\beta}\Big)\boldsymbol{X}\boldsymbol{X}'\Big(\boldsymbol{b}-\boldsymbol{\beta}\Big)' + \Big(\boldsymbol{b}_f-\boldsymbol{\beta}\Big)\boldsymbol{X}_f\boldsymbol{X}_f'\Big(\boldsymbol{b}_f-\boldsymbol{\beta}\Big)' = \\ &\qquad \Big(\boldsymbol{\beta}-\boldsymbol{F}\boldsymbol{A}^{-1}\Big)\boldsymbol{A}\Big(\boldsymbol{\beta}-\boldsymbol{F}\boldsymbol{A}^{-1}\Big)' + \Big(\boldsymbol{b}_f-\boldsymbol{b}\Big)\boldsymbol{H}^{-1}\Big(\boldsymbol{b}_f-\boldsymbol{b}\Big)', \end{aligned} \tag{11.4.13}$$

where

$$\boldsymbol{F} = \boldsymbol{b}\boldsymbol{X}\boldsymbol{X}' + \boldsymbol{b}_f\boldsymbol{X}_f\boldsymbol{X}_f',\ \boldsymbol{A} = \boldsymbol{X}\boldsymbol{X}' + \boldsymbol{X}_f\boldsymbol{X}_f',\ \text{and}\ \boldsymbol{H} = [\boldsymbol{X}\boldsymbol{X}']^{-1} + [\boldsymbol{X}_f\boldsymbol{X}_f']^{-1}. \tag{11.4.14}$$

Then, let

$$\begin{aligned} I_{\sigma^2\boldsymbol{\beta}} &= \int_{\boldsymbol{\beta}} I_{\sigma^2}\, d\boldsymbol{\beta} \\ &= \gamma_o^{\frac{n+n_f}{2}} B\left(\frac{\gamma_o}{2}, \frac{n+n_f}{2}\right) [s_f^2]^{\frac{n_f-p-2}{2}} \\ &\quad \times \int_{\boldsymbol{\beta}} \left[\Big(\boldsymbol{b}_f-\boldsymbol{b}\Big)\boldsymbol{H}^{-1}\Big(\boldsymbol{b}_f'-\boldsymbol{b}\Big) + s^2 + s_f^2 + h\Big(\boldsymbol{\beta}, \boldsymbol{A}\Big)\right]^{-\frac{n+n_f}{2}} d\boldsymbol{\beta} \\ &= \frac{\gamma_o^{\frac{n+n_f}{2}} (\pi)^p B\left(\frac{\gamma_o}{2}, \frac{n+n_f}{2}\right) \Gamma(\frac{n+n_f-p}{2})}{|\boldsymbol{A}|^{\frac{1}{2}}} [s_f^2]^{\frac{n_f-p-2}{2}} \\ &\quad \times \left[\Big(\boldsymbol{b}_f-\boldsymbol{b}\Big)\boldsymbol{H}^{-1}\Big(\boldsymbol{b}_f-\boldsymbol{b}\Big)' + s^2 + s_f^2\right]^{-\frac{n+n_f-p}{2}}, \end{aligned} \tag{11.4.15}$$

where

$$h\Big(\boldsymbol{\beta}, A\Big) = \Big(\boldsymbol{\beta} - \boldsymbol{F}\boldsymbol{A}^{-1}\Big)\boldsymbol{A}\Big(\boldsymbol{\beta} - \boldsymbol{F}\boldsymbol{A}^{-1}\Big)'. \tag{11.4.16}$$

In the same way, let

$$\begin{aligned} I_{\sigma^2\boldsymbol{\beta}\boldsymbol{b}_f} &= \int_{\boldsymbol{b}_f} I_{\sigma^2\beta}\, d\boldsymbol{b}_f \\ &= \frac{\gamma_o^{\frac{n+n_f}{2}} \pi^{\frac{p}{2}} \Gamma\left(\frac{n+n_f-p}{2}\right) B\left(\frac{\gamma_o}{2}, \frac{n+n_f-p}{2}\right)}{|\boldsymbol{A}|^{\frac{1}{2}}} [s_f^2]^{\frac{n_f-p-2}{2}} \\ &\quad \times \int_{\boldsymbol{b}_f} \left[\Big(\boldsymbol{b}_f - \boldsymbol{b}\Big)\boldsymbol{H}^{-1}\Big(\boldsymbol{b}_f - \boldsymbol{b}\Big)' + s^2 + s_f^2\right]^{-\frac{n+n_f-p}{2}} d\boldsymbol{b}_f \\ &= \frac{\gamma_o^{\frac{n+n_f}{2}} \pi^{p} \Gamma\left(\frac{n+n_f-p}{2}\right) B\left(\frac{\gamma_o}{2}, \frac{n+n_f-p}{2}\right)}{|\boldsymbol{A}|^{\frac{1}{2}} |\boldsymbol{H}|^{-\frac{1}{2}}} \\ &\quad \times [s_f^2]^{\frac{n_f-p-2}{2}} [s^2 + s_f^2]^{-\frac{n+n_f-2p}{2}}. \end{aligned} \tag{11.4.17}$$

Finally, let

$$\begin{aligned} I_{\sigma^2\boldsymbol{\beta}\boldsymbol{b}_f s_f^2} &= \int_{s_f^2} I_{\sigma^2\boldsymbol{\beta}\boldsymbol{b}_f}\, ds_f^2 \\ &= \frac{\gamma_o^{\frac{n+n_f}{2}} \pi^{p} \Gamma\left(\frac{n+n_f-2p}{2}\right) B\left(\frac{\gamma_o}{2}, \frac{n+n_f-p}{2}\right)}{|\boldsymbol{A}|^{\frac{1}{2}} |\boldsymbol{H}|^{-\frac{1}{2}}} \\ &\quad \times \int_{s_f^2} [s_f^2]^{\frac{n_f-p-2}{2}} [s^2 + s_f^2]^{-\frac{n+n_f-2p}{2}} ds_f^2 \\ &= \frac{\gamma_o^{\frac{n+n_f}{2}} \pi^{p} \Gamma\left(\frac{n-p}{2}\right) \Gamma\left(\frac{n_f-p}{2}\right) B\left(\frac{\gamma_o}{2}, \frac{n+n_f-p}{2}\right) B\left(\frac{n-p}{2}, \frac{n_f-p}{2}\right)}{|\boldsymbol{A}|^{\frac{1}{2}} |\boldsymbol{H}|^{-\frac{1}{2}} [s^2]^{\frac{n-p}{2}}}. \end{aligned} \tag{11.4.18}$$

Thus, the normalizing constant for the joint distribution of $\boldsymbol{\beta}$, σ^2, $\boldsymbol{b}_f$, and s_f^2 becomes

$$\Psi_3(\cdot) = \frac{\Gamma\left(\frac{\gamma_o+n+n_f}{2}\right) |\boldsymbol{A}|^{\frac{1}{2}} |\boldsymbol{H}|^{-\frac{1}{2}} [s^2]^{\frac{n-p}{2}}}{\gamma_o^{\frac{n+n_f}{2}} \pi^{p} \Gamma\left(\frac{\gamma_o}{2}\right) \Gamma\left(\frac{n-p}{2}\right) \Gamma\left(\frac{n_f-p}{2}\right)}. \tag{11.4.19}$$

The marginal density of $\boldsymbol{\beta}$, $\boldsymbol{b}_f$, and s_f^2, conditional on $\boldsymbol{y}$, is derived by integrating out σ^2 from the above joint density. Thus, we have

$$\begin{aligned} p\Big(\boldsymbol{\beta}, \boldsymbol{b}_f, s_f^2 | \boldsymbol{y}\Big) &= \Psi_4 \times [s_f^2]^{\frac{n_f-p-2}{2}} \Big[s^2 + \Big(\boldsymbol{\beta} - \boldsymbol{F}\boldsymbol{A}^{-1}\Big)\boldsymbol{A} \\ &\quad \times \Big(\boldsymbol{\beta} - \boldsymbol{F}\boldsymbol{A}^{-1}\Big)' + \Big(\boldsymbol{b}_f - \boldsymbol{b}\Big)\boldsymbol{H}^{-1}\Big(\boldsymbol{b}_f - \boldsymbol{b}\Big) + s_f^2\Big]^{-\frac{\gamma_o+n+n_f}{2}} \end{aligned} \tag{11.4.20}$$

where Ψ_4 is the normalizing constant.

Similarly, the marginal density of $\boldsymbol{b}_f$ and s_f^2 is obtained by integrating out $\boldsymbol{\beta}$ over $\mathbb{R}^p$ from (11.4.20). This gives the joint density of $\boldsymbol{b}_f$ and s_f^2, conditional on $\boldsymbol{y}$, as

$$p\Big(\boldsymbol{b}_f, s_f^2|\boldsymbol{y}\Big) = \Psi_5 \times [s_f^2]^{\frac{n_f-p-2}{2}} \times \Big[s^2 + s_f^2 + \Big(\boldsymbol{b}_f - \boldsymbol{b}\Big)\boldsymbol{H}^{-1}\Big(\boldsymbol{b}_f - \boldsymbol{b}\Big)\Big]^{-\frac{n+n_f-p}{2}} \tag{11.4.21}$$

where Ψ_5 is the normalizing constant given by

$$\Psi_5 = \frac{[s^2]^{\frac{n-p}{2}}\Gamma\left(\frac{n+n_f-p}{2}\right)}{|\boldsymbol{H}|^{\frac{1}{2}}\pi^{\frac{p}{2}}\Gamma\left(\frac{n-p}{2}\right)\Gamma\left(\frac{n_f-p}{2}\right)}.$$

The prediction distribution of the FRV, $\boldsymbol{b}_f = \boldsymbol{b}_f(\boldsymbol{y}_f)$, can now be obtained by integrating out s_f^2 from (11.4.21). The integration yields

$$p\Big(\boldsymbol{b}_f\Big|\boldsymbol{y}\Big) = \Psi_6 \times \left[s^2 + \Big(\boldsymbol{b}_f - \boldsymbol{b}\Big)\boldsymbol{H}^{-1}\Big(\boldsymbol{b}_f - \boldsymbol{b}\Big)'\right]^{-\frac{n}{2}}, \tag{11.4.22}$$

where

$$\begin{aligned}\Psi_6 &= \Psi_5 \times B\Big(\frac{n_f - p}{2}, \frac{n}{2}\Big)[s^2]^{\frac{n-p}{2}} \\ &= \frac{[s^2]^{\frac{n-p}{2}}\Gamma\left(\frac{n}{2}\right)}{\pi^{\frac{p}{2}}\Gamma\left(\frac{n-p}{2}\right)|\boldsymbol{H}|^{\frac{1}{2}}}.\end{aligned}$$

The prediction distribution of $\boldsymbol{b}_f$ can be written in the usual multivariate t-distribution form as follows:

$$p\Big(\boldsymbol{b}_f\Big|\boldsymbol{y}\Big) = \Psi_6 \times \left[1 + \Big(\boldsymbol{b}_f - \boldsymbol{b}\Big)\Big[s^2\boldsymbol{H}\Big]^{-1}\Big(\boldsymbol{b}_f - \boldsymbol{b}\Big)'\right]^{-\frac{n}{2}} \tag{11.4.23}$$

in which $n > p$.

Remark 11.4.1. *The multivariate t-distribution, because the density in* (11.4.23) *is a t-density, the prediction distribution of the FRV, $\boldsymbol{b}_f$, conditional on the realized responses, thus follows a multivariate t-distribution of dimension p, with $(n-p)$ degrees of freedom. Thus, $[\mathbf{b}_f|\boldsymbol{y}] \sim M_t^{(p)}(\mathbf{b}, s^2\boldsymbol{H}, n-p)$, where $\mathbf{b}$ is the location vector and $\boldsymbol{H}$ is the scale matrix. It is observed that the degrees of freedom parameter of the prediction distribution of $\boldsymbol{b}_f$ depends on the sample size of the realized sample and the dimension of the regression parameter vector of the model. The above prediction distribution can be used to construct a β-expectation tolerance region for the future regression parameter.*

11.4.2 Distribution of Future Residual Sum of Squares

The prediction distribution of the future RSS from the future regression, $s_f^2(\boldsymbol{y}_f)$, based on the future responses, $\boldsymbol{y}_f$, conditional on the realized responses, $\boldsymbol{y}$, is obtained by integrating out $\boldsymbol{b}_f$ from (11.4.21) as follows:

$$p\left(s_f^2(\boldsymbol{y}_f)\middle|\boldsymbol{y}\right) = \Psi_7 \times \left[s_f^2(\boldsymbol{y}_f)\right]^{\frac{n_f-p-2}{2}} \left[s^2 + s_f^2(\boldsymbol{y}_f)\right]^{-\frac{n+n_f-2p}{2}}. \quad (11.4.24)$$

The density function in (11.4.24) can be written in the usual beta distribution form as follows:

$$p\left(s_f^2\middle|\boldsymbol{y}\right) = \Psi_7 \times \left[s_f^2\right]^{\frac{n_f-p-2}{2}} \left[1 + s^{-2}s_f^2\right]^{-\frac{n+n_f-2p}{2}}, \quad (11.4.25)$$

where

$$\Psi_7 = \frac{\left[s^2\right]^{-\frac{n-p}{2}} \Gamma\left(\frac{n+n_f-2p}{2}\right)}{\Gamma\left(\frac{n-p}{2}\right)\Gamma\left(\frac{n_f-p}{2}\right)}.$$

This is the prediction distribution of the future residual sum of squares based on the future response $\boldsymbol{y}_f$, conditional on the realized responses, from the multiple regression model with multivariate t-error variable. The density in (11.4.25) is a modified form of beta density of the second kind with $(n_f - p)$ and $(n - p)$ degrees of freedom. Obviously, for the existence of the above distribution of s_f^2 we must have $n_f > p$ in addition to $n > p$. Once again we note that the distribution of the FRSS of the future regression model does not depend on the shape parameter, γ_o, of the multiple regression model with multivariate t-errors. Khan (2004) obtained similar results for the multiple regression model with normal errors. In the forthcoming section, we adopt an example from Khan (2004).

11.5 An Illustration

To illustrate how the method works, we consider a real-life data set from Barlev and Levy (1979). The simple regression model fitted to this data is a special case of the model considered here. The data set was used in a study of the relationship between accounting rates on stocks and market returns. It provides information on the two variables for 54 large companies in the USA. Considering the market return to be the response and the accounting rate as the explanatory variables, the fitted model becomes $\hat{y} = 0.084801 + 0.61033x$ with $\bar{x} = 12.9322$, $\bar{y} = 8.7409$, $\sum_{j=1}^{54} x_j^2 = 10293.10893$, $\sum_{j=1}^{54} x_j y_j = 6874.4131$, and $s^2 = 25.864$, the mean squared error. The prediction distribution of the regression parameter involves H, which in this special case becomes $[\sum_{j=1}^{54} x_j^2]^{-1} + [{x_f}^2]^{-1}$.

The top two graphs in Figure 11.1 display the prediction distributions of future responses for future accounting rates 5 and 25, respectively. Both distributions are

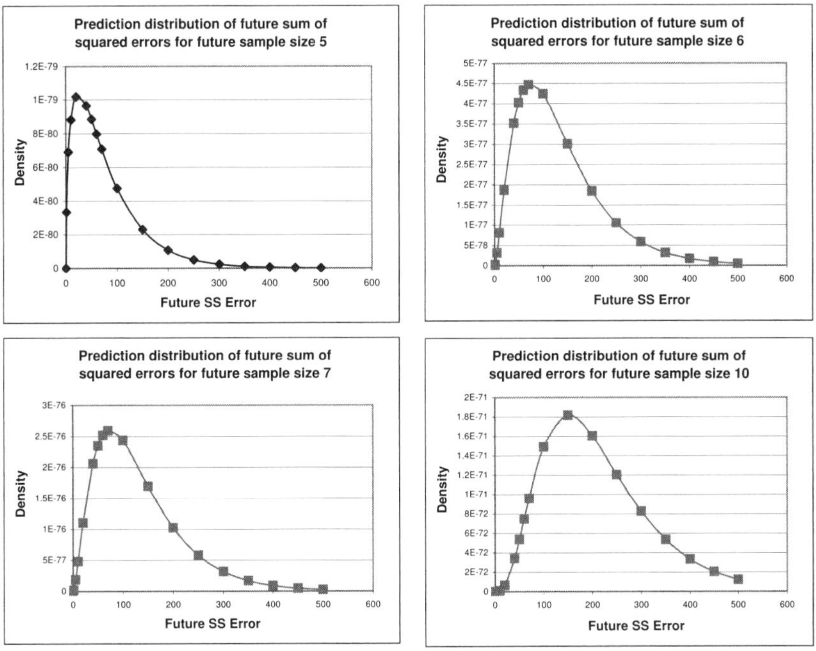

Figure 11.1 Prediction distribution for different future sample sizes

the t-distributions, but the one with a higher future accounting rate has wider spread than the one with a lower value of future accounting rate. The prediction distribution of the future regression (slope) parameter of the regression of future market rate on the future accounting rate is given in the middle two graphs of Figure 1. Both the graphs represent the t-distributions with different parameters. Although the shape of the distribution of both the graphs is roughly the same, the first graph here has a slightly greater spread, but lower pick, than the second graph. The bottom two graphs of Figure 11.1 displays the prediction distribution of the future sum of squared errors for different sample sizes. These last two graphs in Figure 11.1 represent the beta distribution with varying arguments.

11.6 Problems

1. Verify (11.1.4)-(11.1.7).
2. Prove (11.1.8).
3. Prove (11.2.1).
4. Show that the density function given by (11.2.1) does not depend on $d(e)$.

5. Verify (11.2.5)-(11.2.6).
6. Prove (11.2.7)-(11.2.8)
7. Prove (11.3.7)-(11.3.8).
8. Prove (11.4.2).
9. Verify (11.4.3).
10. Verify (11.4.6)-(11.4.7).
11. Prove (11.4.10).
12. Verify (11.4.13).
13. Prove (11.4.20)-(11.4.23).
14. Prove (11.4.24).

CHAPTER 12

STEIN ESTIMATION

Outline

Stein (1956) showed that when the dimension of parameter space is greater than 2 ($p \geq 3$), the best invariant and minimax estimator of the mean of a normal population is inadmissible. The approach by Stein (1956), which is the shrinkage approach, incorporates both uncertain prior information (on the parameter of interest) and the sample observations (from the underlying statistical distribution). To be more specific, comparing to what was proposed in previous chapters, Stein (1981) established conditions under which a more general class of shrinkage estimators dominated the usual minimax estimator under normal theory. See also the papers of James and Stein (1961), Casella (1990), Brandwein and Strawderman (1991), Ouassou and Strawderman (2002), and Xu and Izmirlian (2006), and the monographs of Lehmann and

Statistical Inference for Models with Multivariate t-Distributed Errors, First Edition.

A. K. Md. Ehsanes Saleh, M. Arashi, S.M.M. Tabatabaey.

Casella (1998) and Saleh (2006) for more information. In this chapter, we consider a general class of Stein-type shrinkage estimators under the multivariate t-model.

12.1 Class of Estimators

In this chapter, we will be discussing on a more general class of Stein-type shrinkage estimators, namely the Baranchik-type (due to Baranchik, 1970) class of estimators (BTEs). In the following, we propose our study in two categories. Because the purpose of this chapter is only to consider the performance of BTE from a classical viewpoint, in the first part of the chapter (first category), we consider the usual well-known BTE under the multivariate t-model, and then (second category) we consider its performance in a regression model with diffuse prior information, which is nothing more than a classical point of view. No real Bayesian analysis is taken into account in this chapter compared to the result of Chapter 10.

12.1.1 Without Prior Information

In this subsection, we consider BTEs. We propose conditions under which the underlying class outperforms the sample mean in two different setups:

Setup 1: In this case, we consider BTEs without prior knowledge of the parameter space. In a precise setup, let $\boldsymbol{X} \sim M_t^{(p)}(\boldsymbol{\theta}, \boldsymbol{\Sigma}, \gamma_o)$ for a known positive definite symmetric matrix $\boldsymbol{\Sigma}$. Now, consider the class of estimators given by

$$\boldsymbol{d}_r(\boldsymbol{X}) = \left(1 - \frac{r\left(\|\boldsymbol{X}\|^2\right)}{\|\boldsymbol{X}\|^2}\right)\boldsymbol{X}, \tag{12.1.1}$$

where it is assumed that $r(.) : \mathbb{R}^+ \to \mathbb{R}^+$ is an infinitely continuously differentiable function.

Setup 2: Because, that in real-world problems, taking the scale structure to be known is not practical, we also consider the following extended class, which admits unknown scale situations. To this end, first we suppose that $\boldsymbol{Y} \sim M_t^{(p)}(\boldsymbol{\theta}, \sigma^2\mathbf{V}_p, \gamma_o)$ for an unknown positive-valued scalar σ and a known positive definite symmetric matrix $\mathbf{V}_p$. In addition, according to Arashi and Tabatabaey (2010), assume that the statistic S^2 is distributed independent of $\boldsymbol{Y}$ with density (for a known $f(.)$)

$$f_{S^2}(s) = \frac{1}{\sigma^2} f\left(\frac{s}{\sigma^2}\right), \tag{12.1.2}$$

i.e., it belongs to the class of scale distributions, as such is the t-distribution.

Then the class of estimators under study has the form

$$\boldsymbol{d}_r^*(\boldsymbol{Y}) = \left(1 - \frac{r(F)}{F}\right)\boldsymbol{Y}, \tag{12.1.3}$$

where $F = \frac{\|\boldsymbol{Y}\|^2}{S^2}$.

12.1.2 Taking Prior Information

The precise setup of the problem is as follows: Let $\boldsymbol{Y}_i$ be a $p \times 1$ response vector with model (a typical multivariate location model)

$$\boldsymbol{Y}_i = \boldsymbol{\theta} + \boldsymbol{\varepsilon}_i, \quad 1 \leq i \leq N. \tag{12.1.4}$$

Here, $\boldsymbol{\theta}$ is a $p \times 1$ vector of location parameters and $\boldsymbol{\varepsilon}_i$ is a $p \times 1$ error vector such that

$$E(\boldsymbol{\varepsilon}_i) = \mathbf{0}, \ \ Cov(\boldsymbol{\varepsilon}_i \boldsymbol{\varepsilon}_j) = \boldsymbol{\Sigma}, \quad i, j = 1, \cdots, N, \quad N > p. \tag{12.1.5}$$

It is assumed, in general, $\boldsymbol{\varepsilon} = (\boldsymbol{\varepsilon}_1, \cdots, \boldsymbol{\varepsilon}_N)'$ have a joint p-variate t-distribution. In other words, we have that $\boldsymbol{\varepsilon}_i \sim M_t^{(p)}(\mathbf{0}, \boldsymbol{\Sigma}, \gamma_o)$, $\boldsymbol{\Sigma} > \mathbf{0}$, $i = 1, \ldots, N$.

Now, under a Bayesian framework, it is properly assumed that in distribution, little is known of course, a priori, about the parameters, the elements of $\boldsymbol{\theta}$ and the $\frac{1}{2}p(p+1)$ distinct elements of $\boldsymbol{\Sigma}$. We shall first of all suppose that the elements of $\boldsymbol{\theta}$, and those of $\boldsymbol{\Sigma}$ are approximately independent (see Box and Tiao, 1992, p. 425), i.e.,

$$\pi(\boldsymbol{\theta}, \boldsymbol{\Sigma}) \doteq \pi(\boldsymbol{\theta})\pi(\boldsymbol{\Sigma}). \tag{12.1.6}$$

Using the invariant theory due to Jeffreys (1961), we take

$$\begin{aligned} \pi(\boldsymbol{\theta}) &\propto \text{constant}, \\ \pi(\boldsymbol{\Sigma}) &\propto |\boldsymbol{\Sigma}|^{-\frac{p+1}{2}}, \end{aligned}$$

as the prior knowledge about the parameter space.

The next step is giving results for the marginal posterior distribution of the location parameter given responses.

Lemma 12.1.1. *Assume in the location model* (12.1.4), $\boldsymbol{e}_i \sim M_t^{(p)}(\mathbf{0}, \boldsymbol{\Sigma}, \gamma_o)$, *where* $\boldsymbol{\Sigma}$ *is a symmetric positive definite matrix. Then, w.r.t. the prior distribution given by* (12.1.6), *the posterior distribution of* $\boldsymbol{\theta}$ *is a multivariate t-distribution, i.e.,* $\boldsymbol{\theta}|\boldsymbol{Y} \sim M_t^{(p)}(\bar{\boldsymbol{Y}}, \boldsymbol{S}, N-p)$, *where* $\boldsymbol{Y} = (\boldsymbol{Y}_1, \cdots, \boldsymbol{Y}_N)$ *and*

$$\bar{\boldsymbol{Y}} = \frac{1}{N}\sum_{i=1}^{N} \boldsymbol{Y}_i, \qquad \boldsymbol{S} = \sum_{i=1}^{N} (\boldsymbol{Y}_i - \bar{\boldsymbol{Y}})(\boldsymbol{Y}_i - \bar{\boldsymbol{Y}})'. \tag{12.1.7}$$

Proof: Using Proposition 1 of Ng (2002), one can directly obtain

$$f(\boldsymbol{\theta}|\boldsymbol{Y}) \propto \left| \sum_{i=1}^{N} (\boldsymbol{Y}_i - \boldsymbol{\theta})(\boldsymbol{Y}_i - \boldsymbol{\theta})' \right|^{-\frac{N}{2}}, \tag{12.1.8}$$

which is the same as when we take the errors to be normally distributed (Zellner, 1971, p.243).

At this level, by making a conclusion based on the following equality,

$$\begin{aligned}(\boldsymbol{Y}_i-\boldsymbol{\theta})(\boldsymbol{Y}_i-\boldsymbol{\theta})' &= (\boldsymbol{Y}_i-\bar{\boldsymbol{Y}})(\boldsymbol{Y}_i-\bar{\boldsymbol{Y}})'+(\bar{\boldsymbol{Y}}-\boldsymbol{\theta})(\bar{\boldsymbol{Y}}-\boldsymbol{\theta})' \\ &\quad +2(\boldsymbol{Y}_i-\bar{\boldsymbol{Y}})(\bar{\boldsymbol{Y}}-\boldsymbol{\theta})',\end{aligned}$$

we observe

$$\left|\sum_{i=1}^{N}(\boldsymbol{Y}_i-\boldsymbol{\theta})(\boldsymbol{Y}_i-\boldsymbol{\theta})'\right| = |\boldsymbol{S}+N\boldsymbol{A}|, \tag{12.1.9}$$

where $\boldsymbol{A}=(\boldsymbol{\theta}-\bar{\boldsymbol{Y}})(\boldsymbol{\theta}-\bar{\boldsymbol{Y}})'$.

By making use of Corollary A.3.1 of Anderson (2003), we get

$$\begin{aligned}|\boldsymbol{S}+N\boldsymbol{A}| &= \begin{vmatrix}\boldsymbol{S} & -\sqrt{N}(\boldsymbol{\theta}-\bar{\boldsymbol{Y}}) \\ \sqrt{N}(\boldsymbol{\theta}-\bar{\boldsymbol{Y}})' & 1\end{vmatrix} \\ &= \begin{vmatrix}1 & \sqrt{N}(\boldsymbol{\theta}-\bar{\boldsymbol{Y}})' \\ -\sqrt{N}(\boldsymbol{\theta}-\bar{\boldsymbol{Y}}) & S\end{vmatrix} \\ &= |\boldsymbol{S}|\{1+N(\boldsymbol{\theta}-\bar{\boldsymbol{Y}})'\boldsymbol{S}^{-1}(\boldsymbol{\theta}-\bar{\boldsymbol{Y}})\}. \end{aligned} \tag{12.1.10}$$

Therefore, using equations (12.1.8)-(12.1.10), we come to the realization of the following formula:

$$f(\boldsymbol{\theta}|\boldsymbol{Y}) = c(N,p)|\boldsymbol{S}|^{-\frac{N}{2}}\{1+N(\boldsymbol{\theta}-\bar{\boldsymbol{Y}})'\boldsymbol{S}^{-1}(\boldsymbol{\theta}-\bar{\boldsymbol{Y}})\}^{-\frac{N}{2}},$$

where

$$\begin{aligned}c(N,p) &= \left\{\int_{\boldsymbol{\theta}\in\mathbb{R}^p}|\boldsymbol{S}|^{-\frac{N}{2}}\{1+N(\boldsymbol{\theta}-\bar{\boldsymbol{Y}})'\boldsymbol{S}^{-1}(\boldsymbol{\theta}-\bar{\boldsymbol{Y}})\}^{-\frac{N}{2}}d\boldsymbol{\theta}\right\}^{-1} \\ &= |\boldsymbol{S}|^{\frac{N}{2}}\left\{\frac{[\pi(N-p)]^{\frac{n}{2}}\,\Gamma\left(\frac{N-p}{2}\right)|\boldsymbol{S}|^{\frac{1}{2}}}{\Gamma\left(\frac{N}{2}\right)[N(N-p)]^{\frac{p}{2}}}\right\}^{-1} \\ &= \frac{N^{\frac{p}{2}}\Gamma\left(\frac{N}{2}\right)|\boldsymbol{S}|^{\frac{N-1}{2}}}{\pi^{\frac{n}{2}}(N-p)^{\frac{N-p}{2}}\Gamma\left(\frac{N-p}{2}\right)}.\end{aligned}$$

This would prove our claim.

In considering this setup, we shall also assume that the loss function is given by

$$L(\hat{\boldsymbol{\theta}};\boldsymbol{\theta}) = N(\hat{\boldsymbol{\theta}}-\boldsymbol{\theta})'\boldsymbol{\Sigma}^{-1}(\hat{\boldsymbol{\theta}}-\boldsymbol{\theta}) \tag{12.1.11}$$

for any estimator $\hat{\boldsymbol{\theta}}$ of $\boldsymbol{\theta}$.

It has been fully known that the Bayes estimator of $\boldsymbol{\theta}$ with respect to the loss (12.1.11) is the posterior mean (Proposition 2.5.1, Robert, 2001) given by

$$\hat{\boldsymbol{\theta}} = \bar{\boldsymbol{Y}}. \tag{12.1.12}$$

As it can be made real from the estimator given by (12.1.12), the Bayes estimator, reduces to the sample mean, under the setup presented above. So there is no need to deal with the Bayesian aspects of $\hat{\boldsymbol{\theta}}$, and we, in fact, are concerned with the sample mean rather than the Bayes estimator.

Then, from the result of subsection 2.6.3,

$$\hat{\boldsymbol{\theta}} \sim M_t^{(p)}(\boldsymbol{\theta}, N^{-1}\Sigma, \gamma_o). \tag{12.1.13}$$

Under the classical viewpoint, we devote a general class of Stein-type shrinkage estimators to the estimator $\hat{\boldsymbol{\theta}}$, given by

$$\boldsymbol{d}_r(\hat{\boldsymbol{\theta}}) = \left[1 - \frac{r\left(\hat{\boldsymbol{\theta}}'\boldsymbol{S}^{-1}\hat{\boldsymbol{\theta}}\right)}{\hat{\boldsymbol{\theta}}'\boldsymbol{S}^{-1}\hat{\boldsymbol{\theta}}}\right]\hat{\boldsymbol{\theta}}, \tag{12.1.14}$$

where $r : \mathbb{R}^+ \to \mathbb{R}^+$ is an absolutely continuous function.

Furthermore, $r \in \mathcal{S}(\mathbb{R}^+, \mu)$ (the Schwartz space or space of rapidly decreasing functions on $\mathbb{R}^+$ with the measure μ), where

$$\mathcal{S}(\mathbb{R}^+, \mu) = \left\{r \in \mathcal{C}^\infty(\mathbb{R}^+, \mu) : \|r\|_{\alpha,\beta} < \infty \quad \forall\, \alpha, \beta\right\},$$

α and β are indices, $\mathcal{C}^\infty(\mathbb{R}^+, \mu)$ is the set of all smooth functions from $\mathbb{R}^+$ to $\mathbf{C}$ (the set of all complex numbers), and

$$\|r\|_{\alpha,\beta} = \|x^\alpha \mathrm{D}^\beta r\|_\infty = \sup\{|x^\alpha \mathrm{D}^\beta r(x)| : x \in \text{domain of } r\}.$$

Here, D^β stands for the βth derivative of r. See Folland (1999) for more details.

The latter condition plays a strategic position in gaining the main result. Note that for every function such as $r(.)$ belongs to $\mathcal{S}(\mathbb{R}^+, \mu)$, we have

$$\int_0^\infty r'(x)d\mu(x) < \infty, \tag{12.1.15}$$

$$\int_0^\infty r^2(x)d\mu(x) < \infty, \tag{12.1.16}$$

where $r'(x)$ is the 1st derivative of $r(x)$ w.r.t. x.

More interesting than the Schwartz space is that the denseness in the space of all functions satisfies the above conditions in (12.1.15) and (12.1.16).

12.2 Preliminaries and Some Theorems

In this section, we give some lemmas to evaluate the risk function of $\boldsymbol{d}_r(\hat{\boldsymbol{\theta}})$. Provided that all expectations exist, we deserve the following lemmas.

Lemma 12.2.1. *If* $\boldsymbol{x} \sim \mathcal{N}_p(\boldsymbol{\theta}, \alpha\boldsymbol{\Sigma})$, $\alpha > 0$, $\boldsymbol{\Sigma}$ *is independent of* $\boldsymbol{S} \sim W_p(\beta\boldsymbol{\Sigma}, n)$, $\beta > 0$, $n = N - 1$, *then*

$$E\left[\frac{\boldsymbol{x}'\boldsymbol{\Sigma}^{-1}(\boldsymbol{x}-\boldsymbol{\theta})r\left(\boldsymbol{x}'\boldsymbol{S}^{-1}\boldsymbol{x}\right)}{\boldsymbol{x}'\boldsymbol{S}^{-1}\boldsymbol{x}}\right] = \beta\alpha(n-p+1)\Bigg\{(p-2)E\left[\frac{r(\boldsymbol{x}'\boldsymbol{S}^{-1}\boldsymbol{x})}{\boldsymbol{x}'\boldsymbol{\Sigma}^{-1}\boldsymbol{x}}\right] + 2E\left[r'(\boldsymbol{x}'\boldsymbol{S}^{-1}\boldsymbol{x})\right]\Bigg\}$$

and

$$E\left[\frac{\boldsymbol{x}'\boldsymbol{\Sigma}^{-1}\boldsymbol{x}r^2\left(\boldsymbol{x}'\boldsymbol{S}^{-1}\boldsymbol{x}\right)}{\left(\boldsymbol{x}'\boldsymbol{S}^{-1}\boldsymbol{x}\right)^2}\right] = \beta^2(n-p+1)(n-p+3)E\left[\frac{r^2(\boldsymbol{x}'\boldsymbol{S}^{-1}\boldsymbol{x})}{\boldsymbol{x}'\boldsymbol{\Sigma}^{-1}\boldsymbol{x}}\right].$$

Proof: Let $\boldsymbol{y} = \boldsymbol{\Sigma}^{-\frac{1}{2}}\boldsymbol{x}$, $\boldsymbol{\theta}^* = \boldsymbol{\Sigma}^{-\frac{1}{2}}\boldsymbol{\theta}$, and $\boldsymbol{A} = \beta^{-1}\boldsymbol{\Sigma}^{-\frac{1}{2}}\boldsymbol{S}\boldsymbol{\Sigma}^{-\frac{1}{2}}$, where $\boldsymbol{\Sigma}^{-\frac{1}{2}}$ is a symmetric square root of $\boldsymbol{\Sigma}^{-1}$. From independence of $\boldsymbol{x}$ and $\boldsymbol{S}$, $\boldsymbol{y} \sim \mathcal{N}_p(\boldsymbol{\theta}^*, \alpha\boldsymbol{I}_p)$ is independent of both $\boldsymbol{A} \sim W_p(\boldsymbol{I}_p, n)$ and $\frac{\boldsymbol{y}'\boldsymbol{y}}{\boldsymbol{y}'\boldsymbol{A}^{-1}\boldsymbol{y}} \sim \chi^2_{n-p+1}$. Also note that $F = \boldsymbol{x}'\boldsymbol{S}^{-1}\boldsymbol{x} = \beta^{-1}\boldsymbol{y}'\boldsymbol{A}^{-1}\boldsymbol{y}$. Therefore, using Stein's identity (Stein, 1981), we get

$$\begin{aligned}
E\left[\frac{\boldsymbol{x}'\boldsymbol{\Sigma}^{-1}(\boldsymbol{x}-\boldsymbol{\theta})r\left(\boldsymbol{x}'\boldsymbol{S}^{-1}\boldsymbol{x}\right)}{\boldsymbol{x}'\boldsymbol{S}^{-1}\boldsymbol{x}}\right] &= E\left[\frac{\boldsymbol{y}'(\boldsymbol{y}-\boldsymbol{\theta}^*)r(F)}{F}\right] \\
&= E\left[\frac{\boldsymbol{y}'(\boldsymbol{y}-\boldsymbol{\theta}^*)r(F)}{\boldsymbol{y}'\boldsymbol{y}}\right]E\left[\frac{\boldsymbol{y}'\boldsymbol{y}}{F}\right] \\
&= \beta(n-p+1)E\left[\frac{\boldsymbol{y}'(\boldsymbol{y}-\boldsymbol{\theta}^*)r(F)}{\boldsymbol{y}'\boldsymbol{y}}\right] \\
&= \alpha\beta(n-p+1)\Bigg\{(p-2)E\left[\frac{r(F)}{\boldsymbol{y}'\boldsymbol{y}}\right] \\
&\quad +2E\left[r'(F)\right]\Bigg\}. \qquad (12.2.1)
\end{aligned}$$

Similarly,

$$\begin{aligned}
E\left[\frac{\boldsymbol{x}'\boldsymbol{\Sigma}^{-1}\boldsymbol{x}r^2\left(\boldsymbol{x}'\boldsymbol{S}^{-1}\boldsymbol{x}\right)}{\left(\boldsymbol{x}'\boldsymbol{S}^{-1}\boldsymbol{x}\right)^2}\right] &= E\left[\frac{\boldsymbol{y}'\boldsymbol{y}r^2(F)}{F^2}\right] \\
&= \beta^2E\left[\frac{r^2(F)}{\boldsymbol{y}'\boldsymbol{y}}\right]E\left[\left(\frac{\boldsymbol{y}'\boldsymbol{y}}{\boldsymbol{y}'\boldsymbol{A}^{-1}\boldsymbol{y}}\right)^2\right] \\
&= \beta^2(n-p+1)(n-p+3)E\left[\frac{r^2(F)}{\boldsymbol{y}'\boldsymbol{y}}\right].
\end{aligned}$$

Lemma 12.2.2. *The risk function of the estimator* $\boldsymbol{d}_r(\hat{\boldsymbol{\theta}})$ *w.r.t. the loss function* (12.1.11) *is given by*

$$\begin{aligned} R(\boldsymbol{d}_r(\hat{\boldsymbol{\theta}});\boldsymbol{\theta}) &= R(\hat{\boldsymbol{\theta}};\boldsymbol{\theta}) - 4(N-p)\int_0^\infty E\left[r'\left(\hat{\boldsymbol{\theta}}'\boldsymbol{S}^{-1}\hat{\boldsymbol{\theta}}\right)\Big| t\right] t^{-2}W_o(t)dt \\ &+\int_0^\infty E\left\{\frac{(N-p)r\left(\hat{\boldsymbol{\theta}}'\boldsymbol{S}^{-1}\hat{\boldsymbol{\theta}}\right)}{t\hat{\boldsymbol{\theta}}'(t^{-1}\boldsymbol{\Sigma})^{-1}\hat{\boldsymbol{\theta}}}\right. \\ &\times\left.\left[N(N-p+2)r\left(\hat{\boldsymbol{\theta}}'\boldsymbol{S}^{-1}\hat{\boldsymbol{\theta}}\right)-2(p-2)\right]\Big| t\right\}W_o(t)dt. \end{aligned}$$

Proof: As far as the representation (2.6.3) is concerned, it is possible to continue this way:

$$\begin{aligned} R(\boldsymbol{d}_r(\hat{\boldsymbol{\theta}});\boldsymbol{\theta}) &= NE\left[(\boldsymbol{d}_r(\hat{\boldsymbol{\theta}})-\boldsymbol{\theta})'\boldsymbol{\Sigma}^{-1}(\boldsymbol{d}_r(\hat{\boldsymbol{\theta}})-\boldsymbol{\theta})\right] \\ &= R(\hat{\boldsymbol{\theta}};\boldsymbol{\theta}) - 2NE\left[\frac{\hat{\boldsymbol{\theta}}'\boldsymbol{\Sigma}^{-1}(\hat{\boldsymbol{\theta}}-\boldsymbol{\theta})r\left(\hat{\boldsymbol{\theta}}'\boldsymbol{S}^{-1}\hat{\boldsymbol{\theta}}\right)}{\hat{\boldsymbol{\theta}}'\boldsymbol{S}^{-1}\hat{\boldsymbol{\theta}}}\right] \\ &+NE\left[\frac{\hat{\boldsymbol{\theta}}'\boldsymbol{\Sigma}^{-1}\hat{\boldsymbol{\theta}}r^2\left(\hat{\boldsymbol{\theta}}'\boldsymbol{S}^{-1}\hat{\boldsymbol{\theta}}\right)}{\left(\hat{\boldsymbol{\theta}}'\boldsymbol{S}^{-1}\hat{\boldsymbol{\theta}}\right)^2}\right] \\ &= R(\hat{\boldsymbol{\theta}};\boldsymbol{\theta}) - 2N\int_0^\infty E\left[\frac{\hat{\boldsymbol{\theta}}'\boldsymbol{\Sigma}^{-1}(\hat{\boldsymbol{\theta}}-\boldsymbol{\theta})r\left(\hat{\boldsymbol{\theta}}'\boldsymbol{S}^{-1}\hat{\boldsymbol{\theta}}\right)}{\hat{\boldsymbol{\theta}}'\boldsymbol{S}^{-1}\hat{\boldsymbol{\theta}}}\Big| t\right]W_o(t)dt \\ &+N\int_0^\infty E\left[\frac{\hat{\boldsymbol{\theta}}'\boldsymbol{\Sigma}^{-1}\hat{\boldsymbol{\theta}}r^2\left(\hat{\boldsymbol{\theta}}'\boldsymbol{S}^{-1}\hat{\boldsymbol{\theta}}\right)}{\left(\hat{\boldsymbol{\theta}}'\boldsymbol{S}^{-1}\hat{\boldsymbol{\theta}}\right)^2}\Big| t\right]W_o(t)dt. \qquad (12.2.2) \end{aligned}$$

Conditionally (under normal theory), $\hat{\boldsymbol{\theta}}|t \sim \mathcal{N}_p(\boldsymbol{\theta}, t^{-1}N^{-1}\boldsymbol{\Sigma})$ is independent of $\boldsymbol{S} \sim W_p(t^{-1}\boldsymbol{\Sigma}, n)$. Consequently, by making use of Lemma 12.2.1 for the selections $\alpha = (tN)^{-1}$ and $\beta = t^{-1}$, we reach

$$\begin{aligned} E\left[\frac{\hat{\boldsymbol{\theta}}'\boldsymbol{\Sigma}^{-1}(\hat{\boldsymbol{\theta}}-\boldsymbol{\theta})r\left(\hat{\boldsymbol{\theta}}'\boldsymbol{S}^{-1}\hat{\boldsymbol{\theta}}\right)}{\hat{\boldsymbol{\theta}}'\boldsymbol{S}^{-1}\hat{\boldsymbol{\theta}}}\Big| t\right] &= \frac{N-p}{Nt^2}\left\{(p-2)E\left[\frac{r\left(\hat{\boldsymbol{\theta}}'\boldsymbol{S}^{-1}\hat{\boldsymbol{\theta}}\right)}{\hat{\boldsymbol{\theta}}'\boldsymbol{\Sigma}^{-1}\hat{\boldsymbol{\theta}}}\Big| t\right]\right. \\ &\left.+2E\left[r'\left(\hat{\boldsymbol{\theta}}'\boldsymbol{S}^{-1}\hat{\boldsymbol{\theta}}\right)\Big| t\right]\right\}, \\ E\left[\frac{\hat{\boldsymbol{\theta}}'\boldsymbol{\Sigma}^{-1}\hat{\boldsymbol{\theta}}r^2\left(\hat{\boldsymbol{\theta}}'\boldsymbol{S}^{-1}\hat{\boldsymbol{\theta}}\right)}{\left(\hat{\boldsymbol{\theta}}'\boldsymbol{S}^{-1}\hat{\boldsymbol{\theta}}\right)^2}\Big| t\right] &= \frac{(N-p)(N-p+2)}{t^2} \end{aligned}$$

$$\times E\left[\frac{r^2\left(\hat{\boldsymbol{\theta}}'\boldsymbol{S}^{-1}\hat{\boldsymbol{\theta}}\right)}{\hat{\boldsymbol{\theta}}'\boldsymbol{\Sigma}^{-1}\hat{\boldsymbol{\theta}}}\middle| t\right].$$

Finally, substituting the above expressions in (12.2.2), completes the proof.

It should be noted that, in the above result, the existence of the integrals needs to be checked.

12.3 Superiority Conditions

In this section, we derive the superiority conditions of the class of proposed shrinkage estimators under the two settings considered in section 12.1.1.

12.3.1 Without Taking Prior Information

Under setup 1, we have the following result.

Theorem 12.3.1. *Under the conditions of subsection 12.1.1, the shrinkage estimator* $\boldsymbol{d}_r(\boldsymbol{X})$ *given by* (12.1.1) *has smaller quadratic risk than* $\boldsymbol{X}$ *provided*

(i) *r is a concave function,*

(ii) $0 < r \leq \frac{\gamma_o}{\gamma_o - 2}\left[\frac{2tr(\boldsymbol{\Sigma})}{\lambda_1(\boldsymbol{\Sigma})} - 4\right]$,

where $\lambda_1(\boldsymbol{\Sigma})$ is the largest eigenvalue of $\boldsymbol{\Sigma}$.

Proof: Using the risk difference we get

$$\begin{aligned} D_1 &= R(\boldsymbol{d}_r(\boldsymbol{X}), \boldsymbol{I}_p) - R(\boldsymbol{X}, \boldsymbol{I}_p) \\ &= E[\|\boldsymbol{d}_r(\boldsymbol{X}) - \boldsymbol{\theta}\|^2 - \|\boldsymbol{X} - \boldsymbol{\theta}\|^2] \\ &= E\left[\frac{r^2(\|\boldsymbol{X}\|^2)}{\|\boldsymbol{X}\|^2} - 2\frac{r(\|\boldsymbol{X}\|^2)}{\|\boldsymbol{X}\|^2}\boldsymbol{X}'(\boldsymbol{\theta} - \boldsymbol{X})\right] \\ &= \int_0^\infty E_N\left[\frac{r^2(\|\boldsymbol{X}\|^2)}{\|\boldsymbol{X}\|^2} - 2\frac{r(\|\boldsymbol{X}\|^2)}{\|\boldsymbol{X}\|^2}\boldsymbol{X}'(\boldsymbol{\theta} - \boldsymbol{X})\right] W_o(t)dt\, dt, \end{aligned} \tag{12.3.1}$$

where $E_N(.)$ represents obtaining the expectation w.r.t. $\mathcal{N}_p(\boldsymbol{\theta}, t^{-1}\boldsymbol{\Sigma})$.

In order to show that $D_1 \leq 0$, it is enough to show that

$$E_N\left[\frac{r^2(\|\boldsymbol{X}\|^2)}{\|\boldsymbol{X}\|^2} - 2\frac{r(\|\boldsymbol{X}\|^2)}{\|\boldsymbol{X}\|^2}\boldsymbol{X}'(\boldsymbol{\theta} - \boldsymbol{X})\right] \leq 0. \tag{12.3.2}$$

Applying Stein's identity, we get

$$\begin{aligned} & E_N\left[\frac{r(\|\boldsymbol{X}\|^2)}{\|\boldsymbol{X}\|^2}\,\boldsymbol{X}'(\boldsymbol{\theta}-\boldsymbol{X})\right] \\ = \;& t^{-1}E_N\left\{\frac{r(\|\boldsymbol{X}\|^2)}{\|\boldsymbol{X}\|^2}\,tr(\boldsymbol{\Sigma}) - 2\frac{r(\|\boldsymbol{X}\|^2)}{(\|\boldsymbol{X}\|^2)^2}\,\boldsymbol{X}'\boldsymbol{\Sigma}\boldsymbol{X} \right. \\ & \left. +2\frac{r'(\|\boldsymbol{X}\|^2)}{\|\boldsymbol{X}\|^2}\,\boldsymbol{X}'\boldsymbol{\Sigma}\boldsymbol{X}\right\}. \end{aligned} \tag{12.3.3}$$

Therefore, using equations (12.3.2) and (12.3.3) we obtain

$$E_N\left[\frac{r^2(\|\boldsymbol{X}\|^2)}{\|\boldsymbol{X}\|^2} - 2\frac{r(\|\boldsymbol{X}\|^2)}{\|\boldsymbol{X}\|^2}\,\boldsymbol{X}'(\boldsymbol{\theta}-\boldsymbol{X})\right] = A + B, \tag{12.3.4}$$

where

$$\begin{aligned} A &= E_N\left\{\frac{r(\|\boldsymbol{X}\|^2)}{\|\boldsymbol{X}\|^2}\left[r(\|\boldsymbol{X}\|^2) - 2t^{-1}tr(\boldsymbol{\Sigma}) + 4t^{-1}\frac{\boldsymbol{X}'\boldsymbol{\Sigma}\boldsymbol{X}}{\|\boldsymbol{X}\|^2}\right]\right\}, \\ B &= -4t^{-1}E_N\left[\frac{r'(\|\boldsymbol{X}\|^2)}{\|\boldsymbol{X}\|^2}\,\boldsymbol{X}'\boldsymbol{\Sigma}\boldsymbol{X}\right]. \end{aligned}$$

The function $r(.)$ is concave, thus using Lemma 1 from Casella (1990), $r(.)$ is non-decreasing, and $r'(\|\boldsymbol{X}\|^2) \geq 0$. Therefore, $B \leq 0$.

On the other hand,

$$\frac{\boldsymbol{X}'\boldsymbol{\Sigma}\boldsymbol{X}}{\|\boldsymbol{X}\|^2} = \frac{\boldsymbol{X}'\boldsymbol{\Sigma}\boldsymbol{X}}{\boldsymbol{X}'\boldsymbol{X}} \leq \lambda_1(\boldsymbol{\Sigma}). \tag{12.3.5}$$

Note that $A \leq 0$ and, therefore, $D_1 \leq 0$, iff for every $\boldsymbol{X} \in \mathbb{R}^p$ and $t \geq 0$, we have

$$r(\|\boldsymbol{X}\|^2) \leq t^{-1}\left[2tr(\boldsymbol{\Sigma}) - 4\frac{\boldsymbol{X}'\boldsymbol{\Sigma}\boldsymbol{X}}{\|\boldsymbol{X}\|^2}\right]. \tag{12.3.6}$$

But using (12.3.5), the inequality in (12.3.6) satisfies if we take

$$r(\|\boldsymbol{X}\|^2) \leq t^{-1}\left[\frac{2tr(\boldsymbol{\Sigma})}{\lambda_1(\boldsymbol{\Sigma})} - 4\right]$$

and average over the distribution $W_o(.)$ given by (2.6.4).

It should again be noticed that the above result requires the existence of expectations.

Under setup 2, we have the following result

Theorem 12.3.2. *Under the conditions of subsection 12.1.1, the shrinkage estimator $\boldsymbol{d}_r^*(\boldsymbol{X})$ given by (12.1.3) has a smaller quadratic risk than $\boldsymbol{X}$ provided* *(i)* r *is a concave function,*

(ii) $0 < r \leq \frac{\gamma_o}{\gamma_o - 2}\left[\frac{2tr(\mathbf{V}_p)}{\lambda_1(\mathbf{V}_p)} - 4\right]\frac{E_{\sigma=1}(S^2)}{E_{\sigma=1}(S^4)}$.

Proof: By making use of the risk difference, we have

$$\begin{aligned}
D_2 &= R(\boldsymbol{d}_r^*(\boldsymbol{X}), \boldsymbol{I}_p) - R(\boldsymbol{X}, \boldsymbol{I}_p) \\
&= E\left\{\frac{\|\boldsymbol{d}_r^*(\boldsymbol{X}) - \boldsymbol{\theta}\|^2}{\sigma^2} - \frac{\|\boldsymbol{X} - \boldsymbol{\theta}\|^2}{\sigma^2}\right\} \\
&= E\left\{\frac{\|\boldsymbol{X}\|^2 r^2(F)}{\sigma^2 F^2} - \frac{2r(F)}{\sigma^2 F}\boldsymbol{X}'(\boldsymbol{\theta} - \boldsymbol{X})\right\} \\
&= \int_0^\infty E_{N^*}\left[\frac{S^4 r^2\left(\frac{\|\boldsymbol{X}\|^2}{S^2}\right)}{\sigma^2\|\boldsymbol{X}\|^2} - \frac{2S^2 r\left(\frac{\|\boldsymbol{X}\|^2}{S^2}\right)}{\sigma^2\|\boldsymbol{X}\|^2}\boldsymbol{X}'(\boldsymbol{\theta} - \boldsymbol{X})\right]W_o(t)dt,
\end{aligned}$$

where $E_{\mathcal{N}^*}$ represents obtaining the expectation w.r.t. $\mathcal{N}_p(\boldsymbol{\theta}, \sigma^2 t^{-1}\mathbf{V}_p)$. Now, applying Stein's identity and making use of conditional independency between $\boldsymbol{X}$ and S^2, we can obtain

$$\begin{aligned}
D_2 = &\int_0^\infty E_{S^2}\left(E_{N^*}\left\{\frac{r\left(\frac{\|\boldsymbol{X}\|^2}{s^2}\right)}{\|\boldsymbol{X}\|^2}\left[s^4 r\left(\frac{\|\boldsymbol{X}\|^2}{s^2}\right) + \frac{s^2\boldsymbol{X}'\mathbf{V}_p\boldsymbol{X}}{t\|\boldsymbol{X}\|^2}\right.\right.\right. \\
&\left.\left.\left.-2\frac{s^2 tr(\mathbf{V}_p)}{t}\right]\right\} \mid S = s\right)W_o(t)dt \\
&- \int_0^\infty W(dt)E_{S^2}\left(E_{N^*}\left\{\frac{r'\left(\frac{\|\boldsymbol{X}\|^2}{s^2}\right)}{t\|\boldsymbol{X}\|^2}\boldsymbol{X}'\mathbf{V}_p\boldsymbol{X}\right\} \mid S = s\right)W_o(t)dt,
\end{aligned}$$

wherein the function $r(.)$ is concave, $r'\left(\frac{\|\boldsymbol{X}\|^2}{s^2}\right) \geq 0$ (see Casella, 1990). Similar to equations (12.3.5) and (12.3.6), one can conclude that $D_2 \leq 0$ iff $0 < r(.)$ and for every $\boldsymbol{X} \in \mathbb{R}^p$ and $t \geq 0$ we have

$$\left\{s^4 r\left(\frac{\|\boldsymbol{X}\|^2}{s^2}\right) \mid S = s\right\} \leq \left\{t^{-1}\left[\frac{2s^2 tr(\mathbf{V})}{\lambda_1(\mathbf{V})} - 4\right] \mid S = s\right\}.$$

On averaging over S^2 and $W_o(.)$, we deduce $D_2 \leq 0$ provided

$$r \leq \frac{\gamma_o}{\gamma_o - 2}\left[\frac{2tr(\mathbf{V})}{\lambda_1(\mathbf{V})} - 4\right]\frac{E_{\sigma=1}(S^2)}{E_{\sigma=1}(S^4)}.$$

Accordingly, we may make the following conclusions with some care.

Corollary 12.3.2.1. *When* $\boldsymbol{X} \sim \mathcal{N}_p(\boldsymbol{\theta}, \boldsymbol{I}_p)$, *the condition* ***(ii)*** *of Theorem 12.3.1 changes to* $0 < r \leq 2(p-2)$, *since there is no need to integrate over the distribution*

$W_o(.)$, however, the coefficient $\gamma_o/(\gamma_o - 2)$ substitutes with

$$\int_0^\infty t^{-1}\delta(t-1)dt = 1,$$

*which is the same as in Casella (1990), where $\delta(.)$ is the Dirac delta function. Also, for $\boldsymbol{X} \sim \mathcal{N}_p(\boldsymbol{\theta}, \boldsymbol{\Sigma})$ the conditions of Theorem 12.3.1 are the same as Theorem 5.7 of Lehmann and Casella (1998), replacing condition **(i)** by non-decreasing $r(.)$.*

Corollary 12.3.2.2. *For the case under which $\boldsymbol{X} \sim \mathcal{N}_p(\boldsymbol{\theta}, \sigma^2 \boldsymbol{I}_p)$ and $S^2 \sim \sigma^2\chi^2_n$, based on Theorem 12.3.2, because $\frac{E_{\sigma=1}(S^2)}{E_{\sigma=1}(S^4)} = \frac{1}{n+2}$, we obtain the same results stated in Baranchik (1970).*

Corollary 12.3.2.3. *Let $\boldsymbol{X} \sim M_t^{(p)}(\boldsymbol{\theta}, \boldsymbol{\Sigma}, \gamma_o)$, where $\boldsymbol{\Sigma} > \boldsymbol{0}$ is known. Then $\boldsymbol{d}_r(\boldsymbol{X}) \succeq \boldsymbol{X}$ w.r.t. the loss function $L(\boldsymbol{X};\boldsymbol{\theta}) = N(\boldsymbol{X} - \boldsymbol{\theta})'\boldsymbol{W}(\boldsymbol{X} - \boldsymbol{\theta})$, where $\boldsymbol{W}$ is a known symmetric positive definite weight matrix, provided*

(i) *r is concave,*

(ii) $0 < r \leq \frac{\gamma_o}{\gamma_o - 2}\left[\frac{2tr\left(\boldsymbol{\Sigma}^{\frac{1}{2}}\boldsymbol{W}\boldsymbol{\Sigma}^{\frac{1}{2}}\right)}{\lambda_1\left(\boldsymbol{\Sigma}^{\frac{1}{2}}\boldsymbol{W}\boldsymbol{\Sigma}^{\frac{1}{2}}\right)} - 4\right].$

Corollary 12.3.2.4. *Let $\boldsymbol{X} \sim M_t^{(p)}(\boldsymbol{\theta}, \boldsymbol{\Sigma}, \gamma_o)$, where $\boldsymbol{\Sigma} > \boldsymbol{0}$ is known. Define*

$$\boldsymbol{d}_{r,c}(\boldsymbol{X}) = \left(1 - \frac{(c-2)r\|\boldsymbol{X}\|^2}{\|\boldsymbol{X}\|^2}\right)\boldsymbol{X}, \quad c > 2.$$

Then $\boldsymbol{d}_{r,c}(\boldsymbol{X}) \succeq \boldsymbol{X}$ w.r.t. the quadratic loss function, provided

(i) *r is concave,*

(ii) $0 < r \leq \frac{\gamma_o}{\gamma_o - 2}\frac{2tr(\boldsymbol{\Sigma})}{\lambda_1(\boldsymbol{\Sigma})}.$

Corollary 12.3.2.5. *Setting $r(F)$ equal to a constant c and $r(F) = c/\left(1 + cF^{-1}\right)$ in Theorem 12.3.2, we obtain the generalizations of the estimators given in James and Stein (1961) and Alam and Thompson (1969) under multivariate t-assumptions for $0 < c \leq 2(p-2)\frac{E_{\sigma=1}(S^2)}{E_{\sigma=1}(S^4)}$ and $0 < c \leq (p-2)\frac{E_{\sigma=1}(S^2)}{E_{\sigma=1}(S^4)}$, respectively.*

As an example, consider the linear model

$$\boldsymbol{y} = \boldsymbol{X}\boldsymbol{\beta} + \boldsymbol{\varepsilon},$$

where $\boldsymbol{y}$ is an n-vector of response, $\mathbf{X}$ is an $n \times p$ design matrix with full rank p, $\boldsymbol{\beta} = (\beta_1, \cdots, \beta_p)'$ is a p-vector of regression coefficients, and $\boldsymbol{\varepsilon} = (\epsilon_1, \cdots, \epsilon_n)'$ is the n-vector of errors distributed as $M_t^{(n)}(\mathbf{0}, \boldsymbol{I}_n, \gamma_o)$.

The LS estimator of $\boldsymbol{\beta}$ is

$$\tilde{\boldsymbol{\beta}}_n = \boldsymbol{C}^{-1}\boldsymbol{X}'\boldsymbol{y}, \qquad \boldsymbol{C} = \boldsymbol{X}'\boldsymbol{X}.$$

Define the Stein-type estimator of $\tilde{\boldsymbol{\beta}}_n$ by

$$\tilde{\boldsymbol{\beta}}_n^a = \left(1 - \frac{a}{\tilde{\boldsymbol{\beta}}_n'\tilde{\boldsymbol{\beta}}_n + 1}\right)\tilde{\boldsymbol{\beta}}_n, \quad 0 < a \leq \frac{2(p-2)\gamma_o}{(\gamma_o - 2)}.$$

Theorem 12.3.3. *Relative to the quadratic loss function,* $\tilde{\boldsymbol{\beta}}_n^a \succeq \tilde{\boldsymbol{\beta}}_n$.

Proof: Take

$$r(||\tilde{\boldsymbol{\beta}}_n||^2) = a\frac{||\tilde{\boldsymbol{\beta}}_n||^2}{||\tilde{\boldsymbol{\beta}}_n||^2 + 1}.$$

Then

$$\begin{aligned}\left(1 - \frac{r(||\tilde{\boldsymbol{\beta}}_n||^2)}{||\tilde{\boldsymbol{\beta}}_n||^2}\right)\tilde{\boldsymbol{\beta}}_n &= \left(1 - \frac{a}{||\tilde{\boldsymbol{\beta}}_n||^2 + 1}\right)\tilde{\boldsymbol{\beta}}_n \\ &= \tilde{\boldsymbol{\beta}}_n^a.\end{aligned}$$

Further, $r'(y) = \frac{a}{(1+y)^2} > 0$, which shows that $r(||\tilde{\boldsymbol{\beta}}_n||^2)$ is a concave function. On the other hand, $0 < r(||\tilde{\boldsymbol{\beta}}_n||^2) \leq a$. Using the fact that $\tilde{\boldsymbol{\beta}}_n \sim M_t^{(p)}(\boldsymbol{\beta}, \boldsymbol{C}^{-1}, \gamma_o)$, according to Corollary 12.3.2.3, taking $\boldsymbol{W} = \boldsymbol{C}$, $\tilde{\boldsymbol{\beta}}_n^a \succeq \tilde{\boldsymbol{\beta}}_n$ as soon as

$$\begin{aligned} r &\leq \frac{\gamma_o}{\gamma_o - 2}\left[\frac{2tr\left(\boldsymbol{C}^{-\frac{1}{2}}\boldsymbol{W}\boldsymbol{C}^{-\frac{1}{2}}\right)}{\lambda_1\left(\boldsymbol{C}^{-\frac{1}{2}}\boldsymbol{W}\boldsymbol{C}^{-\frac{1}{2}}\right)} - 4\right] \\ &= \frac{\gamma_o}{\gamma_o - 2}\left[\frac{2tr\left(\boldsymbol{C}^{-\frac{1}{2}}\boldsymbol{C}\boldsymbol{C}^{-\frac{1}{2}}\right)}{\lambda_1\left(\boldsymbol{C}^{-\frac{1}{2}}\boldsymbol{C}\boldsymbol{C}^{-\frac{1}{2}}\right)} - 4\right] \\ &= \frac{\gamma_o}{\gamma_o - 2}\,(2p - 4)\,. \end{aligned}$$

The result follows by noting that $a \leq \frac{2\gamma_o}{\gamma_o - 2}\,(p-2)$.

12.3.2 Taking Prior Information

In this section, we demonstrate the minimaxity of the estimator $\boldsymbol{d}_r(\hat{\boldsymbol{\theta}})$, under some mild conditions made on the function $r(.)$.

Theorem 12.3.4. *Under the assumptions of section 12.1.2, the estimator $\boldsymbol{d}_r(\hat{\boldsymbol{\theta}})$ is minimax under the Schwartz space, providing*

(i) *r is non-decreasing,*

(ii) $r \leq \frac{2(p-2)}{N(N-p+2)}$.

Proof: The estimator $\hat{\boldsymbol{\theta}}$ given by (12.1.12) is minimax, Therefore, in order to show that $\boldsymbol{d}_r(\hat{\boldsymbol{\theta}})$ is minimax it is enough to show that $R(\boldsymbol{d}_r(\hat{\boldsymbol{\theta}});\boldsymbol{\theta}) - R(\hat{\boldsymbol{\theta}};\boldsymbol{\theta}) \leq 0$. But from Lemma 12.2.2 we have $R(\boldsymbol{d}_r(\hat{\boldsymbol{\theta}});\boldsymbol{\theta}) - R(\hat{\boldsymbol{\theta}};\boldsymbol{\theta}) = \boldsymbol{A} + \boldsymbol{B}$, where

$$\begin{aligned}
\boldsymbol{A} &= -4(N-p)\int_0^\infty E_N\left[r'\left(\hat{\boldsymbol{\theta}}'\boldsymbol{S}^{-1}\hat{\boldsymbol{\theta}}\right)\,\middle|\, t\right] t^{-2}W_o(t)dt \\
\boldsymbol{B} &= +\int_0^\infty E_N\left\{\frac{(N-p)r\left(\hat{\boldsymbol{\theta}}'\boldsymbol{S}^{-1}\hat{\boldsymbol{\theta}}\right)}{t\hat{\boldsymbol{\theta}}'\left(t^{-1}\boldsymbol{\Sigma}\right)^{-1}\hat{\boldsymbol{\theta}}}\right. \\
&\quad \times \left[N(N-p+2)r\left(\hat{\boldsymbol{\theta}}'\boldsymbol{S}^{-1}\hat{\boldsymbol{\theta}}\right) - 2(p-2)\right]\left.\middle|\, t\right\}W_o(t)dt,
\end{aligned}$$

Wherein $r(.)$ is non-decreasing, $r'\left(\hat{\boldsymbol{\theta}}'\boldsymbol{\Sigma}^{-1}\hat{\boldsymbol{\theta}}\right) \geq 0$. Also, following Srivastava and Bilodeau (1989), we have

$$\int_0^\infty E_N\left[r'\left(\hat{\boldsymbol{\theta}}'\boldsymbol{S}^{-1}\hat{\boldsymbol{\theta}}\right)\,\middle|\, t\right] t^{-2}W_o(t)dt \geq 0. \tag{12.3.7}$$

Therefore, $\boldsymbol{A} \leq 0$ and by making use of (12.1.15), $\boldsymbol{A} < \infty$. Taking $r \leq \frac{2(p-2)}{N(N-p+2)}$, $\boldsymbol{B} \leq 0$ is achieved for finite $\boldsymbol{B}$, which completes the proof. But for demonstrating that $\boldsymbol{B} < \infty$, it is sufficient to show that

$$(i) \quad \int_0^\infty t^{-1}E_N\left(\frac{1}{N\hat{\boldsymbol{\theta}}'\left(t^{-1}\boldsymbol{\Sigma}\right)^{-1}\hat{\boldsymbol{\theta}}}\,\middle|\, t\right)W_o(t)dt < \infty, \tag{12.3.8}$$

$$(ii) \quad \int_0^\infty t^{-1}E_N\left(\frac{r^2\left(\hat{\boldsymbol{\theta}}'\boldsymbol{S}^{-1}\hat{\boldsymbol{\theta}}\right)}{N\hat{\boldsymbol{\theta}}'\left(t^{-1}\boldsymbol{\Sigma}\right)^{-1}\hat{\boldsymbol{\theta}}}\,\middle|\, t\right)W_o(t)dt < \infty.$$

Note that for a fixed t, $N\hat{\boldsymbol{\theta}}'\left(t^{-1}\boldsymbol{\Sigma}\right)^{-1}\hat{\boldsymbol{\theta}}$ has noncentral chi-square distribution with p d.f. and noncentrality parameter $Nt\boldsymbol{\theta}'\boldsymbol{\Sigma}^{-1}\boldsymbol{\theta}$. In conclusion, from the fact that

$$E_N\left[\frac{1}{N\hat{\boldsymbol{\theta}}'\left(t^{-1}\boldsymbol{\Sigma}\right)^{-1}\hat{\boldsymbol{\theta}}}\right] \leq E\left(\chi_p^{-2}\right) = \frac{1}{p-2} \tag{12.3.9}$$

(12.3.8) (i) is followed by

$$\int_0^{\infty} t^{-1} E_N \left(\frac{1}{N\hat{\boldsymbol{\theta}}' \left(t^{-1}\boldsymbol{\Sigma}\right)^{-1} \hat{\boldsymbol{\theta}}} \middle| t \right) W_o(t) dt \leq \frac{\gamma_o}{(p-2)(\gamma_o - 2)} < \infty$$

and (12.3.8) (ii) is followed by (12.1.16) and using the covariance inequality (see Lemma 6.6, page 370 , of Lehmann and Casella, 1998)

$$\begin{aligned} E_N \left\{ \frac{r^2 \left(\hat{\boldsymbol{\theta}}' \boldsymbol{S}^{-1} \hat{\boldsymbol{\theta}} \right)}{N\hat{\boldsymbol{\theta}}' \left(t^{-1}\boldsymbol{\Sigma}\right)^{-1} \hat{\boldsymbol{\theta}}} \middle| t \right\} &\leq E_N \left\{ r^2 \left(\hat{\boldsymbol{\theta}}' \boldsymbol{S}^{-1} \hat{\boldsymbol{\theta}} \right) \middle| t \right\} E_N \left\{ \frac{1}{N\hat{\boldsymbol{\theta}}' \left(t^{-1}\boldsymbol{\Sigma}\right)^{-1} \hat{\boldsymbol{\theta}}} \middle| t \right\} \\ &< \infty. \end{aligned} \tag{12.3.10}$$

The proof is then complete.

12.4 Problems

1. Verify (12.1.8).
2. Verify $C(N, p)$ in the proof of Lemma 12.1.1.
3. Verify (12.1.12).
4. Prove the estimator $\hat{\boldsymbol{\theta}}$ is minimax.
5. Verify (12.2.1).
6. Verify (12.3.3)-(12.3.4).
7. Obtain the upper bound of r explicitly in Theorem 12.3.2, if (12.1.2) is assumed to be the t-distribution.
8. Prove Corollary 12.3.2.3.
9. Prove Corollary 12.3.2.4.
10. Obtain the upper bound of c explicitly in Corollary 12.3.2.5.
11. Verify (12.3.7).
12. Verify (12.3.9).
13. Verify (12.3.10).

REFERENCES

Aitchison, J. and Dunsmore, I.R. (1975). *Statistical Prediction Analysis.* Cambridge University Press, Cambridge.

Aitchison, J. and Sculthorpe, D. (1965). Some problems of statistical prediction. *Biometrika*, *55*, 469–483.

Akritus, M., Saleh, A. K. Md. Ehsanes and Sen, P. K. (1985). Nonparametric estimation of intercepts after a preliminary test on parallelism of several regression lines, *Biostatistics: Statistics in Biomedical, Public Health and Environmental Sciences*, Ed. P. K. Sen, Elsevier Science, North-Holland, 221–235.

Alam, K. and Thompson, J. R. (1969). Locally averaged risk, *Ann. Inst. Statist. Math.*, **21**, 457–469.

Anderson, T. W. (1984). *An Introduction to Multivariate Statistical Analysis*, John Wiley, New York.

A. K. Md. Ehsanes Saleh, M. Arashi, S.M.M. Tabatabaey.

Anderson, T. W. (2003). *An Introduction to Multivariate Statistical Analysis*, 3rd ed., John Wiley and Sons, New York.

Anderson, T. W. and Fang, K. T. (1990). Inference in multivariate elliptically contoured distribution based on maximum likelihood, in *Statistical Inference in Elliptically Contoured and Related Distribution*, K. T. Fang and T. W. Anderson Ed., 201–216, Allerton Press, New York.

Arashi, M. (2010). Idea of constructing an image of Bayes action, *Statistical Methodology*, **7**, 22–29.

Arashi, M. (2012). Preliminary test and Stein estimators in simultaneous linear equations, *Lin. Alg. Appl.*, **436**(5), 1195–1211.

Arashi, M., Iranmanesh, Anis, Norouzirad, M. and Salarzadeh Jenatabadi, Hashem (2013). Bayesian analysis in multivariate regression models with conjugate priors, *Statistics*, DOI:10.1080/02331888.2013.809720.

Arashi, M., Saleh, A. K. Md. Ehsanes and Tabatabaey, S. M. M. (2010). Estimation of parameters of parallelism model with elliptically distributed errors, *Metrika*, **71**, 79–100.

Arashi, M., Saleh, A. K. Md. Ehsanes and Tabatabaey, S. M. M. (2013) Regression model with elliptically contoured errors, *Statistics*, **47**(6), 1266-1284.

Arashi, M. and Tabatabaey, S. M. M. (2008). Stein-type improvement under stochastic constraints: Use of multivariate Student-t model in regression, *Statist. Prob. Lett.*, **78**(14), 2142–2153.

Arashi, M. and Tabatabaey, S. M. M. (2009). Improved variance estimation under subspace restriction, *J. Mult. Anal.*, **100**, 1752-1760.

Arashi, M. and Tabatabaey, S. M. M. (2010). A note on Stein-type estimators in elliptically contoured models, *J. Statist. Plann. Inf.*, **140**, 1206–1213.

Arashi, M., Tabatabaey, S. M. M. and Hassanzadeh Bashtian, M. (2014) Shrinkage ridge estimators in linear regression, *Comm. Statist. Sim. Comp.*, **43**, 871–904.

Arashi, M., Tabatabaey, S. M. M. and Iranmanesh, Anis (2010). Improved estimation in stochastic linear models under elliptical symmetry, *J. Appl. Prob. Statist.*, **5**(2), 145–160.

Arashi, M., Tabatabaey, S.M.M. and Khan, Shahjahan (2008). Estimation in multiple regression model with elliptically contoured errors under MLINEX loss, *J. Appl. Prob. Statist.*, **3**(1), 23–35.

Arashi, M., Tabatabaey, S. M. M. and Soleimani, H. (2012). Simple regression in view of elliptical models, *Lin. Alg. Appl.*, **437**, 1675–1691.

Arellano-Valle, R., Galea-Rojas, M. and Iglesias, P. (2000). Bayesian analysis in elliptical linear regression models, *Revista Sociedad Chilena de Estadística*, **17**, 59–104.

Baltberg, R. C. and Gonedes, N. J. (1974). A comparison of the stable and Student distributions as statistical models for stock prices: reply, *J. Business*, **47**(2), 244-280.

Baranchik, A. J. (1970). A family of minimax estimators of the mean of a multivariate normal distribution, *Ann. Math. Statist.*, **41**(2), 642–645.

Barlev, B. and Levy, H. (1979). On the variability of accounting income numbers. *J. Accounting Res.*, **16**, 305–315

Box, G. E. P. and Tiao, G. C. (1973). *Bayesian Inference in Statistical Analysis*, Reading, MA, Addison-Wesley.

Box, G. E. P. and Tiao, G. C. (1992) *Bayesian Inference in Statistical Analysis*, John Wiley, New York.

Brandwein, A. C. and Strawderman, W. E. (1991). Generalization of James-Stein estimator under spherical symmetry, *Ann. Statist.*, **19**(3), 1639–1650.

Broemeling, L. D. (1985). *Bayesian Analysis of Linear Models*, Marcel Dekker, New York.

Casella, G. (1990). Estimators with nondecreasing risk: Application of a chi-square identity, *Statist. Prob. Lett.*, **10**, 107–109.

Chaudhry, J. (1997). *Preliminary test approach to shrinkage estimation of parameters in a variety of statistical models*, PhD Thesis, Bowling Green State University, Ohio, USA.

Chib, S., Osiewalski, J. and Steel, M. F. J. (1991). Posterior inference on the degrees of freedom parameter in multivariate-t regression models, *Economics Lett.*, **37**(4), 391–397.

Chu, K. C. (1973), Estimation and decision for linear systems with elliptically random process. *IEEE Trans. Autom. Cont.*, 18, 499–505.

Dey, D., Ghosh, M. and Strawderman, W. E. (1999). On estimation with balanced loss function. *Statist. Prob. Lett.*, **45**, 97–101.

Dirac, Paul (1958). *Principles of Quantum Mechanics* (4th ed.), Oxford, Clarendon Press.

Eaton, M.L. (1983). *Multivariate Statistics - A Vector Space Approach.* Wiley, New York.

Fisher, S. R. (1956). *Statistical Methods and Scientific Inference*, Oliver & Boyd, Edinburgh.

Fisher, S. R. (1960). On some extensions of Bayesian inference proposed by Mr Lindley, *J. Royal Statist. Soc.* Series B, **22**, 299–301.

Folland, G. B. (1999). *Real Analysis: Modern Techniques and Their Applications*, 2nd ed., John Wiley and Sons, New York.

Fraser, D.A.S. (1968). *The Structure of Inference*, Wiley, New York.

Fraser, D.A.S. (1979). *Inference and Linear Models*, McGraw-Hill, New York.

Fraser, D. A. S. and Fick, G. (1975). Necessary analysis and its implementation, *Proc. Symp. Statist. Related Topics*, Carleton Mathematical Notes, **12**, 501–530.

Fraser, D.A.S. and Haq, M.S. (1969). Structural probability and prediction for the multilinear model. *J Royal Statist. Soc.*, B, **31**, 317–331.

Geisser, S. (1993). *Predictive Inference: An Introduction.* Chapman & Hall, London.

Gibbons, D. G. (1981). A simulation study of some ridge estimators, *J. Amer. Statist. Assoc.*, **76**, 131–139.

Gupta, A. K. and Varga, T. (1995). Normal mixture representations of matrix variate elliptically contoured distributions, *Sankhyā*, **57**, 68–78.

Guttman, I. (1970). *Statistical Tolerance Regions: Classical and Bayesian.* Griffin, London.

Han, C. P. and Bancroft, T. A. (1968). On pooling means when variance is unknown, *J. Amer. Statist. Assoc.*, **63**, 1333–1342.

Haq, M. S. and Rinco, S. (1973). β-expectation tolerance regions for a generalized multilinear model with normal error variables. *J. Mult. Anal.*, **6**, 414–421.

Hassanzadeh Bashtian, M., Arashi, M. and Tabatabaey, S. M. M. (2011). Ridge estimation under the stochastic restriction, *Comm. Statist. Theo. Meth.*, **40**, 3711–3747.

Himmelblau, D.M. (1970). *Process Analysis by Statistical Methods*, John Wiley, New York.

Hogg, R. V., McKean, J. and Craig, A. T. (2012). *Introduction to Mathematical Statistics*, 7th ed., Pearson Education, India.

Hoerl, A. E. and Kennard, R. W. (1970). Ridge regression: biased estimation for non-orthogonal problems, *Technometrics*, **12**, 55–67.

Hoerl A.E., Kennard, R. W. and Baldwin, K.F. (1975) Ridge regression: Some simulations Analysis, *Comm. Statist.*, **4**, 105–123.

Jammalamadaka, S. R., Tiwari, R. C. and Chib, S. (1987). Bayes prediction in the linear model with spherically symmetric errors, *Economics Lett.*, **24**(1), 39–44.

James, W. and Stein, C. (1961). Estimation of quadratic loss, *Proc. of the Fourth Berkeley Symp. on Math. Statist. Prob.*, **1**, 361–379.

Jeffreys, H. (1961). *Theory of Probability*, Oxford: Clarendon.

Jozani, M. J., Marchand, E. and Parsian, A. (2006). On estimation with weighted balanced-type loss function, *Statist. Prob. Lett.*, **76**, 773–780.

Judge, G. G. and Bock, M. E. (1978). *The Statistical Implication of Pre-test and Stein-rule Estimators in Econometrics*, North-Holland, New York.

Kabe, D. G. (1965). Generalization of Sverdrups lemma and its application to multivariate distribution theory, *Ann. Math. Statist.*, **36**, 671–676.

Khan, S. (2002a). A note on an optimal tolerance region for the class of multivariate elliptically contoured location-scale model, *J. Calcutta Statist. Assoc. Bull.*, **53**, 125–131.

Khan, S. (2002b) Distribution of sum of squares and products for the generalized multilinear matrix-T model. *J. Mult. Anal.*, **83**, 124–140.

Khan, S. (2003). Estimation of the parameters of two parallel regression lines under uncertain prior information, *Biometrical J.*, **45**, 73–90.

Khan, S. (2004). Predictive distribution of regression vector and residual sum of squares for normal multiple regression model, *Comm. Statist. Theo. Meth.*, **33(10)**, 2423–2443.

Khan, S. (2006a). Shrinkage estimation of the slope parameters of two parallel regression lines under uncertain prior information, *Model Assisted Statist. Appl.*, **1**, 195–207.

Khan, S. (2006b). Prediction distribution of future regression and residual sum of squares matrices for multivariate simple regression model with correlated normal responses, *J. Appl. Prob. Statist.*, **1**, 15–30.

Khan, S. and Haq, M.S. (1994). Prediction inference for multilinear model with errors having multivariate Student-t distribution and first-order autocorrelation structure. *Sankhya, Part B: Indian J. Statist,* **56**, 95–106.

Khan, S. and Saleh, A. K. Md. E. (1995). Preliminary test estimators of the mean for sampling from some Student-t populations. *Jour. of Statistical Research*, **29**, 67–88.

Khan, S. and Saleh, A. K. Md. Ehsanes (1997). Shrinkage pre-test estimator of the intercept parameter for a regression model with multivariate Student-t errors, *Biom. J.*, **39**(2), 131–147.

Khan, S. and Saleh, A. K. Md. Ehsanes (1998). Comparison of estimators of means based on p- samples from multivariate student-t population, *Commun. Statist. Theo. Meth.*, **27**(1), 193–210.

Khan, S. and Saleh A. K. Md. Ehsanes (2003). Stein-type estimators for mean vector in two-sample problem of multivariate Student-t populations with common covariance matrix, *International Journal of Statistical Sciences*, **1**, 1–19.

Khan, S and Saleh, A. K. Md. Ehsanes (2005). Estimation of parameters of the multivariate regression model with uncertain prior information and Student-t errors, *Journal of Statistical Research*, **39**(2), 79–94.

Khan, B. U. and Saleh, A. K. Md. Ehsanes (2006). Improved estimation of regression parameters when the regression lines are parallel, *J. Appl. Prob. Statist.*, **1**, 1–13.

Khan, S. and Saleh, A. K. Md. E. (2008). Estimation of slope for linear regression model with uncertain prior information and Student-t error, *Comm. Statist. Theo. Meth.*, **37**, 2264–2581.

Kibria, B. M. G., Kristofer, M. and Shukur, G. (2013). Some ridge regression estimators for zero inflated poisson Model, *J Appl. Statist.*, **40**(4), 721–735.

Kibria, B. M. G., Mansson, K. and G. Shukur (2012). Performance of some logistic ridge regression estimators, *Comp. Econ.*, **40**, 401–414.

Kibria, B. M. G. and Saleh, A. K. Md. Ehsanes (2004). Preliminary test ridge regression estimators with Student's t errors and conflicting test-statistics, *Metrika*, **59**, 105–124.

Kibria, B. M. G. and Saleh, K. Md. Ehsanes (2012). Improving the estimators of the parameters of a probit regression model: A ridge regression approach, *J. Statist. Plann. Inf.*, **142**, 1421–1435.

Kollo, T. and von Rosen, D. (2005). *Advanced Multivariate Statistics with Matrices*, Springer.

Kotz, S. and Nadarajah, S. (2004). *Multivariate t Distributions and Their Applications*, Cambridge.

Kunugi. T., Tamura. T. and T. Naito. (1961), New acetylene process uses hydrogen dilution, *Chen. Eng. Prog.*, **57**, 43–49.

Lambert, A., Saleh, A. K. Md. Ehsanes and Sen, P. K. (1985). On least squares estimation of intercept after a preliminary test on parallelism of regression lines, *Comm. Statist. Theo. Meth.*, **14**, 793–807.

Lawless, J. F. (1978). Ridge and related estimation procedure, *Comm. Statist. A*, **7**, 139–164.

Lehmann, E. L. and Casella, G. (1998). *Theory of Point Estimation*, 2nd ed., Springer, New York.

Magnus, J. R. and Neudecker, H. (1999). *Matrix Differential Calculus with Applications in Statistics and Econometrics*, 2nd Ed., Wiley, Chichester.

Marquardt, D.W. and R.D. Snee (1975), Ridge regression practice, *J. Amer. Statist. Assoc.*, **29**(1), 3–20.

McDonald, G. C. and Galarneau, D. I. (1975). A Monte Carlo evaluation of some ridge-type estimators, *J. Amer. Statist. Assoc.*, **70**, 407–416.

Meng, X. L. (1994). Posterior predictive p-value, *Ann. Statist.*, **22**(3), 1142–1160.

Montgomery, D. C., Peck, A. E. and Vining, G. G. (2001). *Introduction to Linear Regression Analysis*, 3rd Ed., John Wiley, NewYork.

Muirhead, R. J. (2005). *Aspect of Multivariate Statistical Theory*, 2nd Ed., John Wiley, New York.

Najarian, S. Arashi, M. and B. M. G. Kibria (2013). A simulation study on some restricted ridge regression estimators, *Comm. Statist. Sim. Comp.*, **42**(4), 871–890.

Ng, V. M. (2002). Robust Bayesian inference for seemingly unrelated regressions with elliptical errors , *J. Mult. Anal.*, **82**, 409–414.

Ng, V. M. (2010). On Bayesian inference with conjugate priors for scale mixtures of normal distributions, *J. Appl. Prob. Statist.*, **5**, 69–76.

Ng, V. M. (2012). Bayesian inference for the precision matrix for scale mixtures of normal distributions, *Comm. Statist. Theo. Meth.*, **41**, 4407–4412.

Osiewalski, J. and Steel, M. F. J. (1993). Robust Bayesian inference in elliptical regression models, *J. Econometrics*, **57**, 345–363.

Ouassou, I. and Strawderman, W. E. (2002). Estimation of a parameter vector restricted to a cone, *Statist. Prob. Lett.*, **56**, 121–129.

Pearson, K. (1923). On non-skew frequency surfaces, *Biometrika*, **15**, 231–244.

Press, S. J. (1989). *Bayesian Statistics: Principles, Mehtods and Applications*, John Wiley, New York.

Prucha, I. R. and Kalajian, H. H., (1984). The structure of simultaneous equation estimators: A generalization towards non-normal disturbances. *Econometrica*, **52**(3), 721–736.

Ravishanker, N. and Dey, Dipak K. (2001). *A First Course in Linear Model Theory*, Chapman & Hall/CRC, New York.

Robert, C. P. (2001). *The Bayesian Choice: From Decision-Theoretic Motivations to Computational Implementation*, 2nd ed., Springer, New York.

Rohatgi, V. K. and Saleh, A. K. Md. Ehsanes, (2001). *An Introduction to Probability and Statistics*, 2nd ed., John Wiley, New York.

Saleh, A. K. Md. Ehsanes, (2006). *Theory of Preliminary Test and Stein-type Estimation with Applications*, John Wiley, New York.

Saleh, A. K. Md. Ehsanes and Kibria, B. M. G. (1993). Performances of some new preliminary test ridge regression estimators and their properties, *Comm. Statist. A*, **22**, 2747–2764.

Saleh, A. K. Md. Ehsanes and Kibria, B. M. G. (2010). Estimation of the mean vector of a multivariate elliptically contoured distribution, *Calcutta Statist. Assoc. Bull.*, **62**, 247-248.

Saleh, K. Md. E. and Kibria, B. M. G. (2011). On some ridge regression estimators: A nonparametric approach, *J. Nonparametric Statist.*, **23**(3), 819-851.

Saleh, A. K. Md. Ehsanes and Sen, P. K. (1985). Nonparametric shrinkage estimation in a parallelism problem, *Sankhya*, **47**, 156–165.

Sarkar, N. (1992). A new estimator combining the ridge regression and the restricted least squares method of estimation, *Comm. Statist. A*, **21**, 1987–2000.

Singh, R. K. (1988). Estimation of error variance in linear regression models with errors having multivariate student-t distribution with unknown degrees of freedom, *Econ. Lett.* **27**, 47–53.

Singh, R. K., Misra, S. and Pandey, S. K. (1995). A generalized class of estimators in linear regression models with multivariate-t distributed error, *Statist. Prob. Lett.*, **23**(2), 171–178.

Srivastava, M. and Bilodeau, M. (1989). Stein estimation under elliptical distribution, *J. Mult. Anal.*, **28**, 247–259.

Srivastava, M. S. and Khatri, C. G. (1979). *An Introduction to Multivariate Analysis*, North-Holland, Amsterdam.

Srivastava, M. S. and Saleh, A. K. M. Ehsanes (2005). Estimation of the mean vector of a multivariate normal distribution: Subspace Hypothesis, *J. Mult. Anal.*, **96**, 55–72.

Stein, C. (1956). Inadmissibility of the usual estimator for the mean of a multivariate distribution, *Proc. Third Berkeley Symp. Math. Statist. Prob.*, **1**, 197–206.

Stein, C. M. (1981). Estimation of the mean of a multivariate normal distribution, *Annal. Statist.*, **9**(6), 1135–1151.

Sutradhar, B. C. and Ali, M. M. (1989). A generalization of the Wishart distribution for the elliptical model and its moments for the multivariate t model, *J. Mult. Anal.*, **29**, 155–162.

Sverdrup, E. (1947). Derivation of the Wishart distribution of second order moments by straight forward integration of a multiple integral, *Skand. Aktuarietdskr.*, **30**, 151–166.

Tabatabaey, S. M. M. (1995). *Preliminary Test Approach Estimation: Regression Model with Spherically Symmetric Errors*, Unpublished PhD Thesis, Carleton University, Canada.

Teng, C., Fang H. and Deng, W. (1989). The generalized noncentral Wishart distribution, *J. Math. Res. Exposition*, **9**(4), 479–488.

Tsukuma, H. (2010). Shrinkage priors for Bayesian estimation of the mean matrix in an elliptically contoured distribution, *J. Mult. Anal.*, **101**, 1483–1492.

Ullah, A. and Walsh, V. Z. (1984). On the robustness of LM, LR and W tests in regression models. *Econometrica*, **52**, 1055–1066.

Vidal, I. and Arellano-Valle, R. B. (2010). Bayesian inference for dependent elliptical measurement error models, *J. Mult. Anal.*, **101**(10), 2587–2597.

Xu, J. L. and Izmirlian, G. (2006). Estimation of location parameters for spherically symmetric distributions, *J. Mult. Anal.*, **97**(2), 514–525.

Zellner, A. (1971). *An Introduction to Bayesian Inference in Econometrics*, John Wiley, New York.

Zellner, A., (1976). Bayesian and non-Bayesian Analysis of regression model with multivariate Students t eorror terms, *J. Amer. Statist. Assoc.* **71**, 400–408.

Zellner, A., (1994). Bayesian and non-Bayesian estimation using balanced loss functions. In: Berger, J. O. and Gupta, S. S. (Eds.), *Statistical Decision Theory and Methods V*, Springer, New York, 337–390.

Zhong, Z. and Yang, H. (2007). Ridge estimation to the restricted linear model. *Comm. Statist. Theo. Meth.*, **36**, 2099–2115.

AUTHOR INDEX

Statistical Inference for Models with Multivariate t-Distributed Errors, First Edition.

A. K. Md. Ehsanes Saleh, M. Arashi, S.M.M. Tabatabaey.

V

X

Y

Z

SUBJECT INDEX

Statistical Inference for Models with Multivariate t-Distributed Errors, First Edition.

A. K. Md. Ehsanes Saleh, M. Arashi, S.M.M. Tabatabaey.

S

T

V

W

Z